Common Rail Fuel Injection Technology in Diesel Engines

Common Rail Fuel Injection Technology in Diesel Engines

Guangyao Ouyang
Naval University of Engineering
China

in collaboration with

Shijie An
Naval University of Engineering
China

Zhenming Liu
Naval University of Engineering
China

Yuxue Li
Naval University of Engineering
China

WILEY

国防工業出版社
National Defense Industry Press

This edition first published in 2019 by John Wiley & Sons (Asia) Pte Ltd. under exclusive licence granted by National Defense Industry Press for all media and languages (excluding simplified and traditional Chinese) throughout the world (excluding Mainland China) and with non-exclusive license for electronic versions in Mainland China.

This work is funded by B & R Book Program.

Registered Offices
John Wiley & Sons, Inc., 111 River Street, Hoboken, NJ 07030, USA
John Wiley & Sons Ltd, The Atrium, Southern Gate, Chichester, West Sussex, PO19 8SQ, UK

Editorial Office
The Atrium, Southern Gate, Chichester, West Sussex, PO19 8SQ, UK

For details of our global editorial offices, customer services, and more information about Wiley products visit us at www.wiley.com.

Wiley also publishes its books in a variety of electronic formats and by print-on-demand. Some content that appears in standard print versions of this book may not be available in other formats.

Library of Congress Cataloging-in-Publication Data

Names: Ouyang, Guangyao, author. | An, Shijie, author.
Title: Common rail fuel injection technology in diesel engines / Guangyao Ouyang, Naval University of Engineering, Wuhan ; in collaboration with, Shijie An, Naval University of Engineering, Wuhan, Zhenming Liu, Naval University of Engineering Wuhan, Yuxue Li, Naval University of Engineering, Wuhan.
Description: Hoboken, NJ : Wiley, 2019. | Includes bibliographical references and index. |
Identifiers: LCCN 2018057453 (print) | LCCN 2018057853 (ebook) | ISBN 9781119107248 (AdobePDF) | ISBN 9781119107262 (ePub) | ISBN 9781119107231 (hardcover)
Subjects: LCSH: Diesel motor–Fuel injection systems. | Marine diesel motors.
Classification: LCC TJ797 (ebook) | LCC TJ797 .O985 2019 (print) | DDC 621.43/61–dc23
LC record available at https://lccn.loc.gov/2018057453

Cover image: © Aun Photographer/Shutterstock
Cover design by Wiley

Set in 10/12pt WarnockPro by SPi Global, Chennai, India
Printed in Singapore by C.O.S. Printers Pte Ltd

10 9 8 7 6 5 4 3 2 1

Contents

Preface

Since the twenty-first century, the diesel engine is facing the challenge of two factors: energy and environmental protection; thus improving the efficiency and controlling the emissions has become an important problem facing today's diesel engine industry. The needs of society and economy of the diesel engine for future environmental protection put forward a higher request for better technology to enable a lift in demand for the diesel engine. The efficiency and lower emissions are closely related to the combustion process. The most feasible approach is to reform the diesel engine fuel system implementation to improve its performance. The electronically controlled fuel injection technology is an implementation of controlling the fuel injection quantity, injection timing, and fuel injection law, in order to realize the well-organized combustion process and optimize performance of economy, power, and emission under various working conditions.

Development of the electronic control fuel injection technology began in the 1970s, and the actual shipment of electronically controlled high-power diesel engines started to be commissioned in 1980. So far, the electronically controlled fuel injection system has passed through three stages of development: the initial development stage in 1970s, the production utility stage in the 1980s, and the stage of technological development in the 1990s. Currently, the most advanced electronic control fuel injection technology is the high-pressure common rail fuel injection technology.

The first generation of a common rail system was launched in the 1990s, with the second and third generations being rolled out after more than 10 years of research and improvement. The concept of a fourth generation of common rail system has been promoted in recent years. The latest mid-high speed and high-power diesel engines developed abroad, with no exception, adapted the common rail technology, so it is apparent that the common rail technology has become one of the important technical measures to respond to emissions and fuel economy.

Although common rail technology is one of the hot topics in the study of the modern diesel engine technology with abundant successful application examples, the system study is relatively rare, and it especially lacks a domestic research report. The author has been engaged in the research of this field for more than 10 years, involving research and development stages of demonstration, design, and key technology research. The author has reached a series of achievements with arduous effort and suffered from a deficiency of system data; thus he has developed the germination of summing up the research achievements systematically over the years, to offer some reference for colleagues. It is his wish to provide a bit of inspiration and hopes that it might make a contribution to diesel engine technology development.

This book is based on the perspective of system analysis in order to provide a comprehensive introduction to common rail technology. The book is divided into seven chapters: the first chapter analyses the present situation of the common rail system; the second chapter introduces the common rail system modeling and simulation technology; the third and fourth chapters introduce the research of key technology and key parts of the common rail system; the fifth chapter introduces the ECU design technology; the sixth chapter introduces the machine assembling technology of the common rail system; and the seventh chapter introduces the research and development of a new type of common rail system.

The book is available for senior students in relevant colleges for graduate teaching and engineering and technical personnel. Due to a possible limited level of knowledge of the author, the book offers a preliminary view; if there are some inappropriate statements, please correct them.

The author gratefully acknowledges the help of Professor Jiang Deming at Xi'an Jiaotong University, Professor Gao Xiaohong at Wuhan University of Science and Technology, with the guidance and recommendation of Professor Wang Changyi and Professor Tang Kaiyuan at Naval Engineering University. Thanks also for the support of the National Defense Industry Press. Thanks also to colleagues and graduate students for 10 years of hard work, and it is their support that provided the author with the determination and confidence to realize the publication of this book.

The author

Introduction

This book is the academic monograph about related technical aspects of the high-pressure common rail system of a diesel engine and summarizes the author's research achievements in the field of electronic injection and common rail technology in the past decade. This book systematically elaborates the following contents: the development history of high-pressure common rail technology, system simulation and optimization, key parts design and the optimization design of a new type of common rail technology, etc.

This book can be used as a reference book for Graduate teaching and is also available for engineering and technical personnel specializing in design, development, and manufacture of a common rail system of a diesel engine.

1

Introduction

Today, the diesel engine is being developed and perfected; with its advantages of high efficiency and a wide range of power, it has been widely used in industry, agriculture, national defense, and other fields. Predictably, the diesel engine will still occupy an important position in the field of engines for a long time in the future. With increasingly serious energy and environmental problems, people pay more and more attention to fuel economy and emissions of the diesel engine, especially putting emission issues at the top of considerations. Governments have developed increasingly stringent emission regulations since the 1970s, and internal combustion engine researchers and related companies have constantly committed to improve the performance of the diesel engine, in order to obtain better noise emissions and a more economic performance.

There are many factors that influence diesel engine exhaust emissions and fuel economy, which are quite complicated. The most important means of improving the emissions and economic performance of diesel engine performance is to improve its combustion performance. Therefore, transformation of the fuel injection system has become an object of primary concern since it has the greatest influence on the combustion performance. The parameters that impact the performance of the diesel fuel injection system mainly include injection pressure, injection quantity, fuel injection advance angle, etc. Traditional methods are not able to make these parameters in the diesel engine achieve optimal results in the broad scope of work carried out on the diesel engine, but the development of modern electronic technology has provided a broad space in which to improve the performance of the diesel engine.

The biggest impact of the diesel engine fuel injection system on combustion concerns three factors: injection timing, injection duration, and the fuel injection law. The main purpose of an electric controlled diesel engine fuel injection system is to realize the flexible adjustment of the above three factors, which ensure that the diesel engine is running in optimal working conditions.

The diesel engine electronic control injection system is usually composed of sensors, controllers, and actuators, as shown in Figure 1.1. The combustion process in the cylinder of the diesel engine is very complex, and is affected by many factors. The method of setting up a mathematical model, with the aid of all kinds of sensors, to realize the closed-loop control of the burning process is difficult. The basic method that most diesel engine electronic control fuel injection systems now adopt is: adopting engine speed and load as a basic signal reflecting the actual working conditions of the diesel engine; then referring to the fuel injection quantity obtained by experiment in the optimum working condition and the injection timing MAP graph in order to determine the basic

Common Rail Fuel Injection Technology in Diesel Engines, First Edition.
Guangyao Ouyang, Shijie An, Zhenming Liu and Yuxue Li.

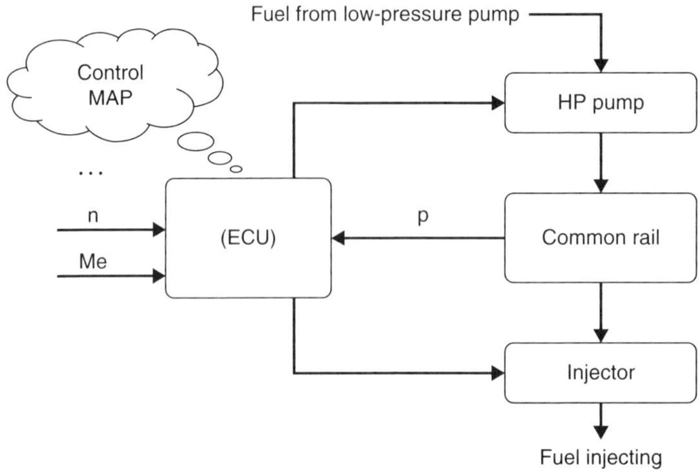

Figure 1.1 The basic compositions of the fuel injection system on an electronic controlled diesel engine.

fuel injection quantity and injection timing; then carrying on the various compensation schemes (such as engine speed, load, water temperature, oil temperature, atmospheric pressure, etc.) in order to determine the cycle fuel injection quantity and injection timing; and then taking the closed-loop feedback control to actuators in the process of working.

After the diesel engine fuel system adopts the electric control system, it has the following features:

(1) The degree of control freedom increases. The electronic control fuel injection system can optimize comprehensive control on the injection parameters in accordance with the different operating conditions.
(2) The control precision improves. For instance, the injection timing control accuracy (CA) is higher than 0.5° CA and the accuracy is four times higher than with mechanical control.
(3) Since the diesel engine fuel injection system has the characteristics of high voltage and high frequency and pulse, it will be able to achieve these objectives and will certainly bring about the complexity of actuator and control and strict requirements on reliability and duration of system that are required.

1.1 The Development of an Electronic Control Fuel Injection System

After decades of development, the diesel engine electronic control fuel injection system has experienced three progressive stages, namely, position type control, time control, and pressure time control.

Table 1.1 Typical representatives of a position type electronic control fuel injection system.

Form	System name	Control characteristics
S T	Pump Zexel (Japan) COPEC	High-speed solenoid valve control injection time, variable inductance displacement control fuel injection quantity
R A	BOSCH (Germany company) EDR	High-speed solenoid valve control gear lever displacement, adjustable CAM phase
I G	(American Caterpillar company) PEEC – ECD-P3	Brushless torque motor control gear lever displacement, linear potentiometer control injection time
H	(Japan Toyota) ECD-p	High-speed solenoid valve control injection time, adjustable CAM phase
T	BOSCH (Germany company) TICS	High-speed solenoid valve control plunger sleeve and gear lever position, the variable stroke
C	(Japan Toyota) ECD-p	High-speed solenoid valve control injection time, adjustable CAM phase
O	(British Lucas) EPIC	The phase of the CAM ring to control injection timing, control the distribution of rotor axial displacement control of oil
L U	(Stanadyne companies in the United States) PCF	Control CAM follower body axial displacement control of oil, the control of fuel injection advance unit control injection time
M N	BOSCH (Germany company) ECD – COVEC	Rotating magnet control sliding sleeve position, solenoid valve control injection time

1.1.1 Position Type Electronic Control Fuel Injection System

A position type electronic control fuel injection system retains the basic structure of a traditional injection system and only replaces the original mechanical control mechanism with electronic components. On the basis of the original mechanical control loop fuel injection quantity and injection timing, the electromagnetic actuator of linear displacement or angular displacement has been adapted to realize electrically controlled fuel injection timing and to improve control accuracy and the mechanical control response speed. Its products involve an array plunger pump electronic control system and a rotor pump distribution electric control system. Typical representative types are shown in Table 1.1.

A position type electronic control injection system adapts electronic control components to replace the original mechanical adjusting mechanism, while the use of electronic control mechanical actuators is to control the process of injection indirectly; thus the control accuracy, response time, is comparatively lower than in other electronic control systems. Since the basic structure of the injection system has not changed, the injection characteristics cannot be greatly changed and so the injection rate is not likely to achieve flexible control.

1.1.2 Time Type Electronic Control Fuel Injection System

As the performance of the diesel engine has further requirements on fuel injection process control, the first generation of fuel injection systems that installed an electronic control device with the original mechanical injector could not meet demand and thus the second generation of electronic control fuel injection system arises at the historic moment to use the electronic control unit (ECU) to control the injection starting point and end point directly. It has changed the traditional injection system of some mechanical structures, switching the original mechanical injector to a high-speed powerful solenoid valve injector, controlling the make and breaks of the electromagnetic valve through a pulse signal, and the action controls the opening and closing of the oil atomizer. The oil pump is completely separated from this mechanism and the control mechanism, and fuel metering is determined by the fuel valve open time length and the size of the injection pressure. Injection timing is controlled by the electromagnetic valve open time, in order to realize flexible control of the fuel injection quantity, injection timing, and the integration of control. It has changed the execution of the first generation of electronic control fuel injection systems, such as a slow response, low control frequency, and unstable control precision, and thus has much greater control of the degrees of freedom and a better control performance, which the first generation of the electric control system cannot reach. The electric control system can be divided into: an electric pump nozzle system, an electric distribution pump system, and an ECU pump or inline pump system. Typical representatives are shown in Table 1.2. Though the performance has improved greatly, it still has the following disadvantages: since the injection pressure is produced by a high-pressure oil pump directly, the injection pressure and fuel injection law is still under the control of the CAM (computer aided manufacture)-shaped line and cannot be adjusted freely.

1.1.3 Pressure–Time Controlled (Common Rail) Type Electronic Control Fuel Injection System

All the electronic control fuel injection systems described above directly adopt the traditional mechanical fuel pump pressure oil and fuel injection mechanism, with its basic principle based on fuel injection pump technology developed by Robert Bosch in 1926.

Table 1.2 Time control type electronic control fuel injection system.

Form	System name	Control characteristics
Pump nozzle	Detroit Diesel Engine Company (DDEC)	Solenoid valve open and close control injection start and end
	British Lucas EUI	As above
	The University of Tokyo in Japan ACE	As above
Distribution pump	Zexel Japanese company ECD-V3	With oil spill control valve adjustment of oil pump
	DS (Stan dyne companies in the United States)	RS with high-speed solenoid valve control oil injection time and quantity
	Germany BOSCH company VP44	As above

In order to meet increasingly stringent needs of emission, noise regulations, and fuel consumption reduction, it must improve the control precision of the fuel injection quantity, injection timing, and fuel injection rate, in order to obtain fine control of each cylinder and adopt the high-pressure jet to get a better atomization effect. All these requirements prompted production of the third generation of electronic control fuel systems, that is, the emergence of the common rail electronically controlled fuel injection system.

The common rail system is characterized by: independent generation of the injection pressure and injection control and pressurization of the fuel in the common rail using a fuel supply pump in which the pressure can be maintained within the scope of 130–160 MPa, although some related research reports claim that it has reached 200 MPa. The opening and closing of the electromagnetic valve control the start and end of the fuel injection process. Thus, it can change the injection pressure according to the engine load and speed, with an operation condition in a wide range of 20–160 MPa, realizing the pilot injection, main injection, multistage spray, etc. It can also change the shape of the fuel injection rate in accordance with the demand, realize a high degree of freedom to control the fuel injection process, greatly improve the combustion efficiency of the diesel engine, and significantly improve emission performance.

The common rail fuel injection system formally entered the stage of practical application in the middle and later periods in the 1990s. This kind of electric control system can be divided into: an electronically controlled medium-pressure common rail fuel injection system (hereinafter referred to as the medium-pressure common rail system) and the electronically controlled high-pressure common rail fuel injection system (hereinafter referred to as the high-pressure common rail system).

1.1.3.1 Medium-Pressure Common Rail System

The fuel pressure in the fuel rail of a medium-pressure common rail system is 525 MPa. The fuel with medium pressure is sprayed into the combustion chamber using fuel injector booster piston pressurization with extremely high pressure. The typical representatives are the Servojet system by the BKM Company and the HEUI system by the Caterpillar Company.

The structure diagram of the HEUI system by the Caterpillar Company is shown in Figure 1.2.This system adapts a pressurization piston with the aid of machine oil pressure to increase the injection pressure and has two public oil ways. One is a high-pressure control oil way (the high-pressure control oil is machine oil), maintaining a certain degree of pressure to push a supercharging piston. Another is the low-pressure fuel oil supply, which provides fuel for the fuel injector. It controls the fuel injection pressure by adjusting the oil pressure in a high-pressure control oil circuit. Fuel injection quantity and injection timing are controlled by the solenoid valve open time length and opening moment.

The main characteristics of the system are as follows:

A. A very high injection pressure may be obtained by changing the proportion of pressure of the piston and plunger of the cross-section area.
B. High pressure only exists in the necessary part (booster amplifier, high-pressure tubing, etc.).
C. It does not need a high-pressure oil pump.

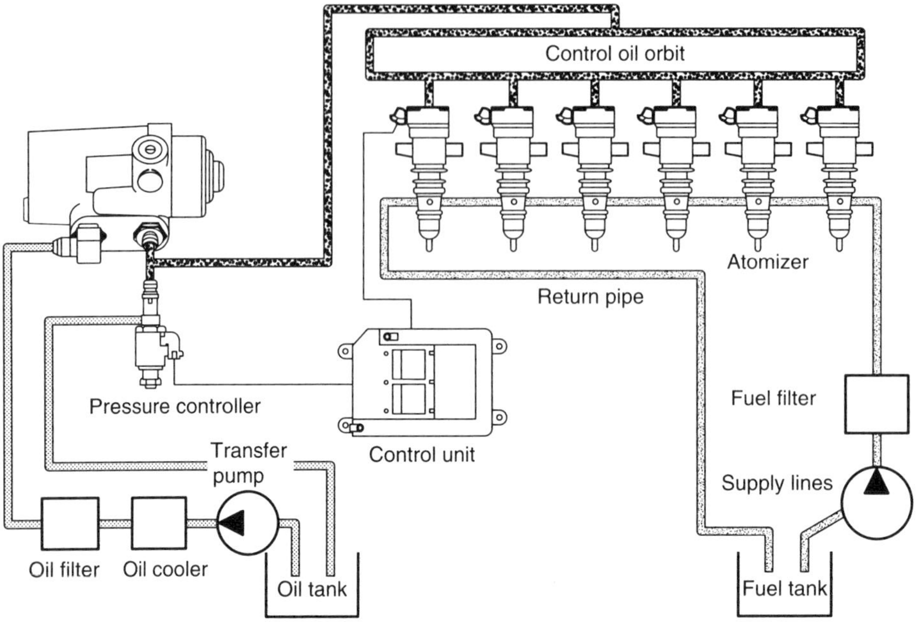

Figure 1.2 Schematic diagram of the liquid pressure type electronic control fuel injection system.

D. The injection shape is affected. It must adopt a large-flow electromagnetic valve (such as where the pressure ratio is 7 : 1 and the circulating oil quantity is more than seven times that of each injection volume). Since the response speed of the large-flow electromagnetic valve is comparatively slow, it is not easy to achieve advance injection when the injection time is very short.
E. Its installation size is comparatively large and it needs two sets of oil ways so that the oil duct size is also bigger.
F. It needs fuel valve plunger parts with high precision in order to ensure the separation of the high-pressure oil and jet fuel control.

1.1.3.2 High-Pressure Common Rail System

A high-pressure common rail system with an accumulator type injector and pressurization piston, and the public oil, the oil pressure directly controls at higher stress levels (the common rail pressure remains above 100 MPa), fuel injection quantity, and injection timing by electromagnetic control of a three-way valve or a two-way valve to adjust the use of a three-way valve or two-way valve control nozzle change of back pressure in order to change the fuel injection quantity and injection timing.

The main characteristics of the system are:

(1) There is freedom to adjust the injection pressure (common rail pressure). Using the pressure sensor, detect fuel pressure in the rail, so as to adjust for the oil pump, control the common rail pressure, and adjust the volume of injection.
(2) With engine speed and throttle opening information, etc., on the basis of optimal fuel injection quantity it is calculated using a computer by controlling the fuel injector solenoid valve moment of electric power and direct control of the fuel injection parameters.

(3) There is freedom to adjust the injection rate shapes. According to the needs of the engine, set and control the fuel injection rate shape after injection, multistage spray, etc.
(4) There is freedom to adjust the injection time. According to the parameters such as the engine speed and load, calculate the optimal injection time and control the open and close with the appropriate time, etc., so as to accurately control the fuel injection time.
(5) It requires a high-pressure fuel pump, as the system components for most of the work are under high pressure and thus may easily fail. Overall, a high-pressure common rail system can be realized in a traditional injection system that cannot otherwise achieve this function.

Its advantages are:

(1) Wide application fields (for cars and light trucks, each cylinder power can be up to 30 kW, while for heavy trucks and motorcycles and marine diesel engines, every cylinder power needs about 200 kW).
(2) A higher injection pressure; the current is currently up to 180 MPa and will soon be more than 200 MPa.
(3) Injection starting point, where the end point of injection can be easily changed.
(4) It can implement pilot injection and main injection, and after injection can be realized according to the discharge requirements, such as five to seven times that of a multistage injection.
(5) It has an injection pressure corresponding to the actual working condition. The establishment of the injection pressure is with no interdependent relationship between the fuel injection and the common rail pipe, and is always full of fuel injection at a certain pressure. The fuel injection quantity is determined by computer through calculation, but is less constrained by the other conditions.
(6) Injection timing and injection pressure are stored in the ECU (MAP) to calculate the characteristic curve of the spectrum; then the electromagnetic valve control is installed on each engine cylinder injector (injection units).

It is because of the advantages of using the high-pressure common rail system that several companies and research institutions at home and abroad are devoting a great effort to its study.

1.2 High-Pressure Common Rail System: Present Situation and Development

1.2.1 For a Common Rail System

In the 1980s research work on the high-pressure common rail system began and in the late 1990s the first generation of common rail system products were introduced.

A typical high-pressure common rail system is mainly composed of a high-pressure pump, electric control, common rail injector tube, current limiter, pressure limiting valve, rail pressure sensor, low pressure pump, filter and fuel tank, and sensors, as shown in Figure 1.3.

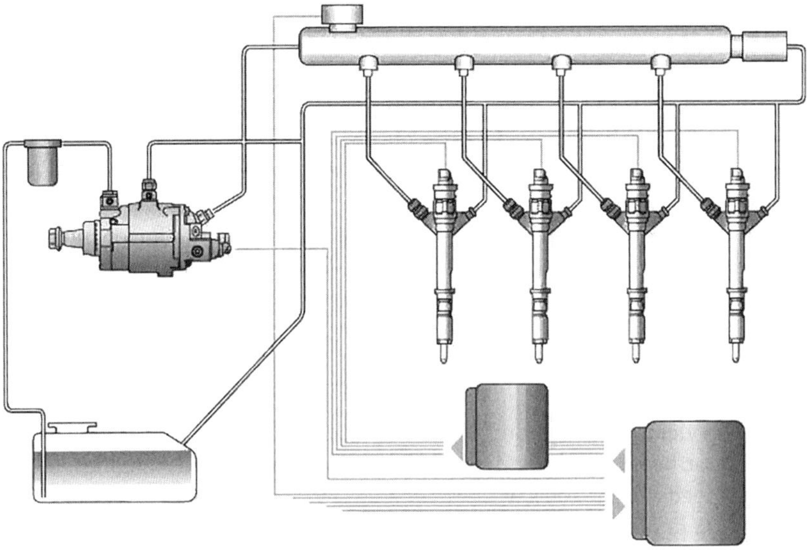

Figure 1.3 The typical schematic of the high-pressure common rail system.

Fuel from the tank passes through at a low pressure to the high-pressure oil pump and then to the radial piston pump, which has three functions: fuel will flow into the high-pressure oil rail, a part of this fuel oil will pass through the injector and is sprayed into the combustion chamber, and a small part will control the injector needle valve after the flow back into the tank. On the high-pressure oil rail there is a pressure sensor; the system will measure the fuel rail pressure compared with the preset value in the ECU, and if the measured value and book value are not consistent, the high-pressure oil rail pressure regulator on the overflow valve will open or close, allowing the fuel back to the fuel tank. Fuel injection timing and fuel quantity control, according to the measured results of each sensor, allow the ECU control high-speed solenoid valve to open and close. The system of the high-pressure oil pump for the three parts of the rotary piston pump has a control input control solenoid valve, when the engine load is low, by closing a feed to reduce the power consumption of the high-pressure oil pump. Fuel injection timing uses the function in the electronic control injector solenoid valve to control the pulse time and the fuel injection quantity uses the function in the electronic control injector solenoid valve to control the pulse width.

Due to the superiority of the high-pressure common rail system, many companies at home and abroad have studied its development and used the characteristics of the common rail system.

1.2.1.1 Germany BOSCH Company of the High-Pressure Common Rail System

So far, BOSCH Company is planning and designing four generations of the high-pressure common rail system. The first-generation batch was on the market in 1997 and with an injection pressure of 135 MPa was mainly used in cars. The second generation of mass production started in 2000, raising the maximum system pressure to 160 MPa, and started using the fuel control function of the high-pressure pump and solenoid valve injector, and improved the injection cycle by pre-injection, main injection, and many multistage jet injections; it is mainly suitable for engine power under 55 kW/l.

In May 2003, BOSCH Company began to produce innovative piezo inline technology of the third generation of the common rail system. In the first two generations of the common rail system, BOSCH Company mainly paid attention to improve the injection pressure, while the third generation of the common rail system's center of gravity shifted to technical complexity and precision, temporarily to keep the pressure at 160 MPa. The special feature of the third generation of the common rail system is that it uses a fast switch compact piezo inline injector. A piezoelectric actuator is built into the fuel injector on the shaft and is very close to the injector nozzle needle valve. The new fuel injector reduces about 75% of the moving parts and quality. An electromagnetic valve actuator compared the injector of the common rail system, its advantage being: a more accurate supply fuel and injection of the fuel in the combustion chamber, as well as better atomization and mixing. A fuel injector higher switching speed means that the time interval between the two jets is reduced, so the injection process has a more flexible control. The result is that the diesel engine is quieter and the fuel burn is cleaner, more efficient, and gives more engine power. From 2003 to 2008, five years, BOSCH injection pressure of the third generation of the common rail system of the company had two versions in order to achieve the 200 MPa high-pressure jet.

BOSCH Company developed heavy commercial vehicles in the fourth generation of the common rail system. The system configuration of the new type of injector had a pressure conversion device and a pressure transducer that could be triggered independently. Figure 1.4 is a BOSCH fourth-generation N4 interchange type automobile engine high-pressure common rail system. This system has the following characteristics: the system uses two levels of pressurization; in the second level within the fuel injector pressure amplifier, the injection pressure can reach 230–250 MPa; it can realize multiple

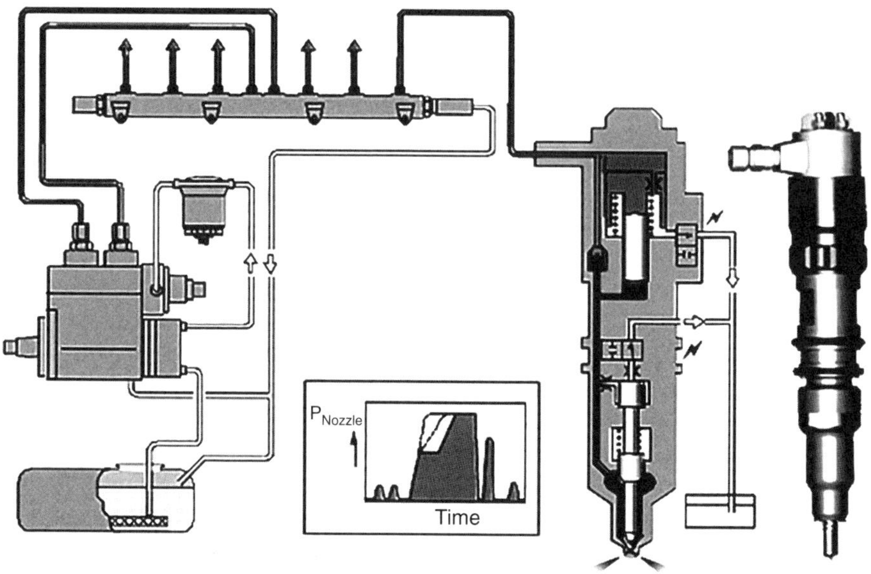

Figure 1.4 BOSCH N4 common rail injector.

injections; the injector with a two-solenoid valve can be used to control the fuel injection rate shape; it is a highly flexible control; and it can make each operation condition point of emission a minimum.

1.2.1.2 The Delphi DCR System of the Company

Delphi is the most representative of the advanced Multec DCR diesel common rail injection systems. The main components of the Multec DCR diesel common rail injection system are a common high-pressure oil rail, high-pressure fuel pressure regulator, ECU for a high-pressure fuel pump, fuel injector, and fuel filter and sensor, etc.

The Multec DCR diesel common rail injection system of injection pressure also has nothing to do with the engine speed and load, for even in low-speed running, the system can still maintain enough pressure for the high-pressure fuel injection. The system can produce injection many times and can meet the requirements of the EU emission regulations although the fuel injector design is unique. Multec DCR mainly adopted a balance control and feedback control strategy of an electric solenoid valve structure of fuel injector, which can provide extremely fast response actions and can accurately measure the fuel flow rate. The quick response, compact structure, small. and exquisite injector solenoid valve control only needs a conventional 12 V car battery drive to work normally. Compared with the world's production of any kind of diesel common rail injection system this system is energy saving, which greatly reduces the production cost of the automobile electronic design system and complexity. The whole system uses a modular design that is easy to apply in different forms and different kinds of engine.

In 2004, a new generation of diesel engine driven directly by the Delphi Company common rail fuel injection system (direct acting diesel common rail (DADCR) system) was introduced into the market. Because the new fuel injector system used piezoelectric actuators, the high-pressure pipe line was not required, which greatly saved the energy waste caused by the high-pressure oil return.

1.2.1.3 Denso High-Pressure Common Rail Injection System of the Company

Denso Co. Ltd. is one of the earliest research and development companies to produce a common rail system, and in 1995 took the lead in the world production of commercial vehicles using the common rail system, the first generation of which entered mass production in 1998. Their product was used in Japan's big four commercial vehicle manufacturing companies. Shortly thereafter, a passenger car using the common rail system began in cooperation and development with Toyota, and in June 1999, production commenced for export of cars to the European market. From that development and experience of the first generation common rail system, the system was further developed to produce the second generation system, with a practical application appearing in June 2002.

The second generation of the common rail system included a fuel injection device where the high-pressure injection was introduced many times with a burning cycle and high-precision injection quantity control, which is critical technology for the development of the second generation of common rail systems focusing on the following two aspects: one is a highest injection pressure of 180 MPa and the second is a high-precision multiple injection capacity.

1.2.2 High-Power Marine Diesel Common Rail System

Relative to the automotive diesel common rail system, the marine diesel common rail system includes new features, mainly including:

(1) A loop supply of a large amount of oil. Under the condition of a small cycle fuel injection quantity, using the electromagnetic valve can realize accurate control of the fuel injection process. Under the condition of a larger circulation of oil, how can control of the fuel injection process ensure the stability of the system pressure; this needs careful study.
(2) A large marine diesel engine power requires high system security. If the diesel engine fuel system fails there will be serious consequences, so the system should include very effective safety protection measures.
(3) As opposed to a relatively stationary diesel engine, in a marine diesel engine the electronic control system electromagnetic interference is stronger, creating a very bad working environment. The narrow space, dampness, corrosive gases such as those produced by environmental conditions of the common rail system signal acquisition, signal processing, electromagnetic compatibility of ECU, etc., make higher requirements necessary.
(4) The vehicle diesel engine has a vehicle load characteristic and its operation with is by a throttle control. In comparison the load characteristics of the marine diesel engine run as a power plant or as the propulsion characteristics of the host, completely by the ECU control strategy during operation control of the diesel engine speed.
(5) As used in a marine diesel engine, the fuel quality is worse than for an automotive machine, so the requirements to ensure the fuel system can run reliably an inferior fuel is used under high pressure.

Because the marine diesel engine has the characteristics above, in order to improve system security, common rail pressure fluctuation should be reduced and the characteristics of the marine common rail system need to be adapted with some significant differences made from that of the common rail system.

1.2.2.1 System Structure

A typical marine common rail system is that of L'Orange for the MTU Company MTU8000 series diesel engine production of the common rail system and that of Wartsila Company for Sulzer RTA – flex marine diesel engine production of the common rail system. Figure 1.5 gives the Wartsila ship common rail system schematic diagram (the system can be applied to the 4–18 marine diesel engine cylinder). The whole system consists of a high-pressure oil pump, high-pressure oil rail, accumulator, electronically controlled injector, control oil, electric control system unit, and the composition of the high-pressure oil pipe. The high-pressure oil pump is used to press the fuel into the high-pressure oil rail connecting each accumulator to each other by high-pressure tubing to offer jet fuel to two injectors. Electronically controlled injector fuel is injected and controlled by an electromagnetic valve oil circuit control. L'Orange for MTU Company MTU8000 series diesel engine production of the common rail system and structure of the Wartsila common rail systems are similar, in that in the system each accumulator is only responsible for providing a fuel injector, jet fuel, and fuel injector directly controlled by an electromagnetic valve, without the need of setting

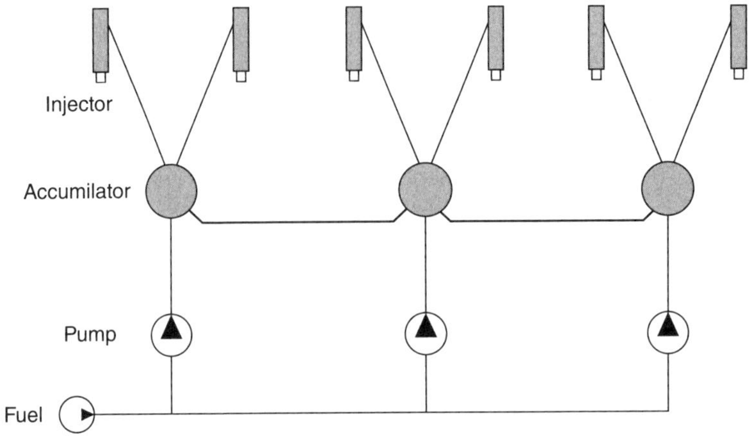

Figure 1.5 Schematic diagram of the Wartsila common rail system.

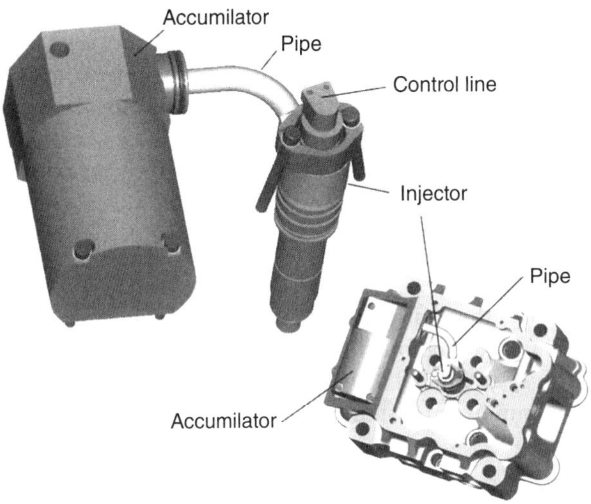

Figure 1.6 MTU8000 injector and accumulator.

up a control hydraulic system to control it. Figure 1.6 shows the accumulator and injector arrangement.

1.2.2.2 High-Pressure Oil Pump

Because the system does not require only high-pressure fuel in the fuel injection stage to be provided, the high-pressure oil pump adopts multiple bumps on the oil supply CAM method, which can effectively reduce the peak torque and improve the high-pressure oil pump volume efficiency. To control the amount of fuel into the accumulator, a high-pressure oil pump inlet is equipped with a rotary solenoid valve to control the oil. Because of the high-pressure oil pump, the oil valve quality has a great influence on the system, so the high-pressure oil pump is used for monitoring the oil valve status of the thermocouple.

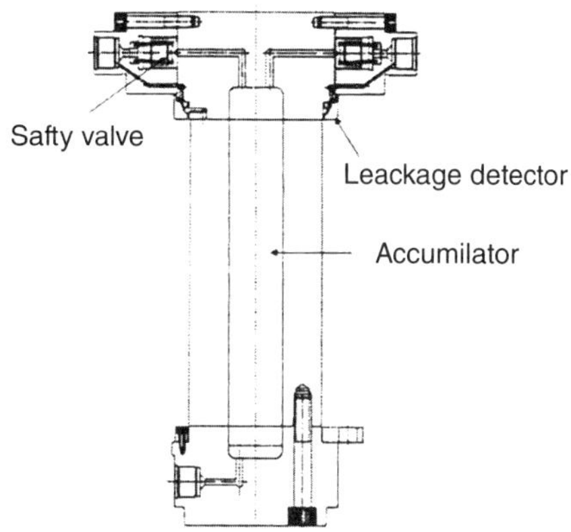

Figure 1.7 Schematic diagram of the accumulator.

1.2.2.3 Accumulator

Because of the large amount of cycle injection, simply increasing the high-pressure oil rail volume will cause it to become very large; this is not conducive to the safety of the system. Marine common rail systems are usually adopted for this accumulator, where the electronic control injector fuel supply goes through each accumulator and high-pressure oil rails connect each accumulator, to ensure that the system pressures balance. Each is equipped with a traffic safety valve and an accumulator when a fuel injector failure occurs, which is used to cut off the fuel. The accumulator structure is shown in Figure 1.7.

1.2.2.4 Electronically Controlled Injector

In order to ensure the safety of the injector when large amounts of fuel are injected, the Wartsila incorporation of an injector was adopted, as shown in Figure 1.8 of the structure of the slide valve control injection, but are of no use for the common rail system as it usually adopts the solenoid valve control mode directly. If the system is in a state of no injection, the slide valve is placed in the closed position of the upper level, the valve will be a high-pressure fuel rail and injector injection oil and control oil circuit partition, and the injector needle valve by the action of a needle valve spring is in the closed position. When the system begins the injection, the electromagnetic valve is opened, control oil makes a slide down, the high-pressure oil rail and control oil and spray oil circuit are connected, because the control piston area is much bigger than the needle bearing area, which makes the needle valve subject to a greatly downward force while still in the closed position. When the closing of the valve in the bottom position of the injector fuel injection channel is connected to the high-pressure oil rail, control of the plunger cavity and the high-pressure oil rail cavity partition, the forces acting on the needle valve in the control column insert a force that is much bigger and makes the needle valve lift the fuel injection process. When the system is asked to stop the injection, the control solenoid valve closes with the control oil valve upside down, the control plunger cavity with the needle valve port with the high-pressure oil rail connected, the control column inserts a force is much larger than the forces acting on the needle valve, and makes the needle valve shut down rapidly. When the valve to the upper position cuts off the high-pressure oil rail contact control oil and spray oil, the fuel injector is then in a state of stop injection.

When adopting slide valve control, the injector and the nozzle flow during fuel injection are not partitioned and the high-pressure oil rail in the fuel injector failure cannot

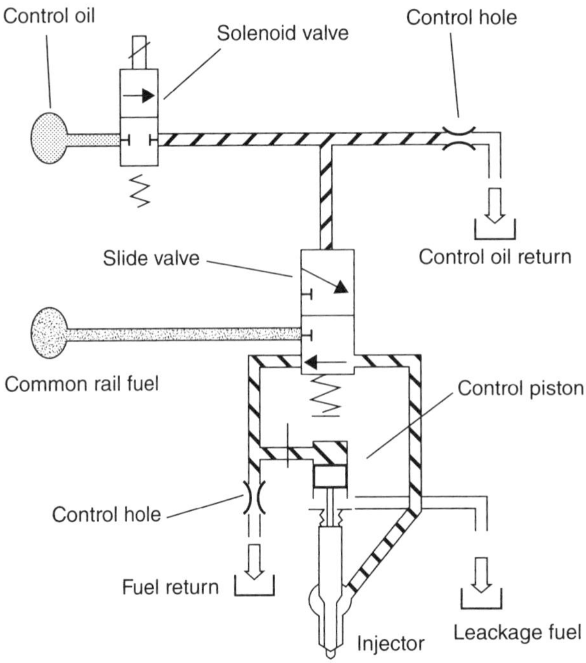

Figure 1.8 Schematic diagram of the injector control.

be completely shut down. This can guarantee that the high-pressure fuel will not cause a serious accident by being sprayed into the cylinder, improves the system security and reliability, and reduces the leakage. The slide valve control has another advantage in that it can ensure that slide valve failure does not occur and that it is in a half open position so that the injector needle valve can effectively shut down. If the slide valve is in the open position, the control plunger and the lower part of the needle valve can close down at the same time by the effect of fuel pressure. However, when the high-pressure oil rail due to the action of the control plunger area is greater than the lower part of the needle valve action area, this guarantees that the needle valve closes down using great force, so as to improve the security of the system.

L'Orange designed both binary and single electronic control injectors for the marine diesel common rail system. The binary structure of the fuel injector design of the electric control fuel injector by Wartsila Company was structurally similar, but more complicated. L'Orange Company considers that these two systems have their own characteristics: a binary system has the possibility of a small fuel leakage to the cylinder and improves security, which can remove the current limiter of each cylinder of advantages, but also has a complex structure, low speed, a lower fuel injection rate control ability, and the disadvantage of the multistage jet not being easy to realize; the single cycle system has a simple structure, high control precision, and fast dynamic response characteristics, and thus can be applied to a high-speed diesel engine and can predict the multistage injection ability. In view of the above reasons, it is still used in MTU4000 and MTU8000 series diesel engines and a vehicle electronically controlled common rail system is similar to that of a single injector system.

2

Common Rail System Simulation and Overall Design Technology

Diesel engine fuel injection system performance, such as the fuel injection pattern, fuel tank, and fuel injection velocity distribution, directly affects fuel atomization of the diesel engine cylinder combustion and emission products generated by one of the determinants and thus has a direct influence on the engine power, economy, and emissions performance. Therefore, the fuel injection system of the engine performance improvement is an important means to effectively increase its applications. A high-pressure common rail system, as a new electronically controlled injection system, which has also been developed in foreign countries in recent years, must, before entering the product development stage and with domestic research just in its infancy, produce many technical difficulties that require further careful study. With the use of modern computing technology, the high-pressure common rail system simulation is very effective in understanding high-pressure common rail system properties of the method.

Through the simulation of a high-pressure common rail system, the following objectives can be achieved:

(1) For the entire system components can be designed to provide an initial range of structural parameters: narrow the scope of the entire system hardware and software design, reduce system design time, save design costs.
(2) Through the system simulation the geometric parameters of the fuel injection system characteristics can be studied, such as an injection characteristics analysis system to study the system under various operating conditions of the injection characteristic factors.
(3) For the system simulation, a variety of dynamic processes can be studied and an injection process to show the internal flow and other characteristics.
(4) Through the system simulation, a quantitative analysis can be made of structural changes in various systems of spray characteristics.
(5) Upon completion of testing on the basis of verification, to further improve the accuracy of the simulation model, part of the system structure parameter optimization techniques can be designed to reduce the workload of purposes.

2.1 Common Rail System Basic Model

In the 1970s a fuel system simulation began. Most of the early simulation models were of a larger simplification, with the system ignoring some factors. Further research and

Common Rail Fuel Injection Technology in Diesel Engines, First Edition.
Guangyao Ouyang, Shijie An, Zhenming Liu and Yuxue Li.

development of computer technology, the fuel system, fuel density changes, changes in fuel midrange sound, leaked impact, and the impact of vacuoles followed, which became the deformation of the system considerations. Into the 1990s, sophisticated software technology and international research institutions vigorously developed the emergence of the fuel system, which produced more accurate simulations using a variety of software, including ones specifically designed for diesel engine fuel system simulation. AVL Company HYDSIM software has many applications and the software is modular in design, both on the conventional fuel system simulation and also on the electronic control system simulation.

HYDSIM is software for the hydraulic system and hydraulic–mechanical system dynamic analysis procedures. It is based on the theory of fluid dynamics and a multi-body system based on vibration analysis software; HYDSIM's main field of application is the fuel injection system simulation. The program is mainly for diesel injection systems and the development of simulation. HYDSIM hydraulic and mechanical systems with dynamic analysis cover vast areas and are very useful. For example, the hydraulic–mechanical control device and the drive system dynamics can both simulate vibration.

The HYDSIM AVL workspace is an integrated tool, which is a working space with user-friendly graphical pre- and post-processing capabilities. The HYDSIM model 2D representation provides user-defined systems in general graphics. Each particular physical system components were basically using an icon on a GUI (graphical user interface) screen (including physical component schematic symbols) to represent items. Icon is associated with mechanical, hydraulic, or logically related graphics. GUI controls the model building process, does not allow contradiction connections and other invalid input parameters.

HYDSIM modular modeling of the module according to the mathematical model and function can be divided into: a cam, a piston, the volume, pipes, pumps, valves, solenoid body orifices, needle and other models; in addition, AVL can develop user-defined special components.

2.1.1 The Common Rail System Required to Simulate a Typical Module HYDSIM

According to a high-pressure common rail fuel injection system, a component feature class will be divided into containers, valves, piping class, the gap flow, etc. and were found in an HYDSIM corresponding module. The following are explanations.

2.1.1.1 Container Class

A container class physical model is shown in Figure 2.1. HYDSIM's container module does not consider geometry. The container module type element is connected to the piston. The procedure will be considered in the calculation of the movement of the piston, which causes volume change.

The mathematical model of the container class module is as follows:

$$\dot{P}(t) = \frac{E}{V(x)} \left\{ \sum_{i=1}^{n_i} \dot{Q}_i + \sum_{i=1}^{n_i} A_i \dot{x}_i - \sum_{j=1}^{l_1} \dot{Q}_j - \sum_{j=1}^{l_2} A_j \dot{x}_j \right\} \tag{2.1}$$

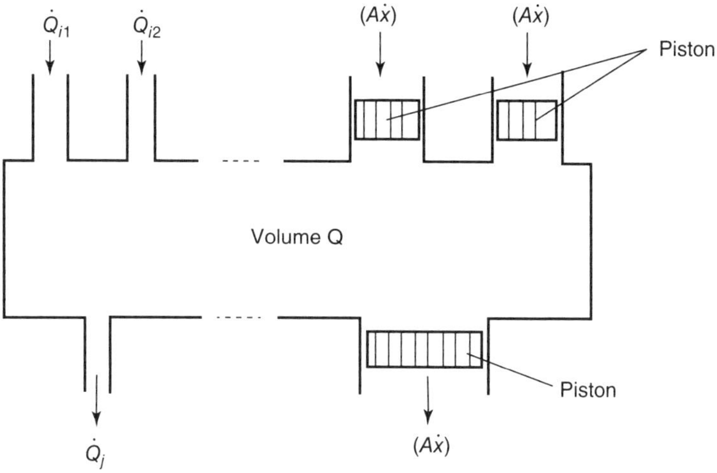

Figure 2.1 Schematic of the container module.

where

P = pressure in the vessel
E = modulus of elasticity of fuel
V = volume of the container
Q = import flows
A = control piston cross-sectional area
X = control piston displacement

The container class module input parameters are: fuel characteristics, the initial volume of the cavitation initial conditions, the heat exchange with the ambient conditions, the volume of the initial pressure, and temperature.

The container class module output parameters are: hydraulic, the actual volume of the inlet flow, and outlet flow.

2.1.1.2 Valves

In the fuel injection system there are many organizations such as valves, such as needle, delivery valve, relief valve, pressure limiting valve, one controlled by a hydraulic pressure plunger, and so on.

The valve module mathematical model is

$$m\frac{d^2x}{dt^2} + kx = \sum_i P_i D_i - A_x\left(\frac{dx}{dy} + u\right)^2 - F_0 \tag{2.2}$$

where

x = valve lift
m = quality of the valve
k = spring constant
u = fuel flow rate
A_x = drag coefficient of the valve head member
F_0 = initial spring force
P_i = acting on the hydraulic pressure on the valve
D_i = P_i acts on the valve area

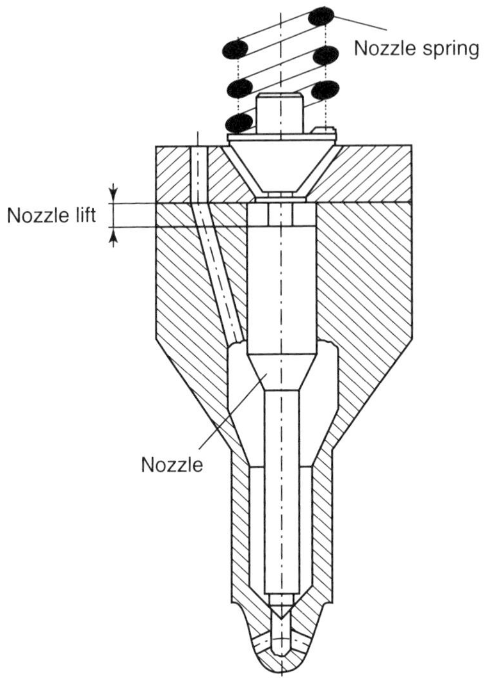

Figure 2.2 Model of the injector.

The needle physical model is shown in Figure 2.2.

The needle model input parameters are mass, dry friction, the maximum needle lift, diameter of the needle guide, needle valve sealing surface diameter, seat stiffness and damping, stiffness and damping of the needle stopper end, the valve seat sealing calculation of the surface pressure distribution at the (constant, discrete), needle initial state (coordinates, velocity), and so on.

The needle model output parameters are needle displacement, velocity and acceleration acting on the needle valve of the hydraulic force, and mechanical force.

The physical model of the pressure limiting valve is shown in Figure 2.3.

The pressure limiting valve input parameters are fluid properties, the spool moving mass, ball diameter, valve maximum lift, seat angle, fluid damping, the pressure limiting threshold, the valve inlet and outlet diameters, spring stiffness, and seat and limit department of stiffness and damping.

The pressure limiting valve output parameters are spool displacement, the spool speed, flow, minimum opening cross-sectional area, flow loss coefficient of the valve seat, and the valve of the force.

The physical model of the piston is shown in Figure 2.4.

The piston module input parameters are select the piston type (rigid or elastomer), the moving mass, dry friction, the input and output cross-sectional area of the piston, the spring stiffness, stiffness and damping at the limit, and the initial state of the piston.

Piston module output parameters are piston displacement, velocity, acceleration, and the force of the piston.

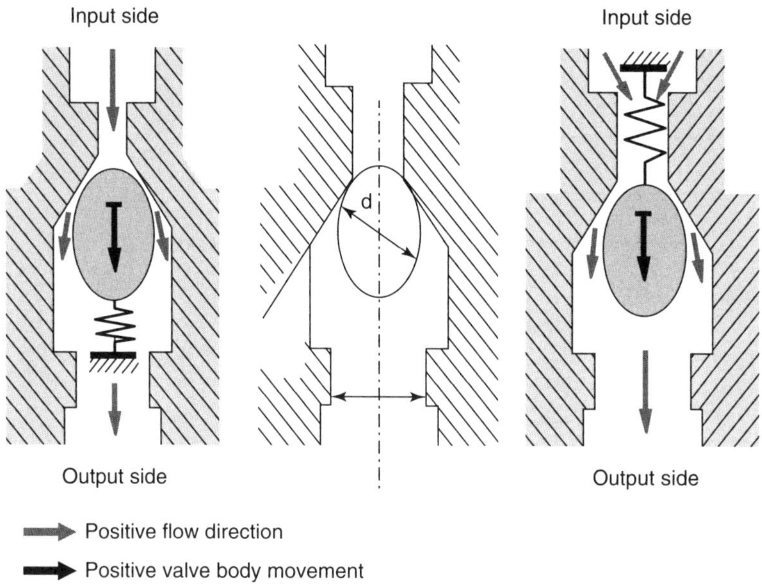

Figure 2.3 Model of the pressure limited valve.

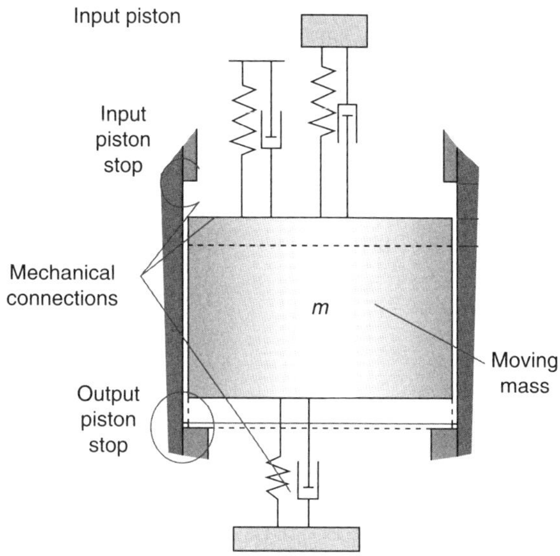

Figure 2.4 Model of the piston.

2.1.1.3 Runner Class Module

The flow class module physical model is shown in Figure 2.5. The flow mathematical model for the class module is

$$\frac{\mathrm{d}m}{\mathrm{d}t} = \mu A_{jk}\sqrt{2|P_j - P_k|} \tag{2.3}$$

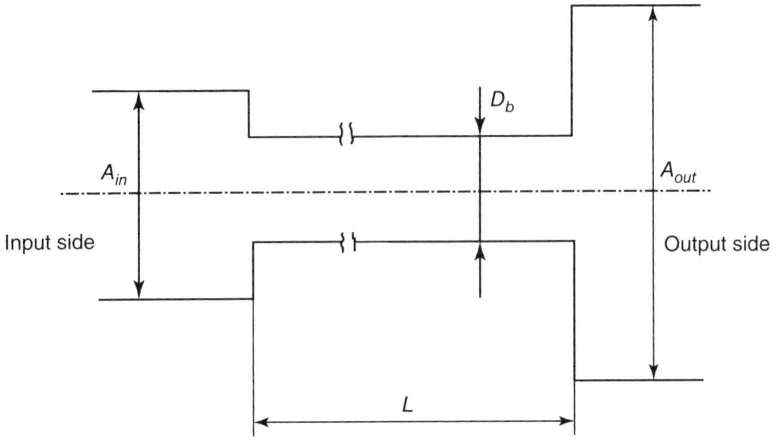

Figure 2.5 Model of the runner.

where

μ = flow coefficient

dm/dt = mass flow through the flow channel

A = cross-sectional area of the flow path

The flow class module input parameters are fluid properties, tube length, diameter, flow resistance, the initial pipe flow state.

The flow class module output parameters are inlet and outlet static pressure, flow rate, and total flow.

2.1.1.4 Annular Gap Class Module Physical Model Shown in Figure 2.6

The annular gap flow calculation formula is

$$Q_{leakout} = \frac{\pi d\delta^3}{12\eta L}(p_{in} - p_{out}) \tag{2.4}$$

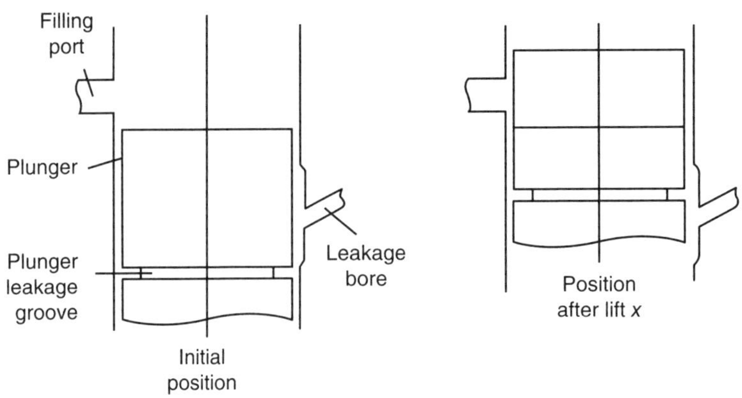

Figure 2.6 Module of the annular gap class.

where

d = piston diameter
δ = piston clearance with the mating surface
L = control piston with the mating surface of the seal length
p_{in} = pressure of the inlet end
p_{out} = outlet pressure

The annular gap module input parameters are fluid properties, the piston diameter, the initial sealing surface length, and width.

The annular gap module output parameters are flow, total flow, the actual gap width, and length.

2.1.2 The Relevant Parameters During the Simulation Calculations

2.1.2.1 Fuel Physical Parameters

The established fuel, the iMPact of fuel, calculated fuel injection process physical parameters are the speed of sound, density, elastic modulus, and viscosity. These parameters are dependent on the fuel temperature and pressure, on the density of the fuel taking the empirical formula, the speed of sound, and the elastic modulus.

Fuel sonic:

$$a = 1390 + 4.228 \times p - 2.9 \times T + 0.0051 \times T \times p \qquad (\text{m/s}) \tag{2.5}$$

Fuel density:

$$\rho = (0.843 + 0.0005916 \times p - 0.000665 \times T + 35.7 \times T \times p) \times 10^3 \;\; (\text{kg/m}^3) \tag{2.6}$$

Fuel modulus of elasticity:

$$E = a^2 \times \rho \tag{2.7}$$

Fuel viscosity:

$$\gamma = \gamma_0 \times K_D^{p/9.81\times10^4} \qquad (\text{m}^2/\text{s}) \tag{2.8}$$

where

K_d = $0.9789 + 0.26 \times 10^{-6} \times \rho_{20}$
γ_0 = atmospheric pressure fuel oil viscosity
γ_{20} = atmospheric pressure at 20 °C density of the fuel

2.1.2.2 Fuel Flow Resistance

According to the calculation of basic assumptions, in order to stabilize the flow resistance of the flow of the process instead of the unsteady flow of the flow resistance per unit mass of fluid flow resistance gives

$$f = \frac{1}{2}\frac{\lambda}{d_T}v^2 \tag{2.9}$$

Table 2.1 Fuel friction coefficient.

Re	Re < 2300 (laminar flow)	Re > 2300 (smooth pipe)	Re > 2300 (rough pipe)
λ	$\lambda = 64/\mathrm{Re}$	$\lambda = (100 \times \mathrm{Re})^{0.25}$	$\lambda = \left(1.14 + 2\ \lg \frac{d_T}{\Delta}\right)^{-2}$

where

γ = fuel friction coefficient
d_T = pipe diameter
v = tubing fuel flow rate

In the current state of various different formulas used to calculate the specific formula, see Table 2.1, where

Re = Reynolds number, $\mathrm{Re} = vd_T/\gamma$
Δ = pipe roughness
γ = fuel kinematic viscosity

2.1.2.3 Partial Loss of Fuel Flow

After the sudden expansion of a pipe, tube, or orifice there is a sudden contraction when the flow loss and partial loss of pressure can be calculated:

$$\Delta p = \xi \times \rho \times v^2/2 \tag{2.10}$$

where

ξ = local loss coefficient

2.1.2.4 Rigid Elastic Volume Expansion and Elastic Compression

In a high-pressure common rail system, the fuel pressure is very high, considered to be due to rigid elastic deformation of the fuel system. In this system, the main consideration is due to the high-pressure pump piston chamber, high-pressure fuel rail, high-pressure tubing. and injector needle elastic deformation.

On the high-pressure fuel pump plunger chamber, the high-pressure fuel rail and high-pressure tubing can be considered as a thick-walled cylindrical elastic expansion and deformation:

$$\Delta V = \Delta p \times V \times \frac{(1-2\gamma) \times 3 \times r^2}{E \times (R^2 - r^2)} \tag{2.11}$$

where

r = thick cylinder diameter
R = thick cylinder diameter
V = thick cylinder volume
Δp = thick cylinder internal pressure changes
E = elastic modulus of steel, $E = 210$ GPa
γ = Poisson's ratio of steel

The injector needle can be considered as a solid shaft needle elastic compression deformation:

$$\Delta V = V \times \Delta P \times \frac{(1 - 2\gamma) \times 3}{E}$$

2.2 Common Rail System Simulation Model

2.2.1 High-Pressure Pump Simulation Model

The typical high-pressure common rail high-pressure pump system structure shown in Figure 2.7, according to its structure, can be modeled as centralized by the cam control volume parts, the high-pressure pump oil pressure process, and the high-pressure fuel pump eccentric cam oil pressure, which is determined by the shape rate and the amount of oil pressure. The high-pressure plunger pumps are to be considered during the movement of leakage. The high-pressure pump HYDSIM simulation model is shown in Figure 2.8.

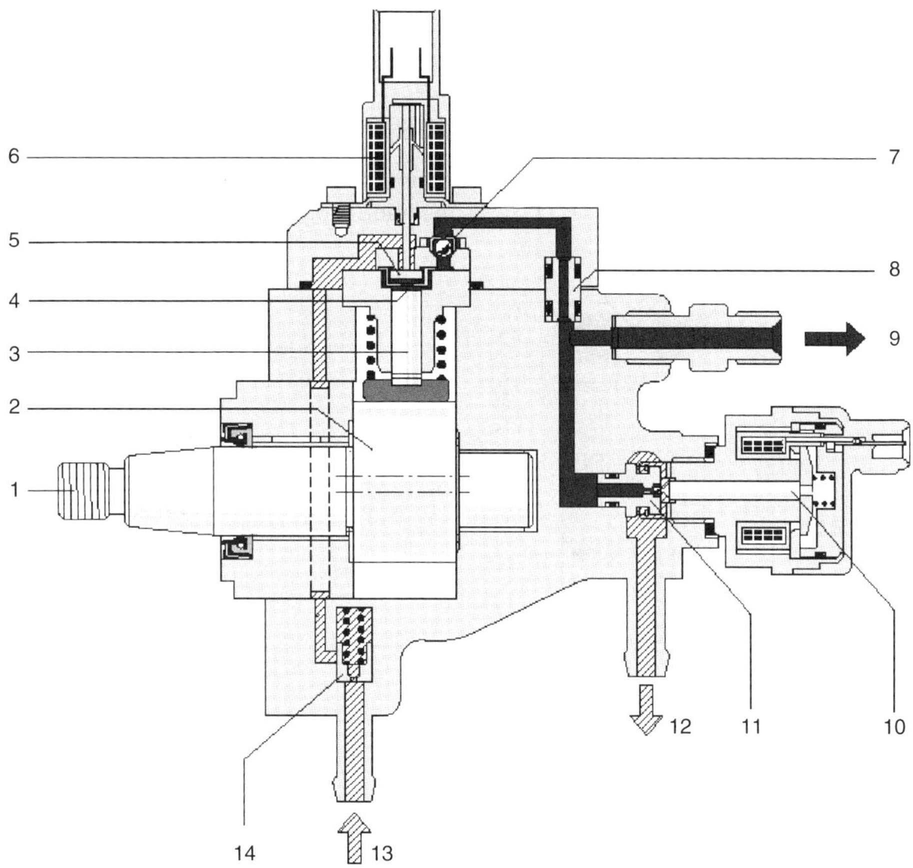

Figure 2.7 Scheme of the high-pressure pump.

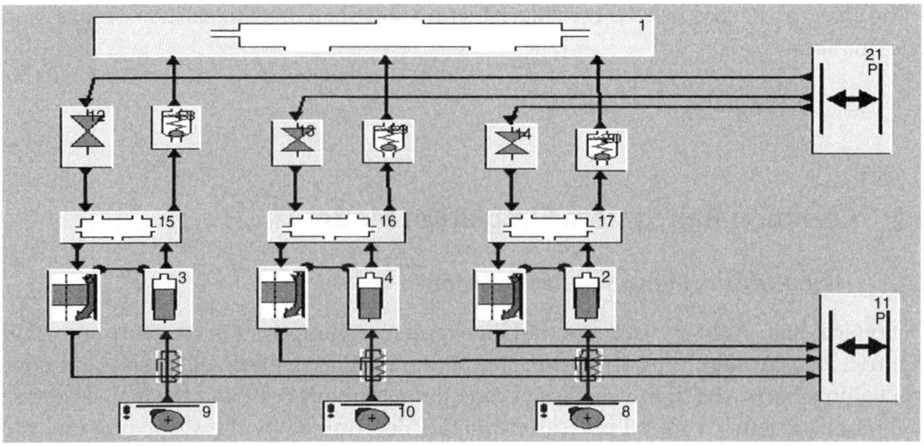

Figure 2.8 Simulation model of the high-pressure pump.

In the cam block diagram 8,9,10, 2,3,4 is shown as the plunger module, 5,6,7 as leakage modules for the high-pressure pump plunger cavity, 15,16,17, 12,13,14 for the low-pressure oil inlet valve, 18,19,20 for the delivery valve, a high-pressure oil collector, 21 is the low-pressure source, and 11 is back to the tank.

2.2.2 Injector Flow Restrictor Simulation Model

The typical flow restrictor high-pressure common rail system structure is shown in Figure 2.9 and the current limiter HYDSIM simulation program is shown in Figure 2.10.

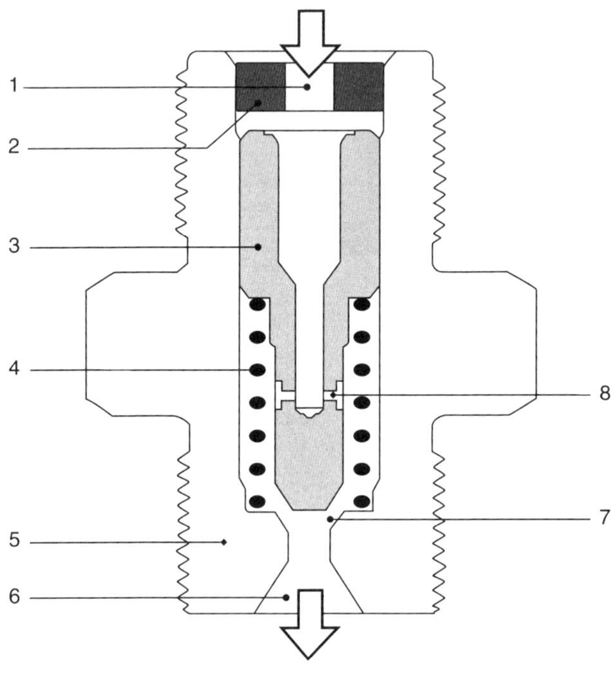

Figure 2.9 Structure of the flow limiter.

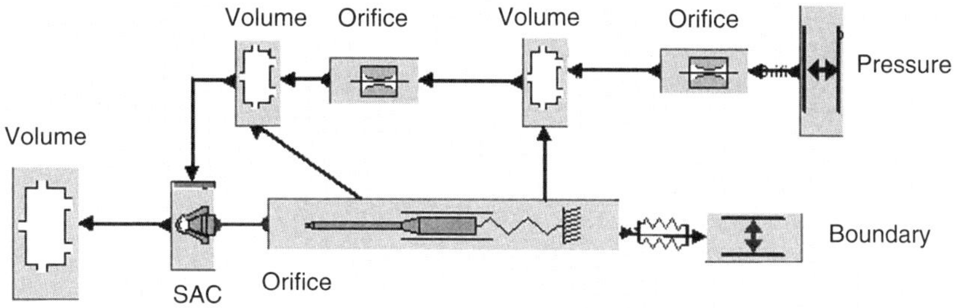

Figure 2.10 Model of the flow limiter.

In Figure 2.10, 1 is imported, the volume concentration of the front chamber 2, 3 is an orifice, the volume concentration of the rear chamber 4, 5 is a control valve, the outlet 6, 8 is the inlet injector, and volume 9 is the control valve chamber.

2.2.3 Simulation Model Electronic Fuel Injector

A typical structure of an electronically controlled fuel injector is shown in Figure 2.11. According to its structure, the electronically controlled injector HYDSIM simulation program is shown in Figure 2.12. Procedures used in the container are the class module, class module piping, valves, class module, class module boundaries, class modules, and the solenoid gap leakage modules.

2.2.4 Overall Model Common Rail System

Upon completion of a high-pressure common rail system, components are modeled on the various components' overall portfolio in order to complete the entire high-pressure common rail fuel injection system modeling and simulation. System simulation model components are shown in Figure 2.13. Using this model, the software used an HYDSIM common rail system simulation, with the system simulation results shown in Figure 2.14.

The purpose of a high-pressure common rail system simulation is to analyze the structure of the system components' injection process parameters on the system pressure and flow characteristics for the system, which are designed to provide reasonable data and reduce the test workload of the high-pressure common rail system design.

Using the established common rail system simulation model, the system pressure pumps, high-pressure tubing, flow restrictor, electronically controlled fuel injection system, and the impact of structural parameters were analyzed and the spray characteristics of the system structure parameters were discussed.

In the common rail system design, to ensure that the common rail pressure fluctuations varied minimally, usually the same number of injection times and fuel, and the fuel injection timing and the combination of time, were used to stabilize the effect of the fuel rail pressure fluctuations.

A high-pressure pump plunger chamber, a high-pressure pump delivery valve chamber, and a high-pressure fuel rail pressure are shown in Figure 2.15. It can be seen that the oil pressure pump piston, the piston chamber, the valve chamber, and the pressure within the fuel rail pressure fluctuations are due to the very large volume of high-pressure fuel rail relative to the high-pressure oil pump volume, so the high-pressure fuel rail produces the pressure fluctuations. The injector control room

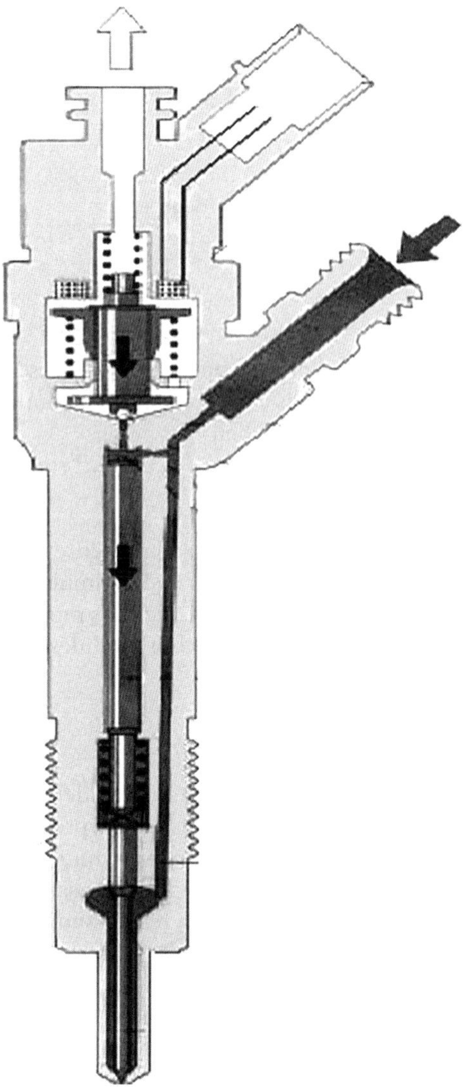

Figure 2.11 Structure of the injector.

pressure, the pressure chamber pressure accumulator pressure, and the outlet pressure variation limit are shown in Figure 2.16.

2.3 Influence Analysis of the High-Pressure Common Rail System Parameters

2.3.1 Influence Analysis of the High-Pressure Fuel Pump Structure Parameters

In the high-pressure common rail fuel injection system, because the high-pressure pump pushes the fuel into fuel rail, the structure of the high-pressure pump will directly affect the pressure in the high-pressure rail.

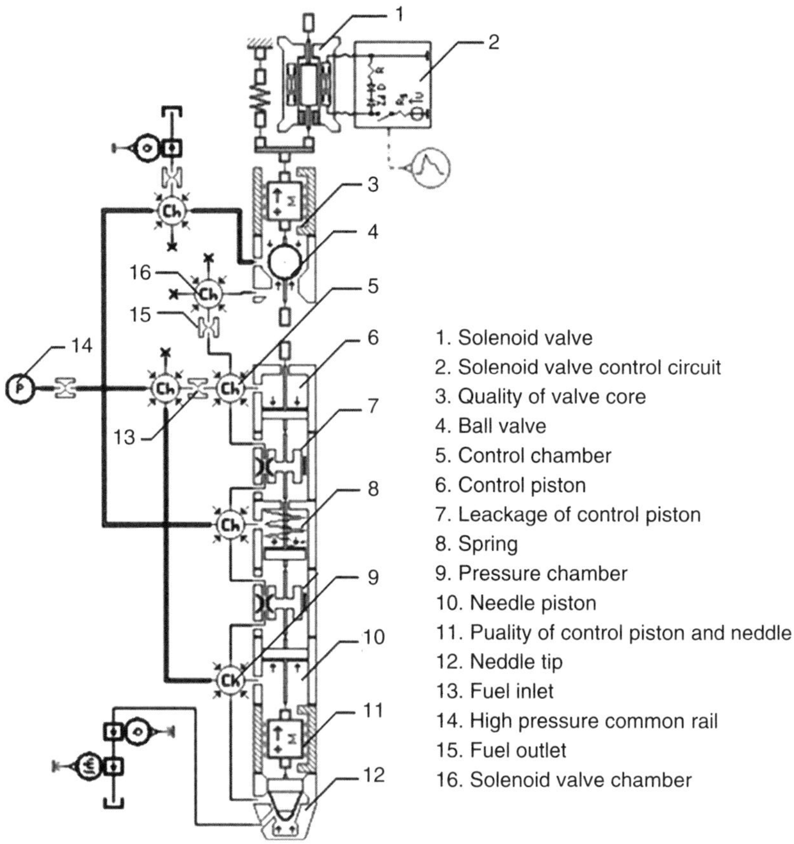

Figure 2.12 Model of the injector.

2.3.1.1 Frequency of the Fuel Supply Pump

To ensure comparability, during the study concerning the impact of the fuel supply frequency, the basic requirement for analysis is the equivalence of supplied fuel by a high-pressure pump in the working cycle. The structure and calculation results of the high-pressure pump are shown in Table 2.2. Figure 2.17 gives the impact of the fueling frequency on the pressure of the high-pressure rail. After every fueling, the fuel injector processes with depressurization in the high-pressure rail, which does not guarantee uniformity of the injection quantity. With the increase of fueling frequency in every working cycle, the range of pressure in the high-pressure rail gradually narrows. For every six times fueling takes place in a cycle, because each injection goes with the fuel supply, the quantity of fuel injection equals the quantity of the fuel supply.

2.3.1.2 Quantity of the Fuel Supply by the High-Pressure Supply Pump

Under the condition of a 7 mm diameter high-pressure pump and six times supplying fuel, the system is stimulated by a different plunger stroke. The change of pressure in the rail is shown in Figure 2.18. When the stroke is less than 8 mm, the sum of the fuel supply in the cycle of the high-pressure pump is less than that of fuel injection and controlled fuel consumption in a working cycle. Therefore, the pressure in the rail is

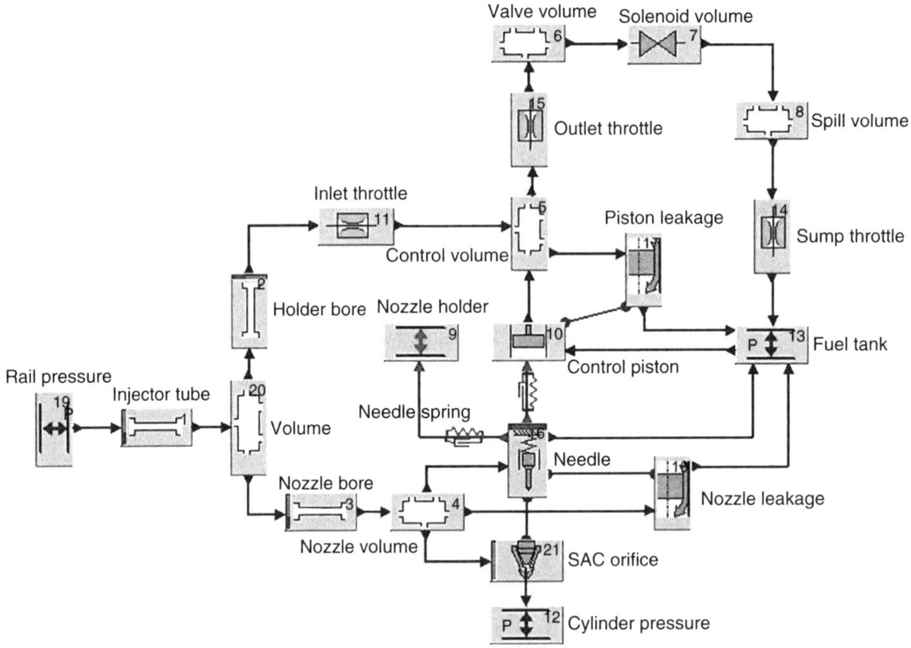

Figure 2.13 Structure of HYDSIM.

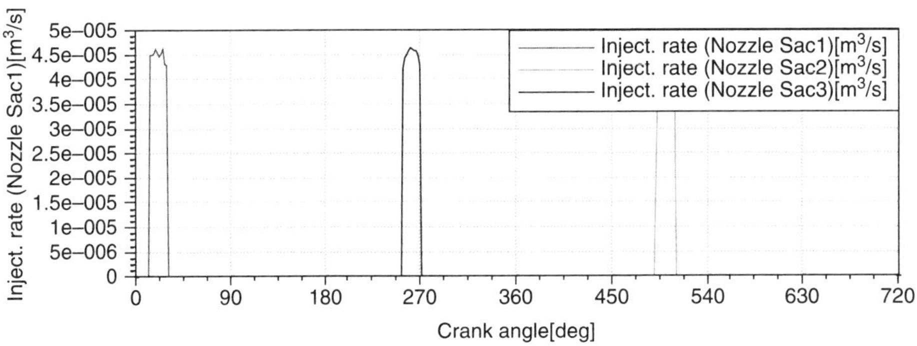

Figure 2.14 System simulation results.

unstable. With an increase in the plunger stroke, the fuel supply increases, and when the fuel in the high-pressure pump rail given from a high-pressure pump is more than the sum of injection and controlled fuel in the fuel injector, the pressure can remain stable. However, when the fuel supply increases, the fuel supplied by the high-pressure pump increases, which causes the pressure fluctuation to increase in the high-pressure rail during the fuel supply, and more fuel flows back to the low-pressure system through an overflow valve. Then more power in the high-pressure pump is consumed, and its economic efficiency is lowered. Therefore, the optimum cyclic fuel supply in a cycle of the high-pressure fuel pump can be appropriately increased and determined on the basis of guaranteeing the fuel consumption of the whole system.

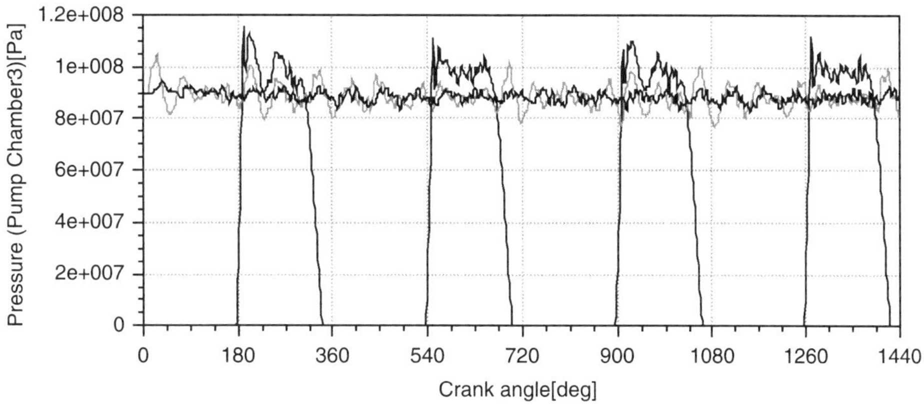

Figure 2.15 Pressure of the pump and common rail.

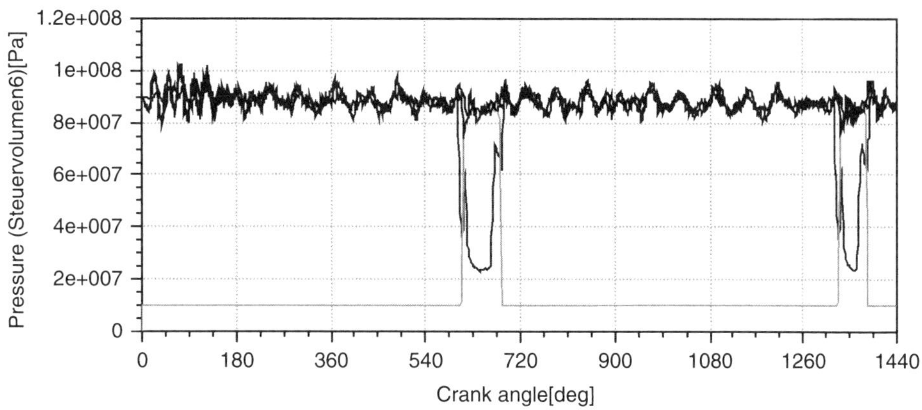

Figure 2.16 Pressure of the injector.

Table 2.2 Influence of fuel supply frequency by the high-pressure pump.

Fuel supply times	1	2	3	6
Plunger diameter (mm)	15	13	11	9
Plunger stroke (mm)	16	11	10	7.5
Amount of fuel supply (mm^3)	2826	2918	2850	2861
Maximum pressure in rail (MPa)	162.2	163.4	170.9	153.1
Minimum pressure in rail (MPa)	112.6	125.5	119.6	133.9

2.3.1.3 Diameter of the Oil Outlet Valve Hole of the High-Pressure Pump

The size of the oil outlet valve hole affects the ability of fuel to flow in the high-pressure pump. The influence analysis of the outlet valve hole of the high-pressure pump on the plunger chamber of the high-pressure fuel pump is shown in Figure 2.19. When the outlet valve is opened, the size of the outlet valve hole has a greater influence on pressure

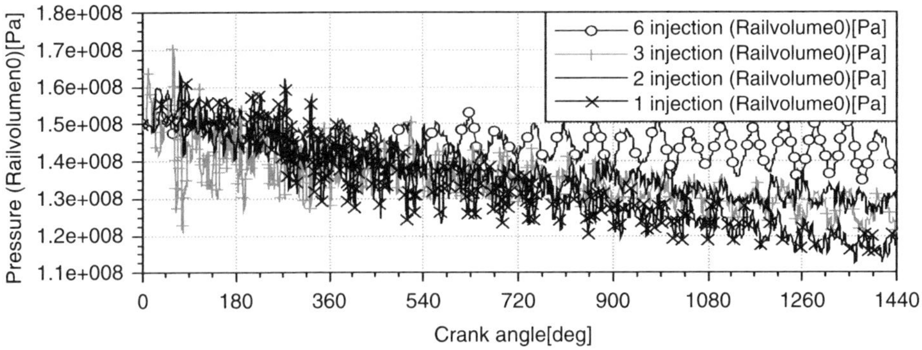

Figure 2.17 Influence of fuel supply times on the pressure of high-pressure rail.

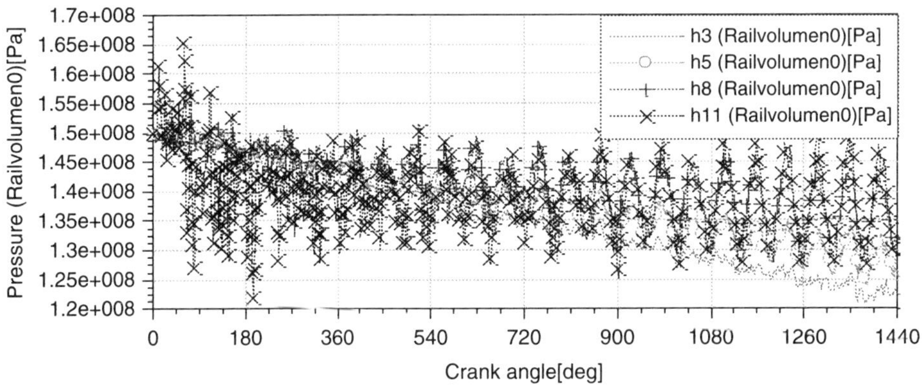

Figure 2.18 Influence of the fuel supply on the high-pressure pump to the pressure in the rail.

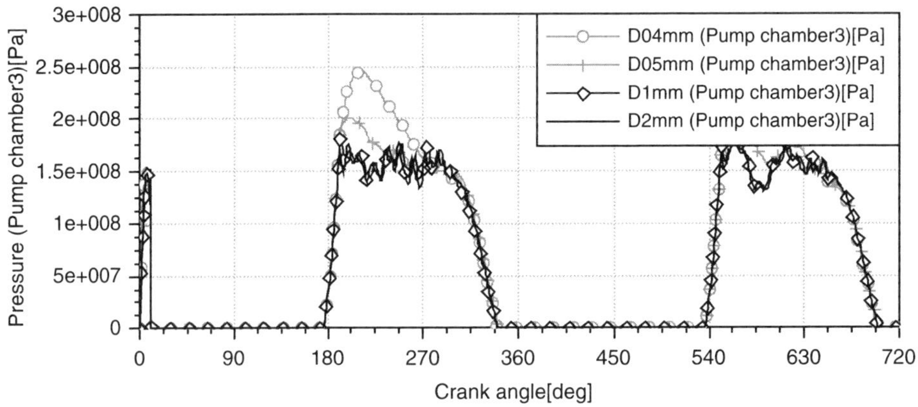

Figure 2.19 Influence of the diameter of the outlet valve diameter on the pressure of the plunger chamber in the high-pressure pump.

change in the plunger chamber than in the oil outlet valve chamber. When the diameter of the outlet valve is small, the pressure of the plunger chamber in the high-pressure pump fluctuates greatly because of the strong throttling effect. If the oil outlet valve hole diameter is very small, the fuel in the high-pressure pump plunger chamber will not be able to flow out freely, and the extremely high pressure in the plunger chamber will force the mechanical load in the high-pressure pump to rise sharply. When the diameter of the outlet valve is large, the influence of the valve hole diameter on the pressure of the plunger chamber in the high-pressure pump is small due to the weak throttling effect. The pressure of the plunger chamber in the high-pressure pump is mainly affected by the pressure wave produced when the oil outlet valve opens, and the fluctuation decreases gradually during the fuel supply process. In conclusion, when the diameter of the outlet valve is small, the pressure in the oil outlet valve chamber fluctuates greatly, but when the diameter of the outlet valve is large, the pressure fluctuation in the oil outlet valve chamber is smaller.

2.3.1.4 Influence of the Pre-tightening Force of the Oil Outlet Valve

The pre-tightening force of the outlet valve in the high-pressure fuel pump directly affects the opening pressure and the ability of the fuel to flow in the outlet valve. Under the condition of different pre-tightening forces of the oil outlet valve, the changes in the high-pressure pump plunger chamber, the oil outlet valve lift, the pressure of the oil delivery valve cavity, and the pressure of the high-pressure oil track are shown in Figures 2.20–2.22. The pre-tightening force of the delivery valve has a great influence on the lift of the high-pressure pump oil outlet valve and the high-pressure pump plunger chamber. As the pre-tightening force of the valve increases, the pressure fluctuation of the high-pressure fuel pump plunger chamber increases; the growing fluctuation mainly occurs because as the valve opening pressure increases, the valve cannot be fully opened and bounces. The pre-tightening force of a high-pressure pump hardly affects the pressure of the oil outlet valve chamber and high-pressure rail when the force is relatively small. When the pre-tightening force is relatively large, due to the instability of opening the oil outlet valve, the oil outlet valve chamber generates a small pressure fluctuation,

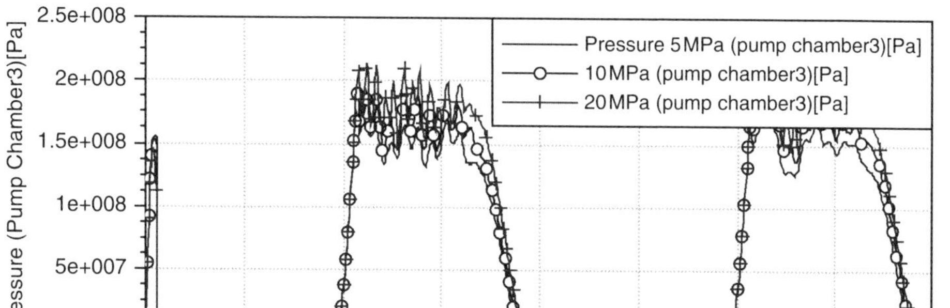

Figure 2.20 Influence of the pre-tightening force of the outlet valve on the pressure of the plunger chamber.

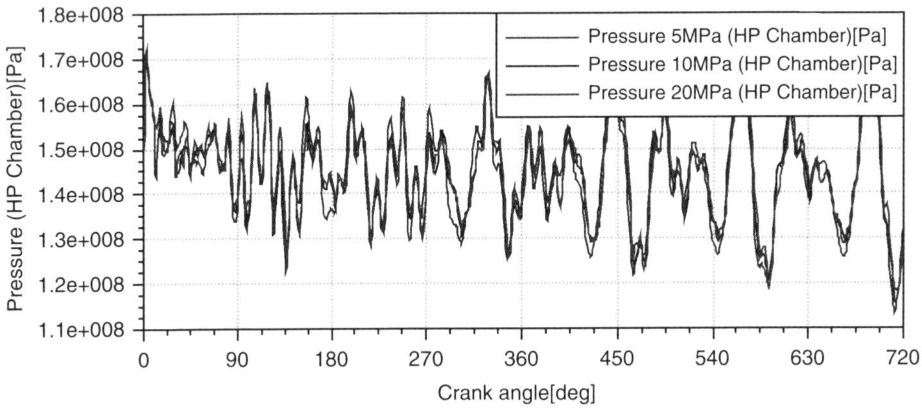

Figure 2.21 Influence of the pre-tightening force of the outlet valve on the pressure of the outlet valve chamber.

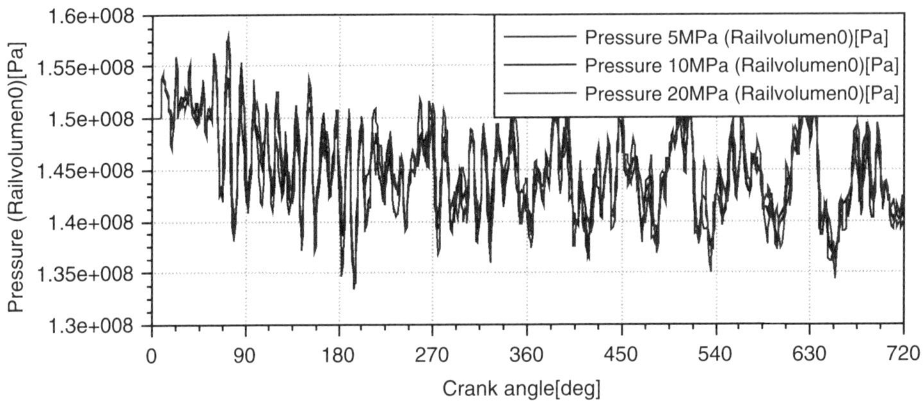

Figure 2.22 Influence of the pre-tightening force of the outlet valve on the pressure of the common rail chamber.

but the fluctuation hardly affects the pressure of the high-pressure rail. Therefore, it is possible to see the pressure stability of the high-pressure rail.

Through the above analysis, the following conclusions are given:

(1) When the number of fuel pumps is equal to the number of cylinders in the diesel engine, the fluctuation of the fuel pressure in the common rail is small.
(2) When the fuel supply quantity of the high-pressure pump is selected, based on the cyclic injection and controlled quantity of the diesel engine, the fuel supple quantity can be appropriately increased to ensure the stability of the whole system pressure.
(3) When the hole diameter of the high-pressure outlet valve is selected, a proper larger size (approximately 0.5–2 mm) can be selected to ensure less throttling loss.
(4) When the pre-tightening force of the oil outlet valve spring is determined, a smaller force can be selected to ensure the stable opening of the oil outlet valve and to reduce the pressure fluctuation in the system.

2.3.2 Analysis of the Influence of the High-Pressure Rail Volume

The high-pressure rail can be used as an accumulator to reduce the pressure fluctuation during the fuel supply of a high-pressure pump and fuel injection of an electronically controlled injector. The volume of the rail must therefore be large enough. In this section, the influence of the relatively large volume of the common rail (with a volume more than 100 times the injection volume) is investigated. When the volume of the common rail is different, the pressure fluctuation of the high-pressure rail and the change of injection are shown in Figures 2.23 and 2.24. With an increase of the rail volume, the pressure fluctuation in the high-pressure rail decreases gradually. However, a decrease of pressure fluctuation is not linearly related to the increase of the high-pressure rail volume: when the volume of the high-pressure rail increases to a certain extent, the decrease of pressure fluctuation decreases gradually. When the pressure is stable and the system runs normally in the high-pressure rail, the change in the high-pressure rail volume hardly affects the injection rate. The influence of pressure fluctuation of the high-pressure rail on the injection rate is mainly reduced by the throttle channels in the high-pressure rail and fuel injector, such as the high-pressure tube, delivery valve, and fuel passage in the

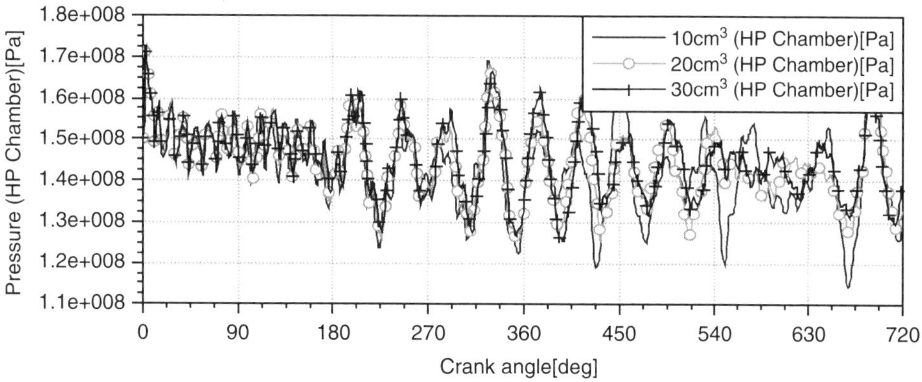

Figure 2.23 Influence of the rail chamber volume on the rail chamber pressure.

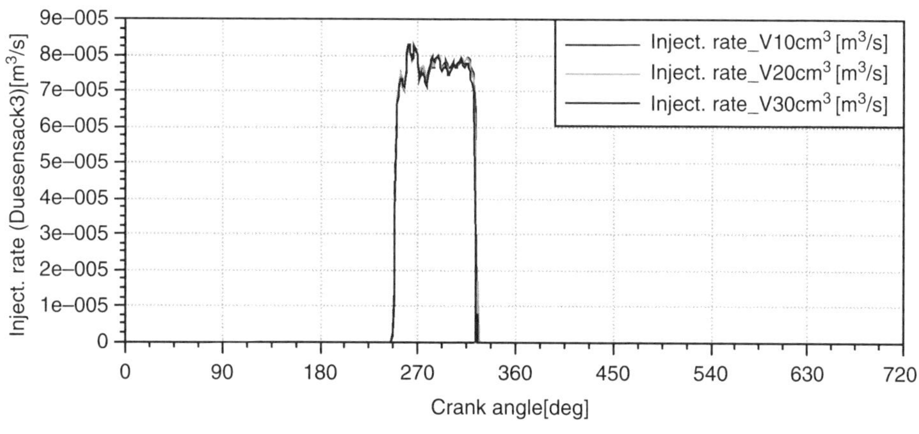

Figure 2.24 Influence of the rail chamber volume on the injection rate.

injector. In addition, the orifice in the delivery valve further reduces the influence of the pressure fluctuation of the high-pressure fuel rail on the injection rate. The increase of the high-pressure rail volume can reduce the pressure fluctuation in the high-pressure rail. However, the increase of the high-pressure rail volume causes the pressure in the high-pressure rail to be established slowly, which is not conducive to starting the diesel engine.

From the above analysis, the following conclusion can be made. When the volume of the high-pressure rail is selected, the quantity of cyclic injections in the diesel engine and the quantity of cyclic fuel supply in the high-pressure pump should be considered comprehensively. The volume should not only ensure a relatively small pressure fluctuation to keep the system stable during fuel supply and injection in the high-pressure rail but should also ensure a quick production of fuel pressure in the rail to improve the startup and variable working conditions of the diesel engine.

2.3.3 Influence of the Injector Structure Parameters

An electronically controlled injector is the most complicated part of a high-pressure common rail system. The structure parameters and performance of the electronically controlled injector directly affect the performance of the whole high-pressure common rail system. Therefore, the electronically controlled injector should be studied in detail.

2.3.3.1 Control Orifice Diameter

The start and stop of fuel injection in the electronically controlled injector is controlled by the fuel pressure in the control chamber, and the size of the orifice directly influences the change of the fuel pressure in the pressure chamber. Therefore, the size of the orifice directly affects the opening and closing of the needle valve of the electronically controlled fuel injector, thus affecting the injection rate.

Under the circumstance where the size of the inlet control orifice remains the same, the influence of the size of the outlet control orifice on the control chamber pressure and the needle valve lift in the electronically controlled injector is as shown in Figures 2.25 and 2.26. When the diameter of the outlet control orifice is relatively small,

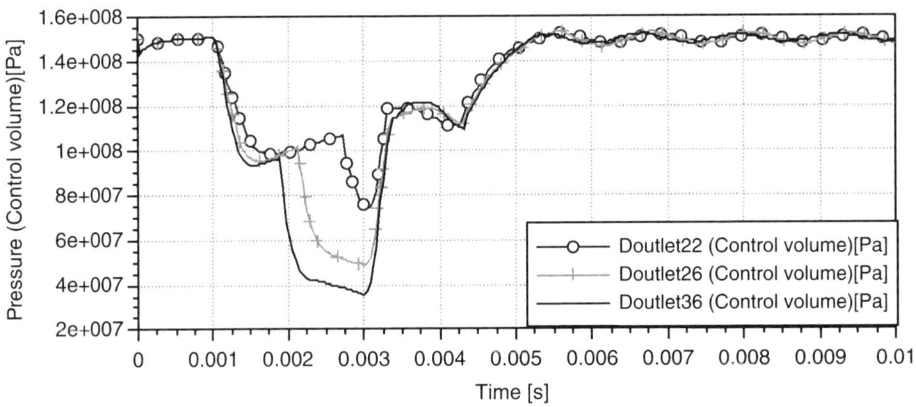

Figure 2.25 Influence of the oil outlet orifice diameter on the pressure in the control chamber.

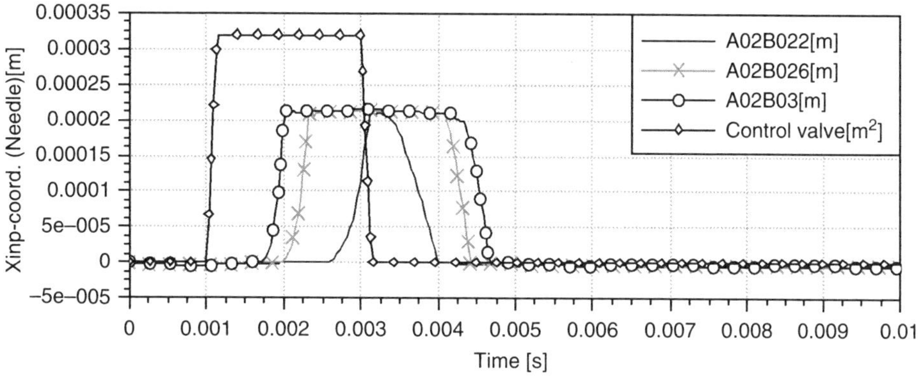

Figure 2.26 Influence of the oil outlet orifice diameter on the lift of the needle valve.

at the moment when solenoid valve opens, the fuel in the control chamber cannot flow out quickly, and the pressure in the control chamber is not effectively reduced, which prevents the needle valve in the injector from opening or not fully opening. With an increase of the diameter of the oil outlet control orifice, at the moment the solenoid valve opens, the pressure in the control chamber decreases rapidly, thus enabling the needle valve to fully open. With the increase of the diameter of the oil outlet control orifice, the speed of pressure drop in the injector control chamber increases, which causes the needle valve to open earlier during the opening process. With the increase of the diameter of the oil outlet control orifice, the pressure in the injector control chamber is greatly reduced. After the solenoid valve is closed, the pressure in the control chamber rises slowly, so that closing the needle valve is delayed during the closing process.

Under the circumstance where the size of the outlet control orifice remains the same, the influence of the size of the inlet control orifice on the control chamber pressure and the needle valve lift in the electronically controlled injector is as shown in Figures 2.27 and 2.28. When the diameter of the inlet control orifice is relatively small, at the moment when the solenoid valve opens, because the pressure in the control chamber is unloaded

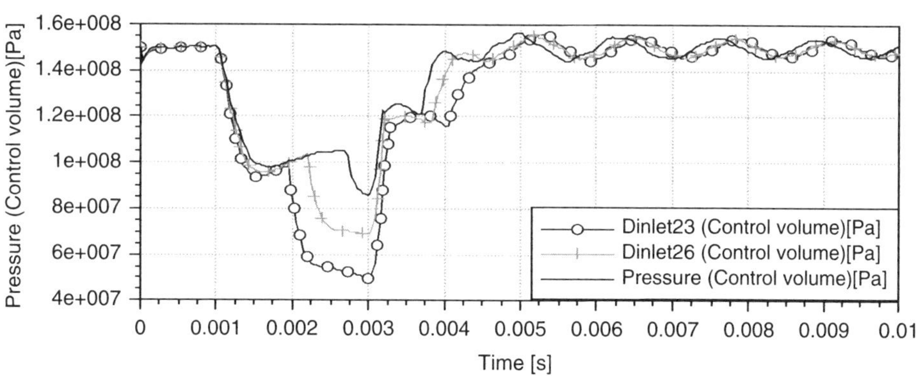

Figure 2.27 Influence of the oil inlet orifice diameter on the control chamber pressure in the control chamber.

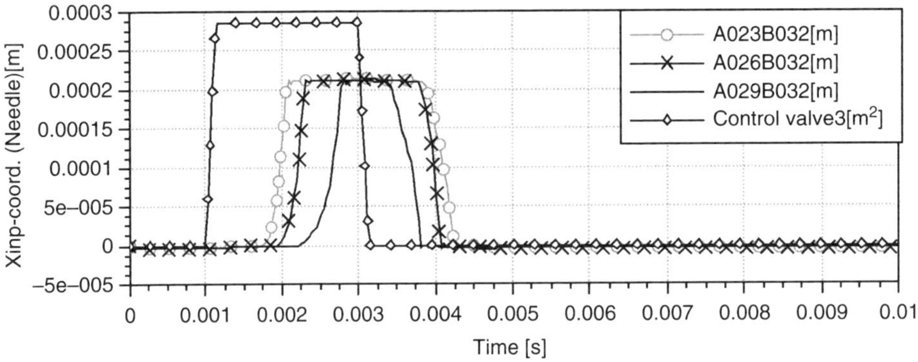

Figure 2.28 Influence of the oil inlet orifice diameter of the control chamber on the needle lift.

rapidly, the needle valve opens quickly. When the solenoid valve is closed, the valve closes slowly because the pressure in the control chamber increases slowly. In addition, when the needle valve remains seated, because of the pressure fluctuation in the needle valve chamber, the needle valve may be raised again, resulting in a secondary injection, which is not conducive to the combustion of diesel. With the increase in the diameter of the oil inlet control orifice, when the solenoid valve opens, the speed of the pressure drop in the control chamber decreases, and the opening speed of the needle valve is reduced. However, when the solenoid valve is closed, the pressure is established rapidly in the control chamber, the needle valve is quickly seated, and no secondary injection is produced. When the diameter of the oil inlet control orifice is large, to a certain extent the solenoid valve is opened, the pressure of the control chamber cannot be effectively reduced, and the needle valve opening process slows down until it cannot be opened at all.

2.3.3.2 Influence of the Control Chamber Volume

The volume of the control chamber has a great influence on the establishment of the control pressure in the control chamber. The influence of the volume of different control chambers on pressure and needle lift in the control chamber is shown in Figures 2.29 and 2.30. When the volume of the control chamber is relatively small, when the solenoid valve is opened, the pressure in the control chamber drops rapidly and the needle valve is lifted; when the solenoid valve is closed, the pressure in the control chamber rises rapidly and the needle valve closes quickly. When the volume of the control chamber is relatively large, when the solenoid valve is opened the pressure in the control chamber drops relatively slowly and the needle valve opens relatively slowly; when the solenoid valve is closed, the pressure established in the control chamber is relatively slow.

2.3.3.3 Influence of the Control Piston Assembly on the Fuel Injector Response Characteristics

In order to study the influence of the control piston parameters, four different piston schemes have been designed, shown in Table 2.3. The calculation result of the response characteristics of the injector to different control piston assembly parameters is shown in Figure 2.31.

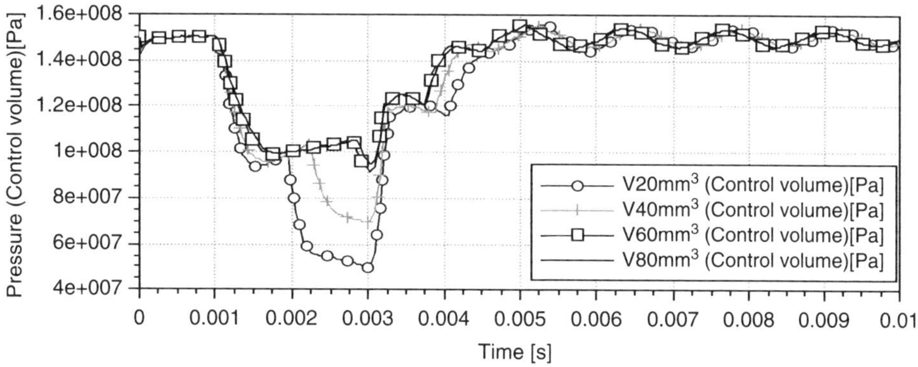

Figure 2.29 Influence of the volume of the control chamber on the pressure in the control chamber.

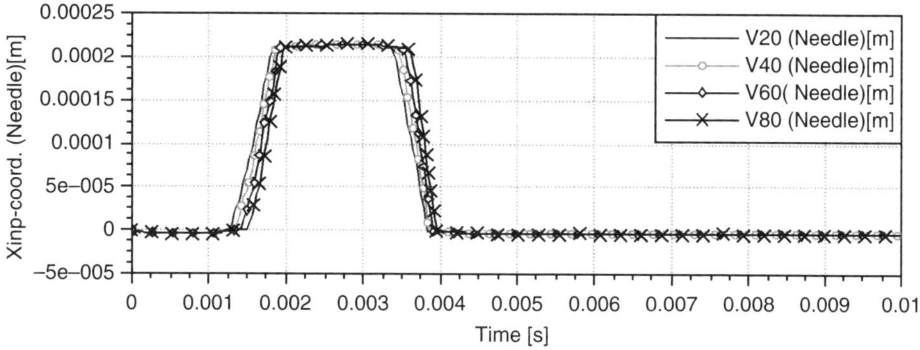

Figure 2.30 Influence of the volume of the control chamber on the lift of the needle valve.

Table 2.3 Control piston scheme.

	D piston	D piston out	Needle guide	Needle seat
A	7	5.5	5	3
B	6	5	4.5	2.8
C	5	4.5	4	2
D	4	4	3.5	1.5

The following are several groups of parameters used for the calculation with units in mm. From the Figure 2.31, reducing the diameter of the control piston can reduce the opening response time of the needle valve, but will extend its closing response time. Considering the matching of the tappet and needle valve, the requirement of strength, and the results of the test, the first group is given priority when selecting the size of the tappet and needle valve.

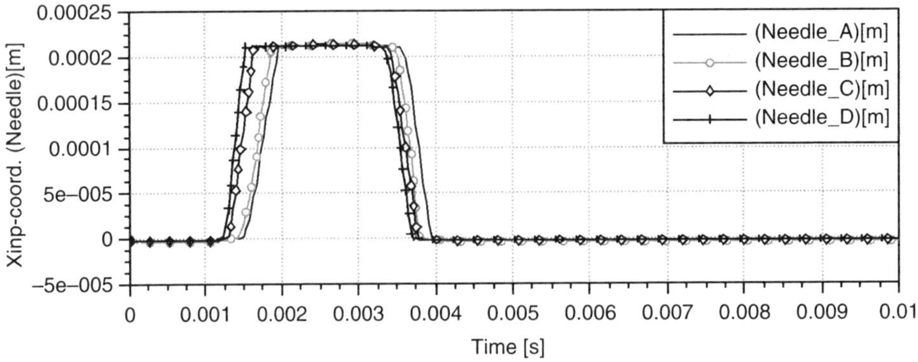

Figure 2.31 The influence of different control piston assembly parameters on the response characteristics of the injector.

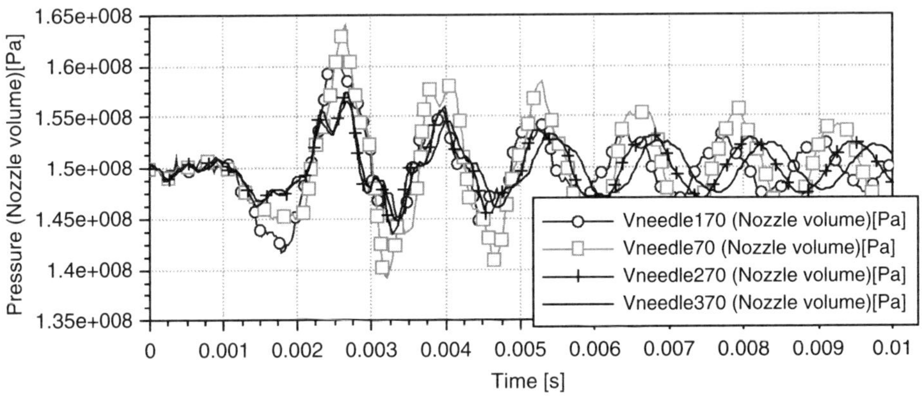

Figure 2.32 Influence of the volume on the pressure of the needle valve chamber.

2.3.3.4 Influence of the Needle Valve Chamber Volume

The influence of the needle valve chamber volume on the pressure and injection rate of the needle valve chamber is shown in Figures 2.32 and 2.33. When the needle chamber volume is relatively small, the pressure in the needle valve chamber fluctuates greatly because of the pressure waves generated by the lift of the needle valve. When the needle valve chamber volume is relatively large, the pressure waves generated by the lift of the needle valve decay rapidly. However, during the fuel injection process, the influence of the needle valve chamber volume on the pressure of the needle valve chamber is not obvious, and the volume hardly affects the injection rate.

2.3.3.5 Influence of the Pressure Chamber Volume

The influence of different pressure chamber volumes on the pressure of the pressure chamber and injection rate is shown in Figure 2.34. It is not difficult to see that the pressure chamber volume hardly affects the injection rate, which is mainly because the volume of the pressure chamber is relatively small and the injection rate is mostly affected by the pressure in the needle valve chamber and the lift of the needle valve. However, the increase in the pressure chamber volume causes the fuel in the pressure chamber to inflate as a result of the high temperature in the combustion chamber after

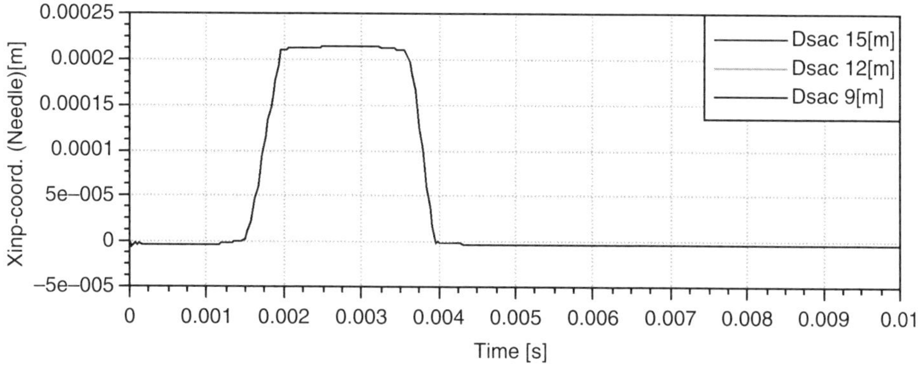

Figure 2.33 Influence of the volume on the lift of the needle valve chamber.

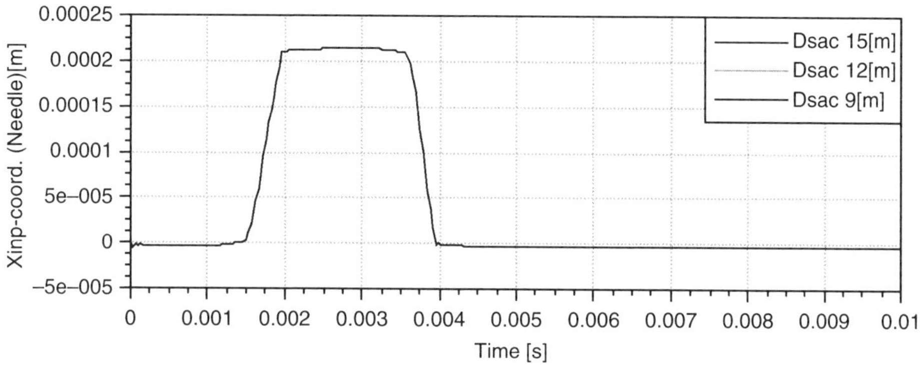

Figure 2.34 Influence of the volume of the nozzle pressure chamber on the lift of the needle valve.

the needle valve is closed. Then the fuel in the pressure chamber enters the combustion chamber again. As this part of the fuel enters the combustion chamber at the end of the combustion period and is poorly atomized, the economic performance and emission performance of the diesel engine become worse. Therefore, as far as possible, the volume of the injector pressure chamber should be minimized.

2.3.3.6 Influence of the Nozzle Orifice Diameter on the Response Characteristics of the Injector

The injector orifice is the passage of fuel injected into the combustion chamber. When the fuel injection pressure and injection duration are constant, the size of the orifice cross-section directly determines the size of the fuel injection. The section area of the orifice should be selected according to the selected high-pressure rail and a maximum injection amount of diesel after the theoretical calculation results, when the shape of the diesel engine combustion chamber and the test results are comprehensively determined.

The nozzle orifice diameters are set to 0.25, 0.2, and 0.15 mm, respectively, and the calculation result of the response characteristics of the injector is shown in Figure 2.35. As shown by the diagram, reducing the diameter of the nozzle orifice shortens the opening response time of the needle valve and prolongs the closing response time of the needle valve. The change of orifice diameter has little influence on the response of the needle valve.

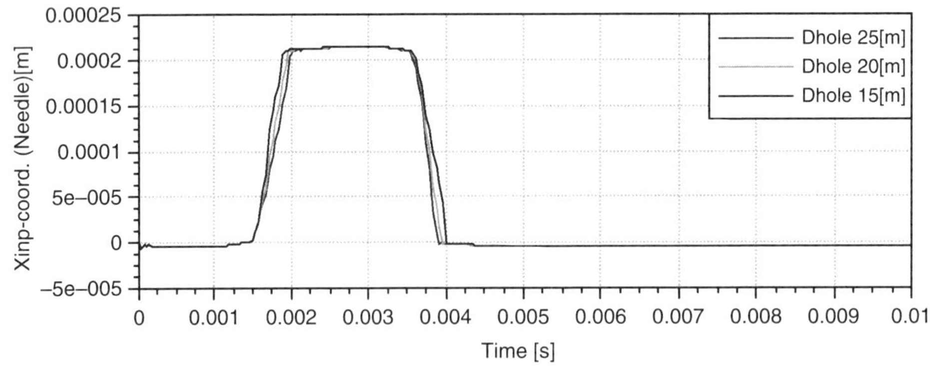

Figure 2.35 Influence of the nozzle orifice diameter on the lift of the needle valve.

Through the above analysis, the following conclusions are given:

(1) The size of the control orifice determines the control pressure setting process of the control chamber and determines the movement regular pattern of the injector needle valve. Therefore, the diameter of two control orifices of the electronic injector should be selected strictly to ensure the rapid opening of the needle valve and to ensure that there is no secondary injection.
(2) The volume of the control chamber affects the setting up of pressure in the control chamber. The volume of the control chamber should be reduced as much as possible to ensure the response speed of the electronic fuel injector.
(3) The size of the control plunger determines the size of the control force at the upper part of the control piston and affects the response speed of the needle valve. Therefore, under the premise of ensuring the opening speed of the needle valve and no secondary injection, when the diameter of the control plunger is determined, the diameter of the needle valve should be considered.
(4) The volume of the needle valve chamber determines the pressure change of the needle valve chamber, and thereby determines the injection pressure and the force acting on the lower part of the needle valve. Therefore, the volume of the needle valve chamber should be selected to be as large as possible to reduce the pressure fluctuation in the needle valve chamber.
(5) The volume of the pressure chamber only slightly affects the injection rate. However, in order to improve the emission performance of the diesel engine, the volume of the pressure chamber nozzle should be selected to be as small as possible.
(6) The cross-sectional area of the nozzle orifice directly affects the amount of cycle fuel injection. Therefore, the area of orifice should be selected based on the pressure of the high-pressure rail and the requirement of the maximum amount of diesel engine injection, combined with the theoretical calculation results, the shape of the diesel engine combustion chamber, and test results.

2.3.4 Influence of the Flow Limiter

2.3.4.1 Influence of the Plunger Diameter

The change in diameter of the flow limiter plunger has a slight influence on the injector needle valve chamber pressure and injection rate. It mainly affects the force on the

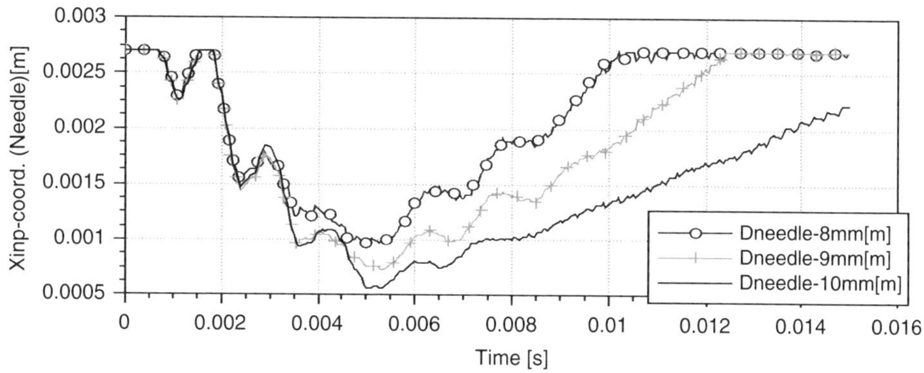

Figure 2.36 Displacement curve of the spool with different diameters and time limits.

plunger, thus affecting its movement. Figure 2.36 shows the influence of the plunger diameter on the stroke of the plunger. With the increase in diameter of the plunger, the force acting on the plunger increases, thus increasing the stroke of the plunger, and when the injector fails, the limited flow of the limiter is correspondingly reduced.

2.3.4.2 Influence of the Flow Limiter Orifice Diameter

The size of the flow limiter orifice affects the throttling effect. The influence of the orifice diameter on the stroke of the plunger is shown in Figure 2.37. When the diameter of the flow limiter orifice is small, the throttling effect is obvious and the pressure difference between the two ends of the plunger is relatively large, which causes the stroke of the plunger to become relatively large. When the diameter of the flow limiter orifice is large, the throttling effect is not obvious and the high-pressure rail is greatly affected by the pressure fluctuation at the injector ends, which causes the plunger stroke to beat relatively quickly. When the diameter of the flow limiter orifice is appropriate, the effect of reducing the pressure fluctuation of the flow limiter is obvious and the plunger pulsation is palpably smaller.

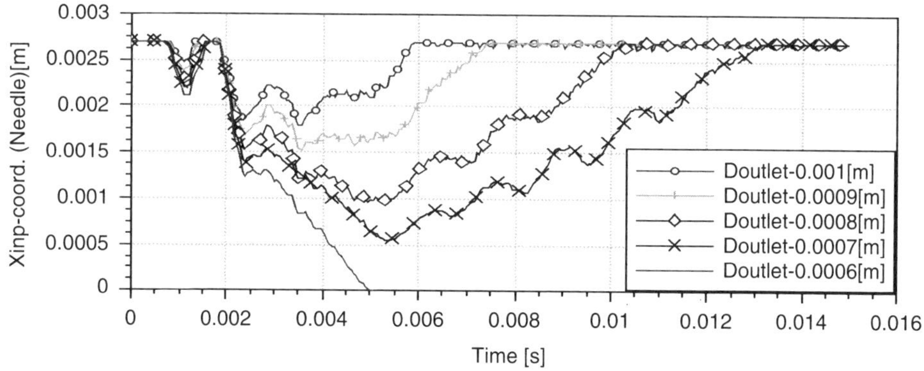

Figure 2.37 Influence of the orifice diameter on the displacement of the flow limiter spool.

2.3.5 Common Rail System Design Principle

Through the above analysis, the following design principles in the common rail system design are given:

(1) The circulation times of the high-pressure fuel pump should be consistent with the number of diesel engine cycle injection times to reduce the pressure fluctuation in the high-pressure rail and to ensure uniformity of the injection quantity in each cylinder. Based on ensuring the cyclic injection quantity of the diesel engine and cyclic control fuel quantity, the fuel supply quantity of the high-pressure pump should be selected appropriately larger in order to ensure the stability of the whole system. An excessive oil supply margin will lead to an increase in pump power consumption and reduce the economic performance of diesel engines. Therefore, the diameter of the oil outlet valve orifice should be selected to be appropriately larger to ensure a smaller throttling loss. The force of the valve spring should be selected appropriately smaller to ensure the stable opening of the oil outlet valve and to reduce pressure fluctuation in the system.
(2) When the volume of the high-pressure rail is selected, the cyclic fuel injection quantity of the diesel engine and the cyclic fuel supply of the high-pressure pump should be taken into consideration. When the volume is selected, it is necessary to ensure that the pressure fluctuation in the high-pressure pump is small during fuel supply and injection to keep the system stable, and that the pressure in the rail can be established quickly to improve the diesel engine startup.
(3) The electronically controlled injector is the most important and complicated component in the high-pressure common rail system, because its structure has a decisive influence on the fuel injection process. When the control orifice is selected, there should be rapid opening of the needle valve but no secondary injection, because the selection directly determines the establishment of the control pressure in the control room and the motion pattern of the injector needle valve. Because the volume of the control chamber affects the establishment of the pressure in the control chamber, the volume of the control chamber should be reduced as much as possible to control the response speed of the electronically controlled injector. When the diameter of the control plunger is selected, the response speed of the needle valve should be controlled with no secondary injection. The volume of the needle valve chamber should be enlarged as much as possible to reduce pressure fluctuations in the needle valve chamber. The volume of the control chamber should be minimized to reduce the emissions of the diesel engine. The cross-section area of the nozzle should be selected according to the pressure of the high-pressure rail and a maximum quantity of fuel circulation in the diesel engine.
(4) The high-pressure rail should be placed near each cylinder to shorten the high-pressure tubing and reduce the flow resistance of the fuel. In addition, high-pressure tubing of a uniform length and a larger inner diameter should be selected to reduce flow resistance of the fuel.
(5) The diameter of the flow limiter plunger and the diameter of the flow limiter orifice should be comprehensively selected according to the limiting flow of the limiter and the definite fluctuation limit of the high-pressure rail.

3

Electronically Controlled Injector Design Technologies

3.1 Electric Control Fuel Injector Control Solenoid Valve Design Technology

As one of the core components of an electric control fuel injector, the solenoid valve has a significant impact on the performance of the injector. An important way of improving the performance injector involves optimizing the design of the solenoid valve.

3.1.1 Solenoid Valve 33 Mathematical Analysis Model

The high-speed solenoid valve for an electric control fuel injector has both the characteristics of the general electromagnetic actuator and its own, which is particularly due to its association with the hydraulic system of the injector. According to intrinsic features of the high-speed solenoid valve, it can be divided for further study into a circuit subsystem, a circuit subsystem, a mechanical subsystem, an hydraulic subsystem, and a thermodynamic subsystem. Interconnected subsystems are shown in Figure 3.1.

For every subsystem, a circuit subsystem should follow the voltage balance equation and a magnetic circuit subsystem should follow the Maxwell equation. In the research of a mechanical motion and hydraulic system, researchers should follow the Darren Bell motion equation and the Bernoulli equation, which follow the heat balance equation during the research of a thermal system. Researchers should fully consider the interrelations of subsystems. As Figure 3.1 shows, the circuit subsystem determines the change of magnetic circuit subsystem by $\mathrm{d}\phi/\mathrm{d}t$, while in turn the change of the magnetic circuit subsystem influences the current change of the circuit subsystem. The electromagnetic force of the magnetic circuit subsystem F_{mag} and hydraulic pressure of the hydraulic subsystem F_h affect the mechanical subsystem, while in turn the hydraulic subsystem and the magnetic circuit subsystem are also related to the displacement x, velocity $\mathrm{d}x/\mathrm{d}t$, and acceleration $\mathrm{d}^2x/\mathrm{d}^2t$ of the solenoid valve armature. The resistance of the solenoid valve coil $R(\theta)$ is a function of the coil temperature θ, while in turn the changes of current in the circuit system affect the coil temperature. Therefore, these subsystems are interrelated and influence each other.

3.1.1.1 Circuit Subsystem

According to the requirement of the electric control fuel injector for a solenoid valve, many kinds of circuit drives have been developed both here and abroad. However, the essence of the main circuit is basically the same, and applies different voltages to the

Common Rail Fuel Injection Technology in Diesel Engines, First Edition.
Guangyao Ouyang, Shijie An, Zhenming Liu and Yuxue Li.

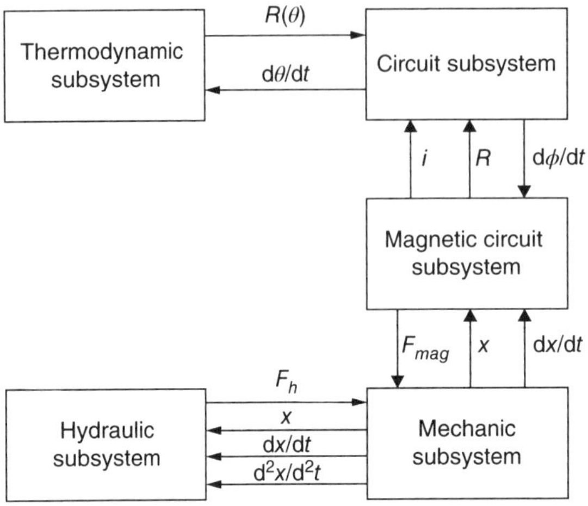

Figure 3.1 Diagram of interconnected subsystems.

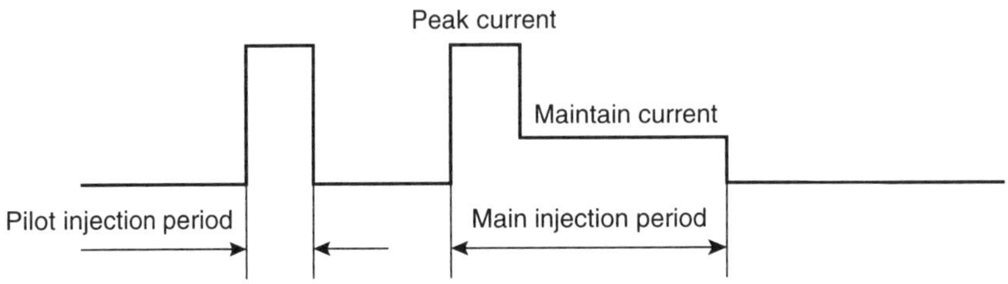

Figure 3.2 Ideal control current of the waveform of a solenoid.

solenoid valve when the solenoid is in the opening stage and is maintaining the opening stage. The circuit drive should provide different amps of current to the solenoid valve in different stages. Figure 3.2 shows the ideal exciting current waveform of the solenoid valve. The period of peak current refers to the situation that once the injection control pulse is released, it is necessary to inject the peak current to the solenoid valve coil immediately so that the solenoid valve can be turned on quickly. The period of maintaining the current refers to the situation that when the solenoid valve is turned on, the magnetic resistance decreases because the air gap between the electromagnet and the armature decreases; maintaining the current at a low level can not only keep the solenoid valve turning on reliably but can also reduce the energy consumption of the solenoid valve and the coil temperature; furthermore, it is beneficial for the ability to turn off the solenoid valve rapidly. The period of current stop refers to the situation where, at the end of the fuel injection, the current drive needs to be quickly cut off to reduce the turning off time of the solenoid valve.

Figure 3.3 is the solenoid drive circuit schematic diagram; the working process is that the direct-current power supply charges the capacitor through the current limiting resistor when the SCR (selective catalytic reducer) is not triggered. When the SCR receives the trigger signal, the capacitor will discharge to the solenoid valve coil in a very short time. The characteristic of this discharge is related to the inductance, the resistance, and the capacitance of the solenoid valve. In the early stage of the discharge, the

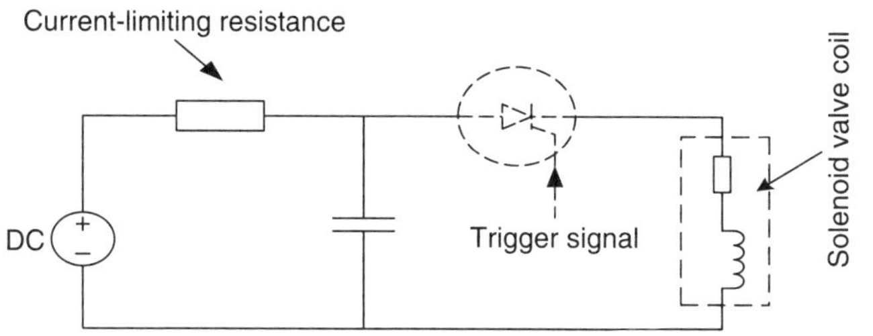

Figure 3.3 Circuit drive.

Figure 3.4 Capacitor discharge pulse source.

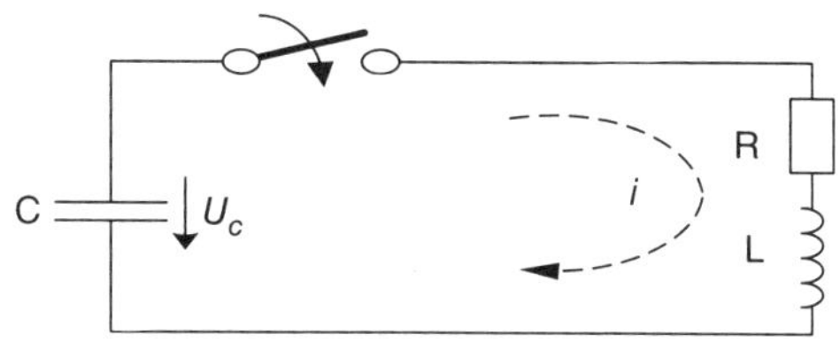

current limiting resistor does not work; the work can be simplified by using a capacitor discharge type pulse source circuit, as shown in Figure 3.4.

Figure 3.4 is an RLC (R stands for electric resistance, L for Lenz, and C for electric capacity) circuit zero input response. According to J.F. Herve's law, the discharge process can be expressed as

$$L\frac{di}{dt} + Ri + U_c = 0 \tag{3.1}$$

Substituting $i = C\frac{dU_c}{dt}$ to the equation above, we get

$$LC\frac{d^2U_c}{dt^2} + RC\frac{dU_c}{dt} + U_c = 0 \tag{3.2}$$

In the equations above:

U_c = capacitor discharge voltage, $U_c(0) = U_0$
i = coil current (A)
C = capacitance (F)
L = inductance (H)
R = resistance (Ω)

This is the two-order constant coefficient differential equation. Setting the dropping voltage on capacitance as $U_c(t)$ and using initial condition $i(0) = 0, U_c(0) = U_0$, we can obtain three solutions under different conditions.

Let $\alpha = (R/2)\sqrt{C/L}$. When the winding of the solenoid value coil is fixed, the values of R and L are determined and the value of α depends on the capacitor.

(1) When $\alpha < 1$, underdamping

$$i(t) = \frac{U_0}{\sqrt{1-\alpha^2}}\sqrt{\frac{C}{L}}\exp[-\alpha(t/\sqrt{LC})]\sin(\sqrt{1-\alpha^2}[t/\sqrt{LC}]) \tag{3.3}$$

At this point, $i(t)$ is the amplitude of the periodic oscillation current and is decreased by an exponential curve; the first peak value I_m and the time of oscillation t_m are, respectively:

$$I_m = U_0\sqrt{\frac{C}{L}}\exp\left[\frac{-\alpha}{(1-\alpha^2)^{1/2}}\tan^{-1}(\sqrt{1-\alpha^2}/\alpha)\right] \tag{3.4}$$

$$t_m = \sqrt{LC}(1-\alpha^2)^{-1/2}\tan^{-1}(\sqrt{1-\alpha^2}/\alpha) \tag{3.5}$$

(2) When $\alpha = 1$, critical damping

$$i(t) = \frac{U_0}{L}te^{(-t/\sqrt{LC})} \tag{3.6}$$

This is a non-periodic pulse current.

(3) When $\alpha > 1$, overdamping

$$i(t) = \frac{U_0}{\sqrt{\alpha^2-1}}\sqrt{\frac{C}{L}}e^{-\alpha(t/\sqrt{LC})}sh\left(\sqrt{\alpha^2-1}\frac{t}{\sqrt{LC}}\right) \tag{3.7}$$

This is also a non-periodic pulse current.

3.1.1.2 Magnetic Circuit Subsystem

According to the structure of a solenoid valve and Ohm's law of a magnetic circuit, the magnetic circuit equations of a solenoid valve can be obtained:

$$iN = R_m\phi \tag{3.8}$$

$$B = \frac{\phi}{S} \tag{3.9}$$

In the equations above:

i = coil current (A)
N = number of windings
ϕ = magnetic flux (Wb)
R_m = total magnetoresistance in a magnetic circuit (H^{-1})
B = magnetic induction (T)
S = cross-section of a magnetic flux (iron core, air gap, armature)

$$R_m = R_{mc} + R_{ma} + R_{mg} = \frac{l_cH_c(\phi_c)}{\phi_c} + \frac{l_a}{\mu_0S_a} + \frac{l_gH_g(\phi_g)}{\phi_g} \tag{3.10}$$

In the equation above:

R_{mc} = reluctance in iron core (H^{-1})
R_{ma} = reluctance in air gap (H^{-1})
R_{mg} = reluctance in armature (H^{-1})
$H_c(\phi_c), H_g(\phi_g)$ = iron core and the magnetic field strength (A/m) of the armature is determined by the magnetization curve B-H of the core and armature material
ϕ_c, ϕ_g = magnetic flux in the iron core and armature (Wb)
l_c, l_a, l_g = core magnetic circuit length, air gap magnetic circuit length, armature magnetic circuit length (m)
S_c, S_a, S_g = sectional area of the iron core, the air gap sectional area, and the sectional area of the armature (m^2)
μ_0 = air permeability (H^{-1}), $\mu_0 = 4\pi \times 10^{-7}$

When the working air gap is large, it is normal that $R_{ma} >> (R_{mc} + R_{mg})$; that is, the magnetoresistance in the magnetic circuit is mainly shown in the air gap.

Therefore, after meeting the work requirement, designing a short stroke solenoid valve is an effective way to improve the response speed. At the same time, in the calculation of magnetic resistance, R_{ma} is the only thing we need to consider. Although the core and armature are made of high permeability materials, $(l_c + l_g) >> l_a$. Therefore, the reluctance of the iron core R_{mc} and the reluctance of the armature R_{mg} must be considered.

When the coil of the solenoid valve is excited by something outside, the voltage applied U_c on the coil should equal the voltage induced on the coil resistance and flux linkage change $N\phi$. That is:

$$U_c = Ri + N\frac{d\varphi}{dt} \tag{3.11}$$

The energy balance equation for it is

$$U_c i\,dt = Ri^2\,dt + Ni\,d\varphi \tag{3.12}$$

An integral in the static period T_{01} from the current is energized as the armature begins to move:

$$\int_0^{To1} U_c i\,dt = \int_0^{To1} Ri^2\,dt + \int_0^{To1} Ni\,d\varphi \tag{3.13}$$

Suppose that the flux remains unchanged when the armature moves; the mechanical work required by the armature is completely converted from stored magnetic energy $\int_0^{To1} Ni\,d\varphi$. Suppose that all the flux passes through the air gap; the work done by the displacement $d\delta$ and suction F_{mag} from the armature in a constant magnetic field is converted from the magnetic energy stored in the air gap:

$$F_{mag}d\delta = dW_{em} \tag{3.14}$$

The energy of the magnetic field in the air gap of the cross-section S_a is

$$W_{em} = \frac{1}{2} \times \frac{B_a^2 S_a \delta}{\mu_0} \tag{3.15}$$

In the equation, B_a is the magnetic induction intensity in the air gap and

$$dW_{em} = \frac{1}{2} \times \frac{B_a^2 S_a}{\mu_0} d\delta \tag{3.16}$$

From Eqs. (3.9), (3.14), and (3.16), we get

$$F_{mag} = \frac{1}{2} \times \frac{B_a^2 S_a}{\mu_0} = \frac{\phi^2}{2S_a\mu_0} \tag{3.17}$$

and

$$B_a = \frac{(iN)\mu_0}{\delta} \tag{3.18}$$

Therefore

$$F_{mag} = \frac{\mu_0 (iN)^2 S_a}{2\delta^2} \tag{3.19}$$

3.1.1.3 Mechanical Circuit Subsystem

According to Newton's law of motion, the equation of motion of the solenoid valve is

$$F_{mag} - k(x + x_0) - F_h - \alpha_x \frac{dx}{dt} = m\frac{d^2x}{dt^2} \tag{3.20}$$

In the equation:

F_{mag} = electromagnetic attraction of solenoid valve (N)
k = spring stiffness (N/m)
x_0 = spring decrement (m)
x = solenoid valve armature displacement (m)
F_h = hydraulic pressure of the fuel on the solenoid valve (N)
m = total mass of the armature, spring, and spool of the solenoid valve (kg)
α_x = head resistance coefficient of the solenoid valve

3.1.1.4 Hydraulic Subsystem

For the hydraulic subsystem of a solenoid valve, the hydraulic force and resistance due to the movement of solenoid valve core and armature are mainly studied.

When the solenoid valve is not energized, the fluid pressure in the solenoid valve is the force acting on the spool from the pressure difference ($p_{con} - p_{sol}$) between the oil in the control cavity and the valve in the solenoid valve, with an action area of $(\pi/4)d_{out}^2$. When the solenoid valve is energized, the fluid pressure in the solenoid valve is still the force acting on the spool from the pressure difference ($p_{con} - p_{sol}$) between the oil in the control cavity and the valve in the solenoid valve, with an action area of $(\pi/4)d_{val}^2$.

Therefore, the oil liquid force acting on the solenoid valve is

$$F_h = \begin{cases} \frac{\pi}{4} d_{out}^2 (p_{con} - p_{sol}); & x = 0 \\ \frac{\pi}{4} d_{val}^2 (p_{con} - p_{sol}); & x > 0 \end{cases} \tag{3.21}$$

In the equation:

F_h = force acting on the spool of the oil in the control cavity (N)
d_{out} = diameter of the control cavity drain port (m)
d_{val} = diameter of the plane valve (m)
P_{con} = pressure in the control cavity (N/m^2)
p_{sol} = pressure in the solenoid valve (N/m^2)
x = spool displacement (m)

3.1.1.5 Thermodynamic Subsystem

For the solenoid, because it works at a high-temperature environment, the change in solenoid valve coil resistance due to the change of temperature cannot be ignored, because it affects the current drive and number of field ampere turns of the solenoid valve. Therefore, it is necessary to perform a thermal equilibrium analysis of the electromagnetic mechanism.

The relationship between the coil resistance and temperature is shown below:

$$R(\theta) = \rho_0(1 + \alpha * \theta)\frac{l}{S} \tag{3.22}$$

In the equation:

ρ_0 = resistivity of a coil conductor (Ω m)
α = coil temperature coefficient
θ = coil temperature (°C)
l, S = length and cross-section of a coil wire (m, m^2)

The steady temperature rise of the coil is shown below:

$$\frac{\mathrm{d}\theta}{\mathrm{d}t} = \frac{i^2 R(\theta)}{K_T(S_w + \alpha_s S_n)} \tag{3.23}$$

where

$R(\theta)$	=	coil resistance (Ω)
K_T	=	heat dissipation coefficient of the coil outer surface
S_w, S_n	=	outer and inner surface areas (m^2)
α_s	=	coefficient of difference of the heat dissipation condition between the inner and outer surfaces of the coil

3.1.1.6 Dynamic Characteristic Synthetic Mathematical Model of the Solenoid Valve

Based on the analysis of the five subsystems mentioned above, because the equations are related to each other, the differential equations of the dynamic process of the solenoid valve are formed as below:

$$\begin{cases} \frac{\mathrm{d}t}{\mathrm{d}t} = -\frac{R(\theta)}{L}t - \frac{U_c}{L} \\ i = C\frac{\mathrm{d}U_c}{\mathrm{d}t} \\ \frac{\mathrm{d}^2x}{\mathrm{d}t^2} = \frac{1}{m}F_{mag} - \frac{1}{m}k(x_0 + x) - \frac{1}{m}F_k - \frac{1}{m}\alpha_s\frac{\mathrm{d}x}{\mathrm{d}t} \\ F_{mag} = \frac{\mu_0(iN)^2 S_n}{2(\delta - x)^2} \\ F_k = f\left(x, {}^{\mathrm{d}x}/_{\mathrm{d}t}, P_{con}, P_{sol}\right) \\ \frac{\mathrm{d}\theta}{\mathrm{d}t} = \frac{t^2 R(\theta)}{K_J(S_w + \alpha_s S_n)} \\ R(\theta) = \rho_0(1 + \alpha^*\theta)\frac{l}{S} \end{cases} \tag{3.24}$$

3.1.2 Solenoid Magnetic Field Finite Element Analysis

The response characteristic of the solenoid valve greatly determines the response characteristics of the fuel injector, because the turning on and turning off time of the solenoid valve determines the injection start and end time of the fuel injector. There are many factors affecting the time response characteristics of the solenoid valve, such as the solenoid valve structure type, solenoid valve structural parameters, the work cycle of the armature, electromagnetic coil parameters, spring structural parameters, the electromagnetic valve material, and so on. Using the ANSYS software, the influence of the structural parameters of the electromagnetic valve on the dynamic response time of the solenoid valve is analyzed using the finite element method by transient analysis in two aspects: electromagnetic attraction and magnetic field distribution.

ANSYS is a large-scale finite element software integrated with a study of the structure, heat, fluid, electromagnetism, and acoustics. In an electromagnetic application, it can analyze many kinds of equipment effectively, conveniently, rapidly, and accurately.

The finite element formulas of ANSYS for magnetic field analysis are derived from the Maxwell equations of the magnetic field. By introducing the scalar magnetic or vector magnetic fields into the Maxwell equations and considering the relationship of other electromagnetic properties, users can develop the equations for finite element analysis. The ANSYS provides various expressions for both linear and non-linear materials, including isotropic, orthotropic, linear permeability, B-H curves of materials, and demagnetization curves for permanent magnets. The post-processing function allows the user to display magnetic field lines, magnetic flux density, and magnetic field strength, and calculates forces, moments, source input energy, terminal voltage, coil currents, and other parameters.

Like other finite element analyses, the electromagnetic field analysis of ANSYS is mainly composed of the following five steps:

(1) Create a physical environment.
(2) Build models, assign features, and mesh.
(3) Add boundary conditions and loads (excitation).
(4) Find the solution.
(5) After processing, check the results. Lines of force, magnetic force and torque, coil resistance or inductance, and so on, are available. List displays, such as a graphics vector display or a contour display, are available, along with the path of the display and calculation of the number of units.

3.1.2.1 Model Establishment and Mesh Creation

Figure 3.5 shows a structural model of the solenoid valve while Figure 3.6 shows a simplified version. The suction plate is far from the core, so that part has a minimal influence on the magnetic circuit, so is not considered here. Because of the axis symmetry structure and the fact that use of the two-dimensional form can completely reflect the three-dimensional situation, in order to reduce the computation burden and improve the precision of the calculation, analysis of the electromagnetic valve magnetic field uses 2D axisymmetric model analysis.

Because of the use of the 2D axisymmetric model for analyzing the magnetic field, the method of the vector potential (MVP) is adopted. The vector potential method is a method based on the node method (the scalar potential method is another node method). In the vector potential method, the degree of freedom on a node basis is more than that of the scalar potential method. These potentials are AX, AY, and AZ, i.e. the magnetic vector potentials in the direction of X, Y, and Z. Three additional degrees of freedom, including current (CURR), electromotive force (EMF), and potential (VOLT), can also be introduced in the analysis of the load voltage or circuit coupling.

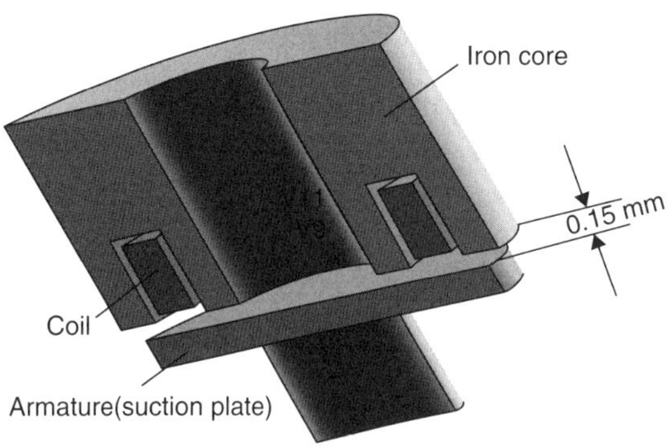

Figure 3.5 3D solid model of the solenoid valve.

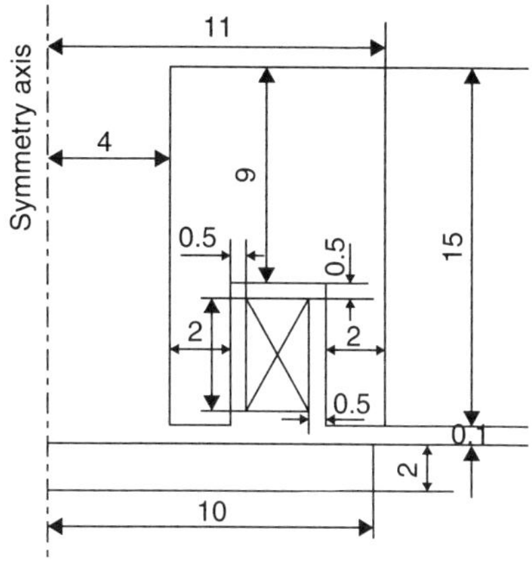

Figure 3.6 2D analysis model of the solenoid.

In the vector potential method, the current supply (wire region) should be used as part of the whole finite element model. Because it has more degrees of freedom on a node basis, it runs at a slower speed.

The vector potential is defined as $B = \nabla \times A$. In the equation, B is the magnetic flux density and A is the magnetic vector potential. A has three components: AX, AY, and AZ. In the 2D plane analysis and axial symmetry analysis, only AZ is not zero. AZ is solved by ANSYS at each finite element node and calculates the field quantity, such as the magnetic flux density.

Assuming that the core and armature materials are made of materials of higher magnetic permeability and that the gap between the core and the armature is very small, the following assumptions are made:

(1) The current density of all wires is distributed uniformly.
(2) The hysteresis effect is neglected.
(3) Neglecting the surrounding magnetic leakage, which is not modeled by the surrounding air, the parallel boundary condition of magnetic lines is needed on the outer surface of the model.
(4) The permeability of the core and armature is isotropic.
(5) The effect of vortex is not considered.

In this section, meshes are divided by quadrilateral 8 nodes of the PLANE53 unit, and are totally divided into 542 units and 1689 nodes. The actual finite element mesh model is shown in Figure 3.7.

3.1.2.2 Loading Analysis

Table 3.1 shows the parameters of the coil structure, material properties, loading voltage, and so on. In the loading analysis, a 2D transient analysis is used. In a transient analysis, boundary conditions and loads are functions of time. ANSYS divides the time-dependent load into multiple time periods, and each period of time relates to certain boundary conditions. Then each time period is analyzed and the loading solution for each time period is called a load step. In each load step, you need to specify how it is loaded, including stepped, ramped, or automatic time stepping. In this section, step and ramp loading are combined.

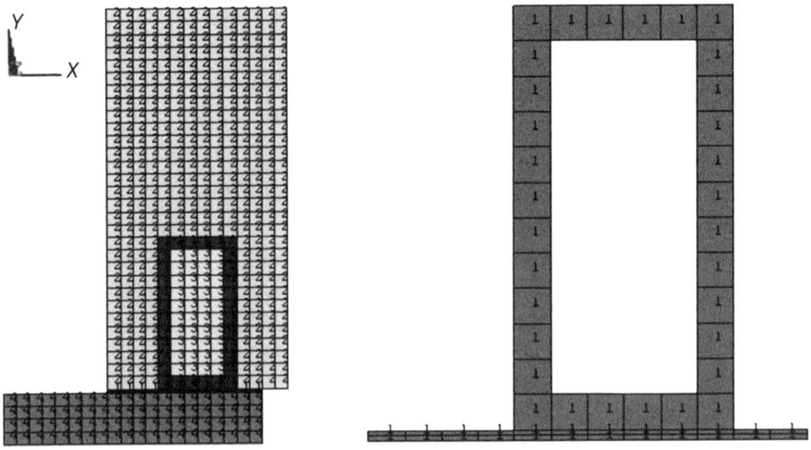

Figure 3.7 Solenoid valve finite element mesh model.

Table 3.1 Finite element analysis of the loading setup.

Coil structure parameter	Way of voltage loading	Core material	Armature material
Wire diameter: 0.2 mm Number of turns: 150 Resistance: 2 Ω	Capacitor discharge variable voltage loading	Material: industrial pure iron Average permeability: B-H curve	Material: industrial pure iron Average permeability: B-H curve

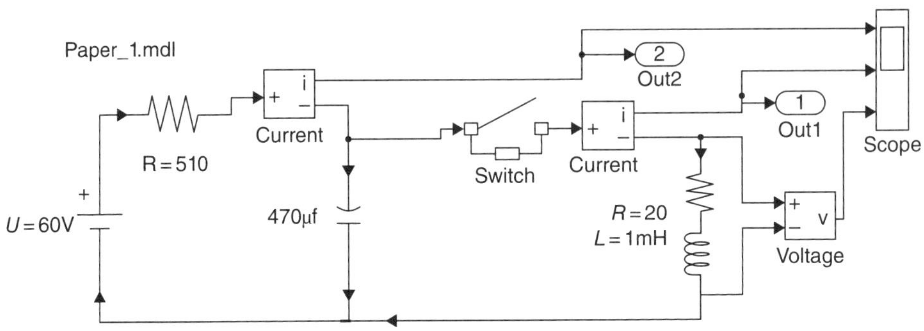

Figure 3.8 Simulation model of the solenoid valve drive circuit in Simulink.

Figure 3.8 is the simulation model of the actual solenoid drive circuit in Simulink and Figure 3.9 shows test results of the current and voltage changes of current when a solenoid is electrified. When properly simplified, we can use the voltage changes in the solenoid valve coil shown in Figure 3.9 to obtain the input conditions for step voltage and ramp loading of the solenoid valve coil when Figure 3.10 is used in the ANSYS analysis.

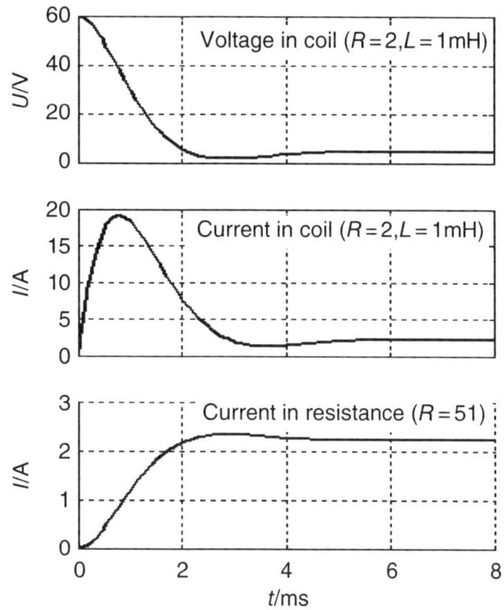

Figure 3.9 Circuit component parameter change diagram.

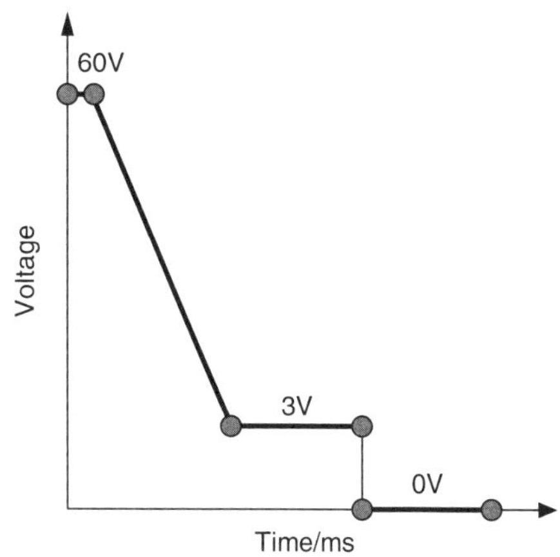

Figure 3.10 Time variation with the coil loading voltage change.

3.1.2.3 Result Display After ANSYS

Figures 3.11–3.13 are, respectively, magnetic force, magnetic flux density in vector form, and nodal magnetic flux density contours after voltage loading for 2 ms. Figures 3.14 and 3.15 are the electromagnetic force output information window and the change of current on the node of the coil unit, respectively.

3.1.3 Solenoid Valve Response Characteristic Analysis

According to formula (3.19), we already know that

$$F_{mag} = \frac{\mu_0 (iN)^2 S_a}{2\delta^2}$$

The change in electromagnetic attraction is related to the gap of the air gap and the change of coil current in the whole process of separation and release. In the actual

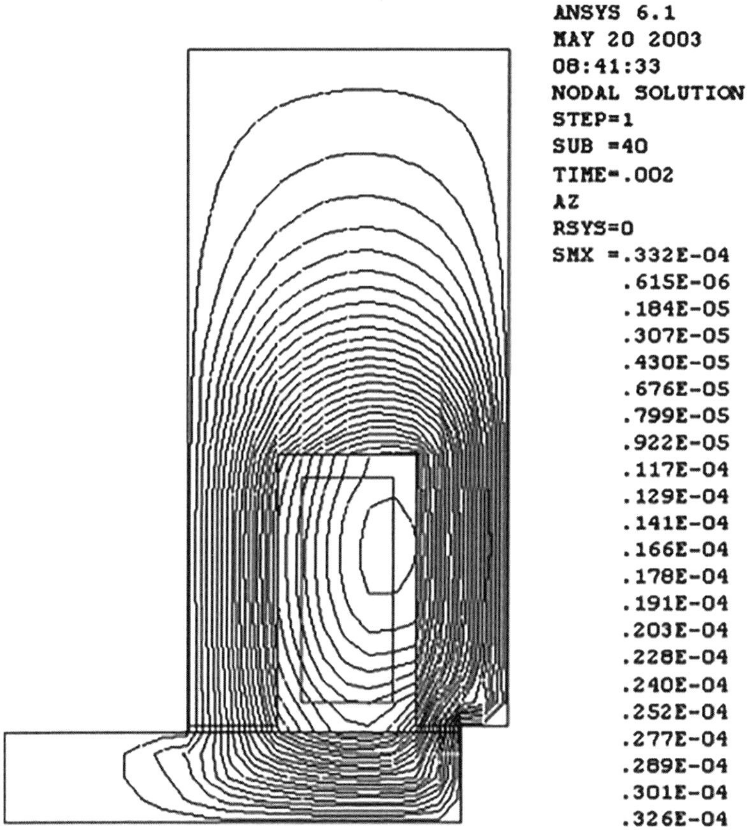

Figure 3.11 Magnetic force.

working process, due to δ, the solenoid valve is 0.15~0.05 mm, its change is small, but the range of current change is relatively large. Figure 3.16 is a variation of the electromagnetic force in the two gaps when the voltage is loaded according to Figure 3.10. As can be seen from the diagram, the change in the smaller range gap has little influence on the change in the electromagnetic force, so dynamic analysis of the solenoid valve can be simplified as follows:

(1) From giving the excitation to the starting pulling phase, $\delta = 0.15$ mm is the actual electromagnetic attraction.
(2) From the start of pulling to the complete pulling phase, the average value of electromagnetic attraction in the two clearances is the actual attraction.
(3) From the beginning of power down to the beginning of the fall phase, $\delta = 0.05$ mm is the actual electromagnetic attraction.
(4) From the beginning of the fall to the complete close phase, the average value of electromagnetic attraction in the two clearances is the actual attraction.

Based on the above simplified hypothesis and using MATLAB to analyze the results of the ANSYS analysis, we get the influence of change of various factors on the response characteristics of the solenoid valve.

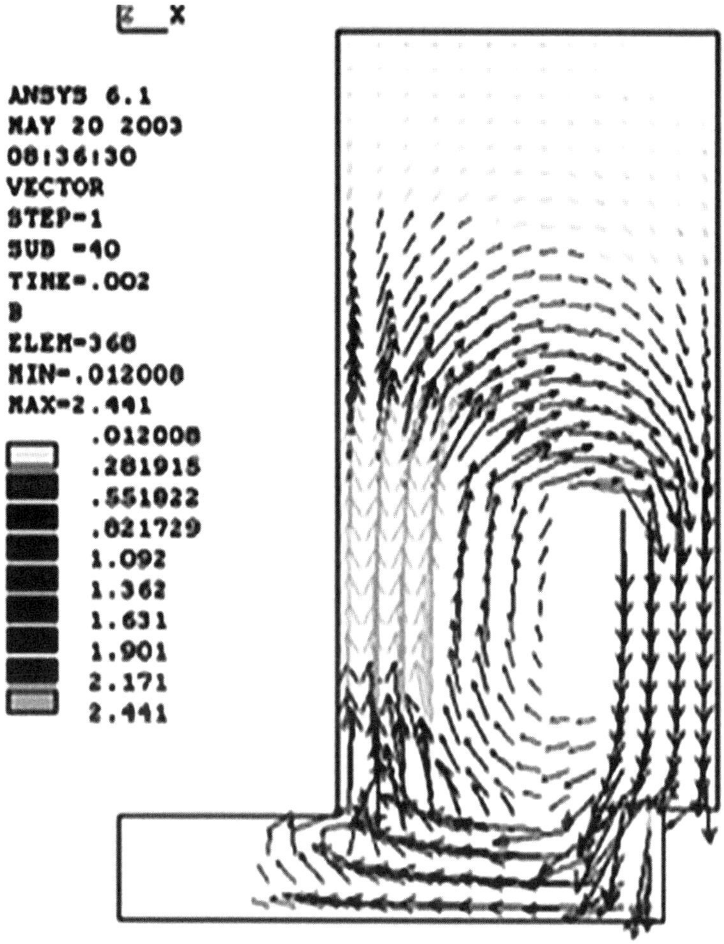

Figure 3.12 Magnetic flux density.

Figure 3.13 Nodal magnetic flux density contours.

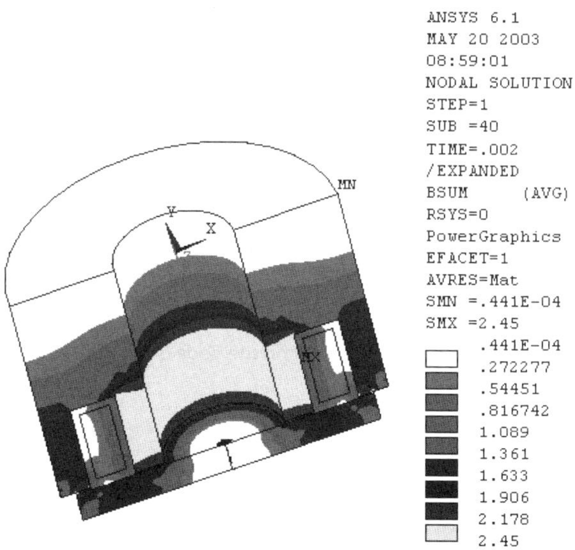

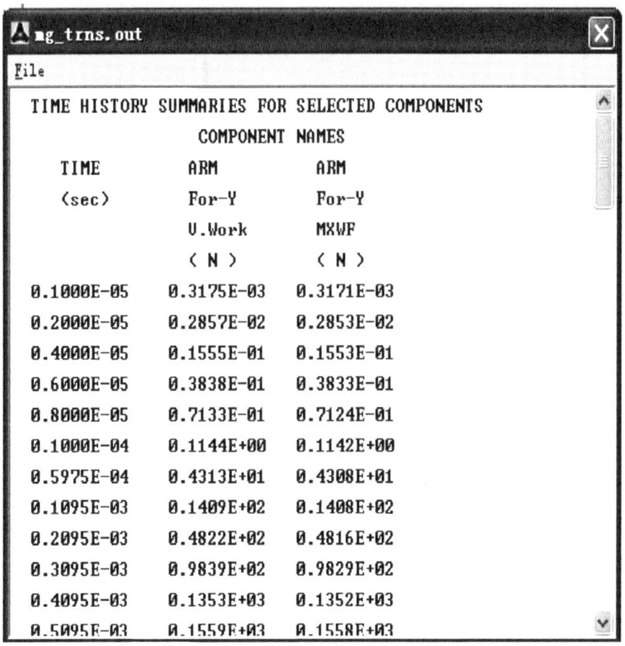
mg_trns.out

File

TIME HISTORY SUMMARIES FOR SELECTED COMPONENTS

COMPONENT NAMES

TIME (sec)	ARM For-Y V.Work (N)	ARM For-Y MXWF (N)
0.1000E-05	0.3175E-03	0.3171E-03
0.2000E-05	0.2857E-02	0.2853E-02
0.4000E-05	0.1555E-01	0.1553E-01
0.6000E-05	0.3838E-01	0.3833E-01
0.8000E-05	0.7133E-01	0.7124E-01
0.1000E-04	0.1144E+00	0.1142E+00
0.5975E-04	0.4313E+01	0.4308E+01
0.1095E-03	0.1409E+02	0.1408E+02
0.2095E-03	0.4822E+02	0.4816E+02
0.3095E-03	0.9839E+02	0.9829E+02
0.4095E-03	0.1353E+03	0.1352E+03
0.5095E-03	0.1559E+03	0.1558E+03

Figure 3.14 Electromagnetic force output information window.

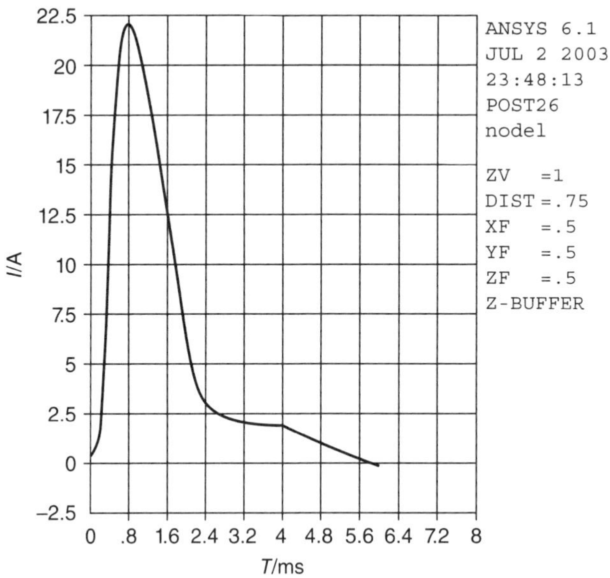

Figure 3.15 Change of current on the node of the coil unit.

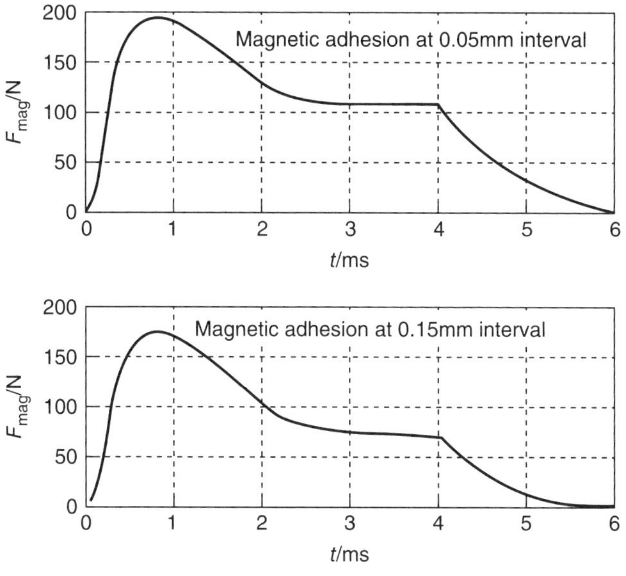

Figure 3.16 Attraction in two kinds of limit clearance.

Figure 3.17 The force of the solenoid valve.

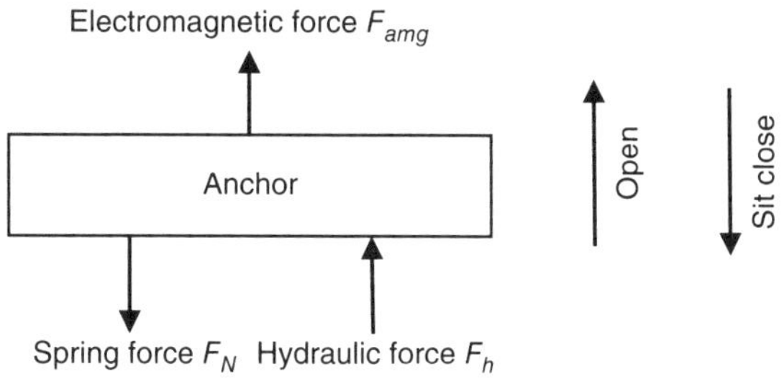

3.1.3.1 The Influence of Spring Pre-load on the Dynamic Response Time of the Solenoid Valve

In addition to the hydraulic pressure, the spring force and the electromagnetic force are the main forces acting on the solenoid valve. In the electromagnetic valve opening process, the electromagnetic force is the active force while the elasticity is the resistance; however, in the electromagnetic valve closing process, the elasticity is the active force while the electromagnetic force is the resistance. The force of the solenoid valve is shown in Figure 3.17.

The pre-load of the spring not only affects the switch response of the solenoid valve but also affects the sealing performance of the solenoid valve when it closes, giving a greater pre-tightening force, a greater sealing performance, a shorter closing response speed, but a worse opening response. Therefore, the pre-tightening force must be considered in order to ensure the tightness of the solenoid valve when it closes, and the comprehensive influence of the opening response and the closing response must be taken into account. Figure 3.18 shows the dynamic response calculation results of the solenoid valve when the spring pre-load is 10–60 N.

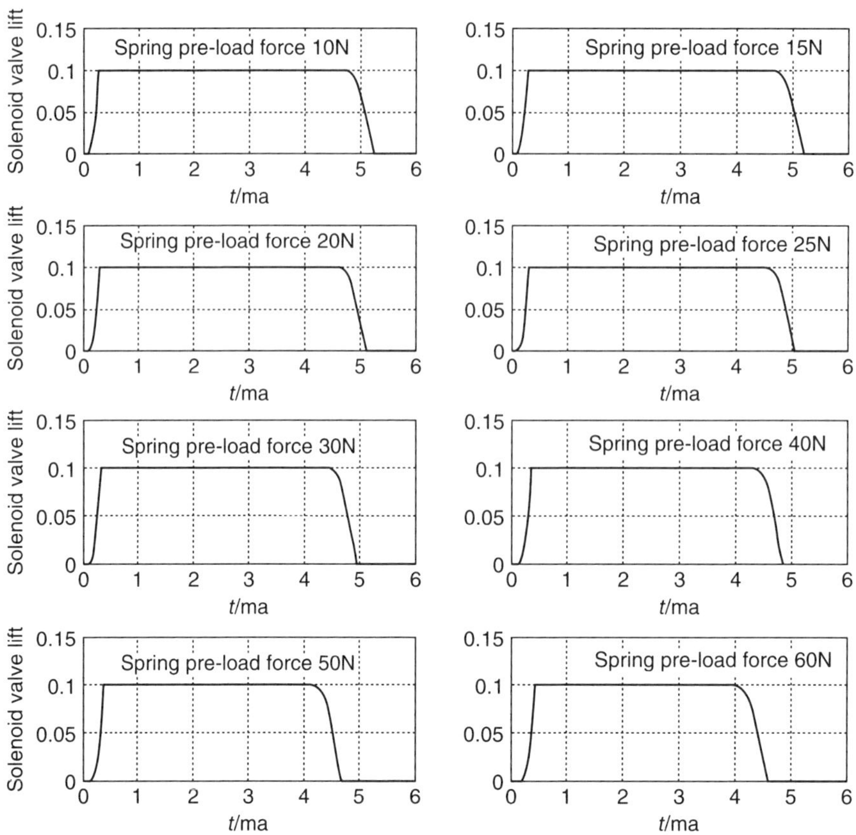

Figure 3.18 Dynamic response of the solenoid valve under different pre-loads.

Figures 3.19 and 3.20 show the effects of different spring pre-loads on the opening time, closing time, and total response time of the solenoid valve. According to Figures 3.19 and 3.20 and Table 3.2, as the pre-load of the spring decreases, the opening response time of the solenoid valve decreases, but the total response time increases; as the spring pre-load increases, the total response time decreases; however, when a certain value is reached, the opening time increases significantly but the total response time decreases less. Furthermore, because the pre-load is too large, the solenoid valve will turn on even when the electronic control unit (ECU) has not yet given the signal; as a result, the injector is not working properly.

According to the above analysis, the maximum peak current can be reached in an extremely short time (0.5 ms after the ECU gives the signal) when the drive circuit is working. Therefore, in a certain range of pre-tightening force, the opening time can be short. A small pre-tightening force will make the closing time too long. If the total response time is the goal of determining the pre-tightening force of the spring, the larger pre-load of the spring should be adopted to ensure that the injector can work properly.

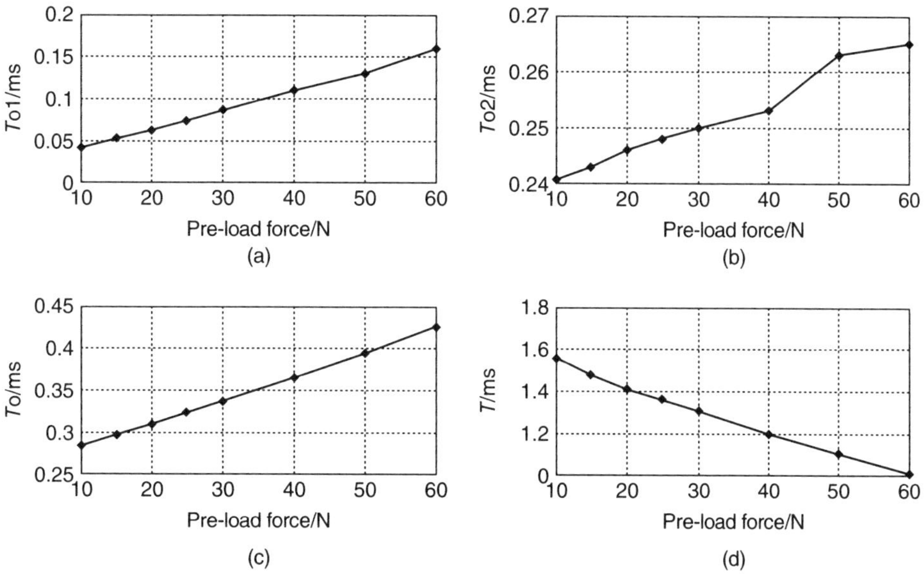

Figure 3.19 The influence of the spring pre-load on opening and the total response time.

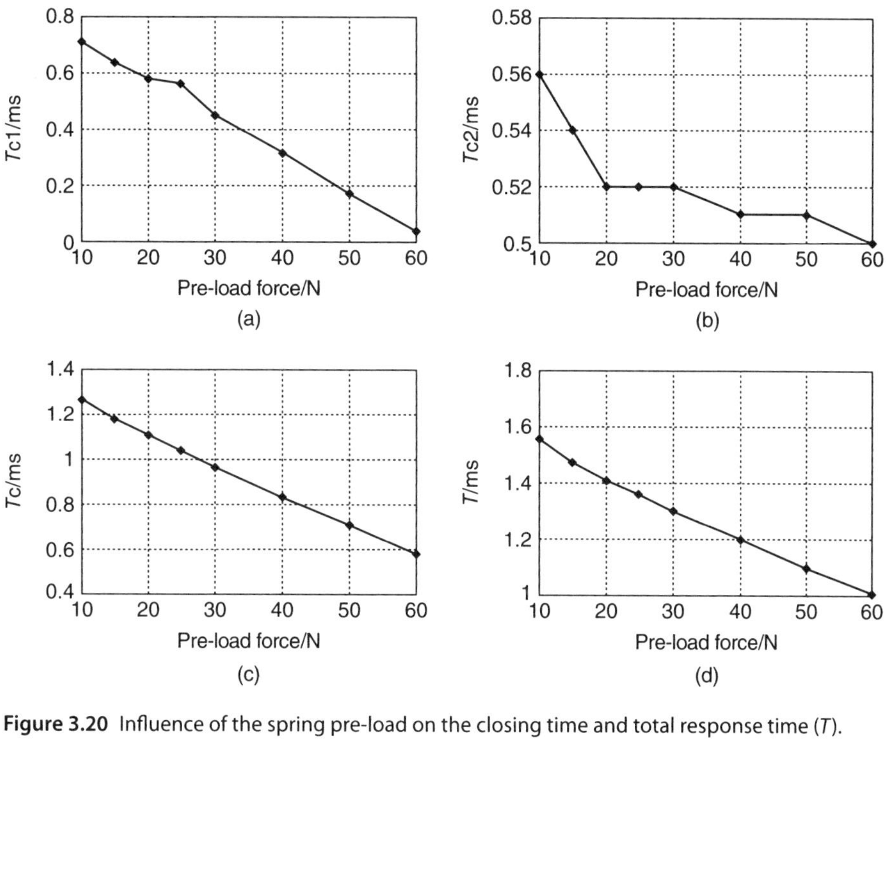

Figure 3.20 Influence of the spring pre-load on the closing time and total response time (T).

Table 3.2 Dynamic response time of the solenoid valve with different spring pre-loads.

Pre-tightening force (nN)	Time (ms)						
	Pick-up time			Break time			Total
	T_{01}	T_{02}	T_0	T_{C1}	T_{C2}	T_C	T
10	0.042	0.241	0.283	0.71	0.56	1.27	1.553
15	0.052	0.243	0.295	0.64	0.54	1.18	1.475
20	0.062	0.246	0.308	0.58	0.52	1.10	1.408
25	0.074	0.248	0.322	0.57	0.52	1.09	1.362
30	0.085	0.25	0.335	0.45	0.52	0.97	1.305
40	0.11	0.253	0.363	0.32	0.51	0.83	1.193
50	0.13	0.263	0.393	0.17	0.51	0.68	1.100
60	0.16	0.265	0.425	0.04	0.50	0.54	0.965

3.1.3.2 The Influence of Spring Stiffness on the Dynamic Response Time of the Solenoid Valve

The study of the influence of the spring pre-load on the response time of the solenoid valve shows that the coil current reaches its peak in a very short period of time during the opening phase, and the electromagnetic force also reaches a large value in a very short time. The magnitude of the pre-load effect on the opening time is not considered relative to the closing time. Therefore, when discussing the total response time, we should pay more attention to the influence of the closing time of the solenoid valve and take a greater number of pre-load values.

Similarly, in discussing the effect of spring stiffness on the response time of a solenoid valve, there is a similar conclusion. Figure 3.21 shows the influence of the spring stiffness obtained by five different spring stiffnesses ($k = 0.1, 1, 3, 4.5, 5$ N/mm) on the response time of the solenoid valve under different pre-load values. From Figure 3.21a and b, it has been found that the magnitude of the different spring stiffnesses has a minimal influence on the opening response time T_{o2} and the closing response time T_{c2}. From Figure 3.21c, it can be found that the magnitude of the spring stiffness has a greater influence on the closing response time T_{c1}. That is, with the increase in spring stiffness, the closing response time will be significantly reduced. Also, from Figure 3.21d, it is found that the influence of the magnitude of the spring stiffness on the total closing response time T is also due to the influence of the closing response time T_{c1}.

3.1.3.3 The Influence of Driving Voltage on the Dynamic Response Time of the Solenoid Valve

Figure 3.22 shows that when the spring pre-load force of the solenoid valve is 40 N and the driving voltages are, respectively, 24, 30, 40, 50, 60, 80, 110, and 120 V, the solenoid valve driving voltage affects the dynamic response time. As shown in the figure, with the increase of driving voltage, the total response time of the solenoid valve is found to reduce with the improvement of the dynamic characteristics; however, when the driving voltage is increased to a certain value, then, with the increase of the driving voltage, the total response time increases. The reason is that due to the influence of magnetic

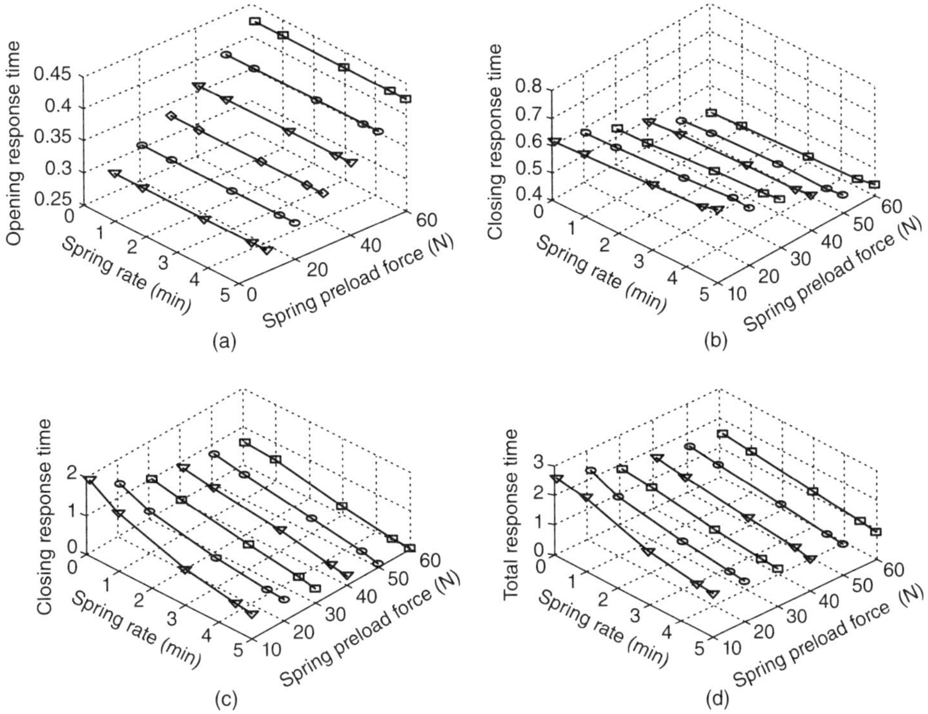

Figure 3.21 The influence of spring stiffness on the response time.

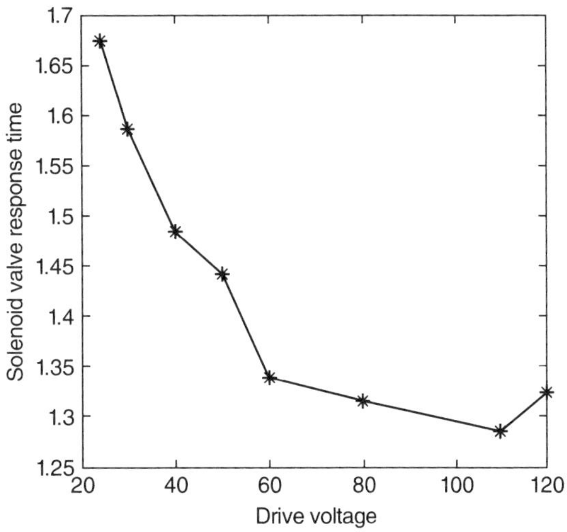

Figure 3.22 Relationship between the driving voltage and dynamic response time of the solenoid valve.

saturation and some other factors, the increase in the driving voltage of the solenoid valve makes the influence of reducing the opening time smaller, and due to the voltage increase while powering off, the closing response time of the solenoid valve significantly increases, resulting in increases in the total response time. The increase in the driving voltage will bring about adverse effects on the power consumption of the drive circuit, and leads to heating of the coil. Under the premise of satisfying the dynamic response characteristics, the driving voltage should somehow be reduced.

3.1.3.4 Influence of Capacitance on the Dynamic Response Time of the Solenoid Valve

When studying the high-speed response characteristics of an opening solenoid valve, the most common concern is the peek current value and its time, as shown in Figure 3.8. Figure 3.22 shows the condition of changing the coil in a solenoid valve with the change in the capacitance C or the change in the driving voltage U.

The capacitance is one of the most important parameters and will be selected in the design of the drive circuit. Figure 3.23a shows that the peak value of the current increases significantly with an increase in the value of the selected capacitor, but the time to reach the peak current also increases.

Furthermore, the next charging time for the capacitor will also be increased. Usually, it will be selected near $\alpha = R/2\sqrt{C/L} = 0.7$ to ensure that the current becomes larger in a short time. Figure 3.23b shows that when the capacitance value is constant and the coil current varies with the driving voltage, the peak value of the current increases with the increase in the driving voltage, but the time to reach the peak current is almost unchanged. Therefore, when the driving voltage is larger, the response will be better, and the influence of the voltage increase on the dynamic response time will be reduced due to the saturation of the magnetic circuit.

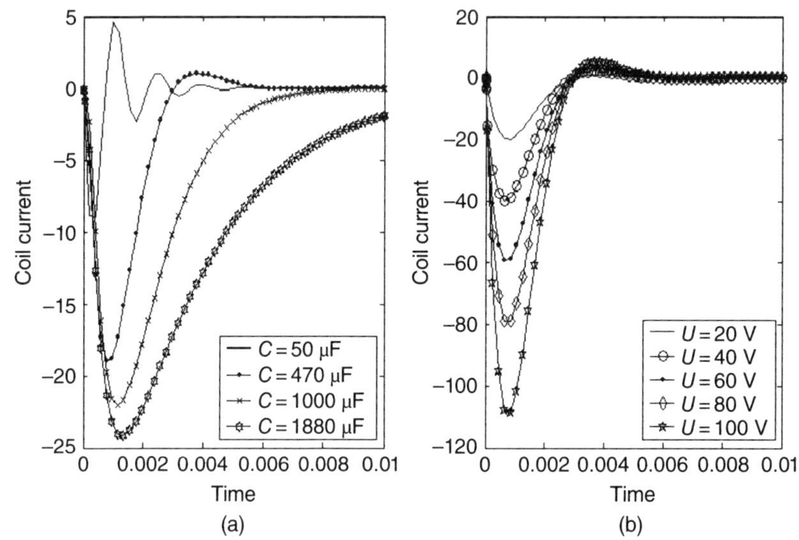

Figure 3.23 Influence of the main parameters of the drive circuit on a current change of the solenoid valve coil.

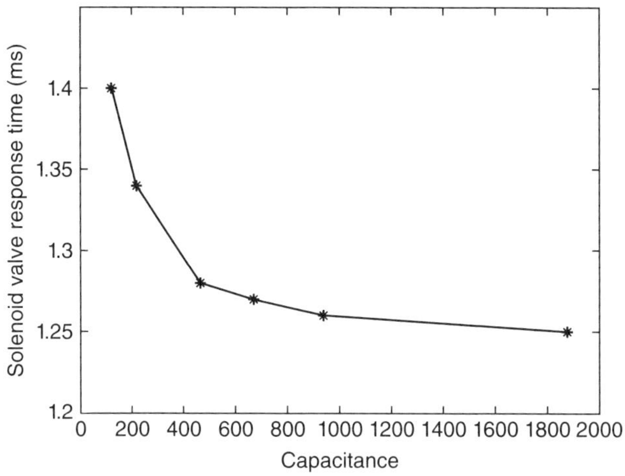

Figure 3.24 Relationship between the capacitance size and dynamic response time of the solenoid valve.

Figure 3.24 shows the response time of the solenoid valve as the capacitance varies. As shown by the diagram, the response time decreases with the increase of the capacitance value; that is, the larger capacitance offers the solenoid valve better response characteristics. However, when the capacitance increases to a certain value, there is little improvement in the response. With the increase of the capacitance, the slope of its current growth remains almost unchanged; although the peak current is large, it appears later; and if the solenoid valve is in the fully open state, then the increased peak has little influence on the open response. Therefore, the situation described in Figure 3.24 occurs (the curve is flat when $C \geq 900\ \mu F$). The increase in the capacitance will inevitably lead to the prolongation of the charging time after its discharge, which is undoubtedly disadvantageous for the high-speed diesel engine with its high injection frequency. When the capacitor value is relatively small, the characteristics of the RLC circuit will show the reverse current phenomenon to that shown in Figure 3.23a, which can reduce the electromagnetic force rapidly and cause the injector to work out of order.

For a fixed solenoid valve, the determination of the capacitor value of the drive circuit can be estimated by the foregoing $\alpha = R/2\sqrt{C/L}$, making α equal to around 0.7.

3.1.3.5 Influence of Structure of the Iron Core on the Response Characteristics of the Solenoid Valve

The number of turns of the coil, the diameter of the wire, and the material property of the wire have a certain influence on the response characteristic of the solenoid valve, while the installation space of the coil depends on the structure of the iron core. In this section, we assumed that the overall dimension remains unchanged, and discussed the influence of the structure of the solenoid valve core on the response characteristics. Specific parameters are shown in Figure 3.25. D1, D2, and H are fixed values; the influence of the sizes of d1, d2, d3, and h on the dynamic response characteristics of the solenoid valve is discussed, and then the window size of the coil is determined.

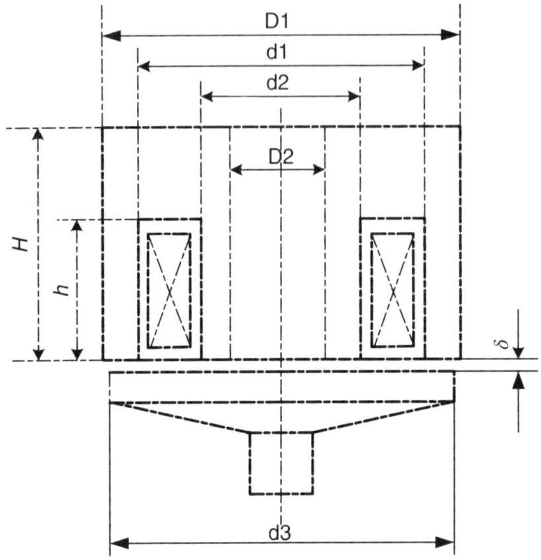

Figure 3.25 Structure diagram of the high-speed solenoid valve.

Table 3.3 presents four structural configuration plans:

(1) The size of the coil window remains the same while changing the size of d1 and d2.
(2) d1 and d2 remain unchanged while changing the size of *h*.
(3) *h* remains unchanged while changing the size of the window.
(4) The structure of the iron core remains unchanged while changing the size of the armature.

Figure 3.26 shows that when the size of the coil window remains unchanged, the influence of the size of d1 and d2 depends on the response characteristics of the solenoid valve; Figure 3.27 shows that when d1 and d2 remain unchanged, the influence of the value of h depends on the response characteristics of the solenoid valve; Figure 3.28 shows that when h remains unchanged, the influence of the size of the window depends on the response characteristics of the solenoid valve; while Figure 3.29 shows that when the structure of the iron core remains unchanged, the influence of changing the size of the armature depends on the response characteristics of the solenoid valve.

Figure 3.26 shows that when using a smaller d1 and d2, that is when the coil window is closer to the axis, the response characteristic of the solenoid valve is better. Also, when considering the structural strength requirements, d2 cannot be too small.

Figure 3.27 shows that when $h = 9$, and $2h/(\mathrm{D1} - \mathrm{D2}) = 1.15$, program B will be used, and the response characteristic of the solenoid valve is better.

Figure 3.28 shows that when h remains unchanged and the size of the coil window adopts the medium plan, that is when $(\mathrm{d1} - \mathrm{d2})/(\mathrm{D1} - \mathrm{D2}) = 0.71$, the response characteristic of the solenoid valve is better.

Figure 3.29 shows that when the size of the armature is chosen to be the medium value, that is when the value of d3 is chosen to be between d1 and D1, the response characteristic of the solenoid valve is better. In order to reduce the mass of the armature, the value can be taken close to d1.

Table 3.3 Four structural configuration plans.

Plan type	Plan description		Conclusion
Window size unchanged, while changing d1 and d2	Plan A	d1, d2 less, d1 = 15, d2 = 9	f = 74.5 t = 0.14
	Plan B	d1, d2 medium, d1 = 16, d2 = 10	f = 69 t = 0.16
	Plan C	d1, d2 bigger, d1 = 20, d2 = 14	f = 62 t = 0.18
D1 and d2 unchanged, while changing h	Plan A	h less, h = 6.2 h/(D1 – D2) = 0.9	f = 68 t = 0.168
	Plan B	h medium, h = 8.2 h/(D1 – D2) = 1.1	f = 74.5 t = 0.14
	Plan C	h biggish, h = 10.2 h/(D1 – D2) = 1.3	f = 61 t = 0.187
	Plan D	h extensive, h = 12.2 h/(D1 – D2) = 1.7	f = 59 t = 0.21
H unchanged, while changing window size	Plan A	less window, (d1 – d2)/(D1 – D2) = 0.4	f = 69 t = 0.17
	Plan B	medium window, (d1 – d2)/(D1 – D2) = 0.7	f = 74.8 t = 0.14
	Plan C	biggish window, (d1 – d2)/(D1 – D2) = 0.9	f = 72 t = 0.16
Changing size of d3	Plan A	less anchor size, d3 = d1	f = 73.2 t = 0.132
	Plan B	medium anchor size, d1 < d3 < D1	f = 74.5 t = 0.14
	Plan C	biggish anchor size, d1 = D1	f = 74.5 t = 0.15

Note that f and t are, respectively, the electromagnetic force F_{mag} (N) and opening response time T_{o1} (ms).

Figure 3.26 When the size of the coil window remains unchanged, the influence of the size of d1, d2 on the response characteristics of the solenoid valve.

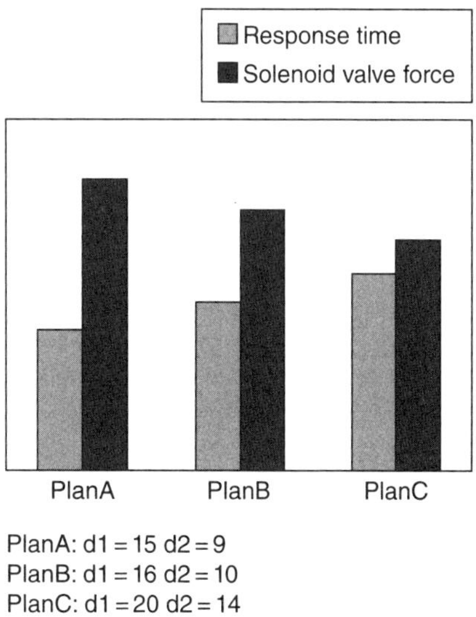

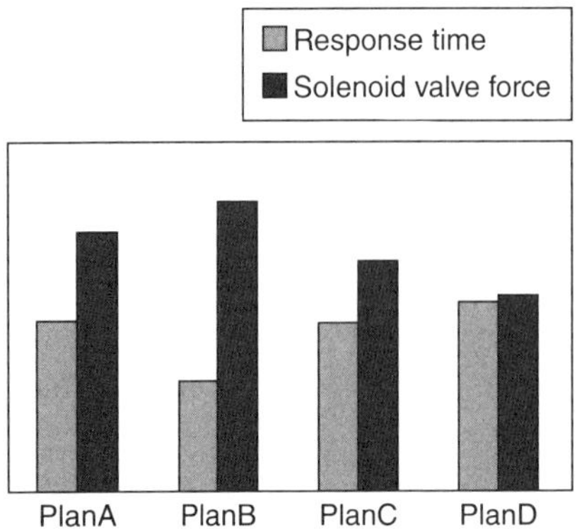

PlanA: 2h/ (D1–D2) = 0.85
PlanB: 2h/ (D1–D2) = 1.15
PlanC: 2h/ (D1–D2) = 1.42
PlanD: 2h/ (D1–D2) = 1.7

Figure 3.27 When d1 and d2 remain unchanged, the influence of the value of *h* on the response characteristics of the solenoid valve.

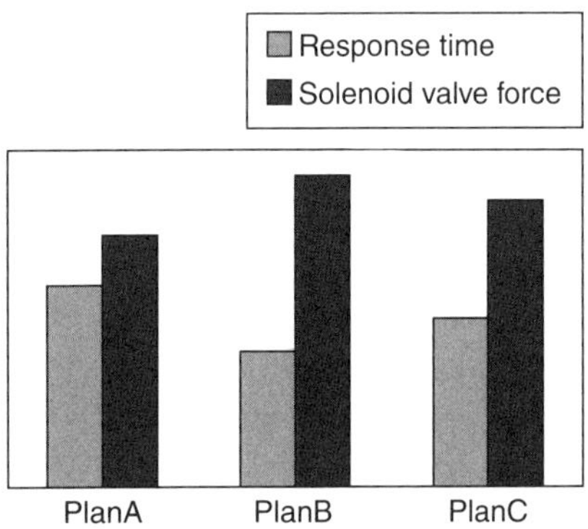

PlanA: Small window (d1–d2)/(D1–D2) = 0.42
PlanB: Medium window (d1–d2)/(D1–D2) = 0.71
PlanC: Big window (d1–d2)/(D1–D2) = 0.86

Figure 3.28 When *h* remains unchanged, the influence of the size of the window on the response characteristics of the solenoid valve.

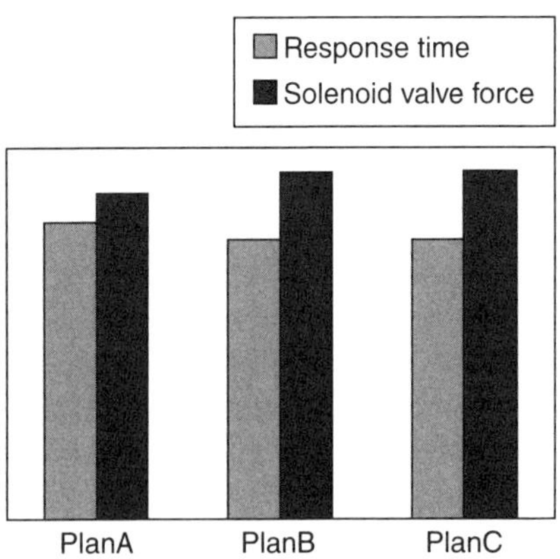

PlanA: Small armature d3 = d1 = 18 mm
PlanB: Medium armature d1 < d3 = 20 mm < D1
PlanC: Big armature d3 = D1 = 22 mm

Figure 3.29 The influence of d3 on the response characteristics of the solenoid valve.

To summarize, when D1 = 22 mm, D2 = 8 mm, and H = 15 mm, as shown in Figure 3.25, and d1 = 16 mm, d2 = 10 mm, d3 = 11 mm, and h = 8 mm, the response characteristic of the solenoid valve is better.

3.1.3.6 Influence of Coil Structure Parameters on the Response Characteristics of the Solenoid Valve

According to the size of the core of Figure 3.6, the relative parameters of the coil, including the diameter of the conductor, the number of turns of the coil, and the resistance of the coil, can be preliminarily determined, and the influence of the number of turns of the coil on the response characteristics is then analyzed. Specific calculations are as follows:

Window area of coil: $Q_{xq} = \frac{1}{2}h(d_1 - d_2) = hb$, where $b = \frac{1}{2}(d_1 - d_2)$

Cross-sectional area of a wire material: $Q_{dx} = \frac{1}{4}\pi d_{dx}^2 N$, where N is the number of turns for the coil and d_{dx} is the diameter of the wire

Filling factor of the coil: $\xi = \frac{Q_{dx}}{Q_{xq}}$, where $\xi = 1$ in this paper

Turn numbers: $N = \frac{2h(d_1-d_2)}{\pi d_{dx}^2}\xi$

Coil resistance: $R = \rho\frac{L}{S} = 2\rho\frac{d_1+d_2}{d_{dx}^2}N$, where ρ is the resistivity of the wire

Coil current: $I = \frac{U}{R}$, where U is the drive voltage of the solenoid valve

Ampere-turns of the coil: $IN = \frac{U}{R}N$

Table 3.4 is obtained, with the premise of using the Figure 3.6 structure, setting the filling coefficient of the coil $\xi = 0.8$, the resistivity of copper wires $\rho = 1.6 \times 10^{-8}(\Omega\ \text{m})$, and using the driving circuit with $V = 100$ V, the spring pre-load is $f = 40$ N, with different diameters of wires.

Table 3.4 shows that when the coil window size of the solenoid valve is constant, the smaller the diameter of the wire, the greater is the number of turns, the greater the resistance, and the greater the inductance when the coil filling factor $\xi = 0.8$. Figure 3.30 shows the influence of the wire diameter of the coil on the response of the solenoid valve. In the smaller diameter range, as the wire diameter increases, the response time will decrease significantly. When the wire diameter of the coil is reduced to a certain value, as shown, when the diameter is 0.3 mm, the response time is minimized. With a further decrease in the diameter of the wires, because now the resistance of the coil inductance decreases rapidly, according to the characteristics of RLC circuits, the capacitance value should remain unchanged, which will cause the coil current to change rapidly, reaching the peak in a very short period of time. When the solenoid valve is not fully open, the current also decreases quickly, so in a certain period of time the total work energy is very small. When the coil diameter increases to a certain value, the solenoid valve is

Table 3.4 The different numbers of turns around a winding, resistances, and inductance with a constant window size of coil.

Coil diameter (mm)	0.1	0.2	0.29	0.41	0.51	0.6
Number of turns	2445	610	290	145	94	68
Resistance (Ω)	203	12.68	2.86	0.72	0.3	0.16
Inductance (H)	14.2	5.6	1.8	0.98	0.2	0.078

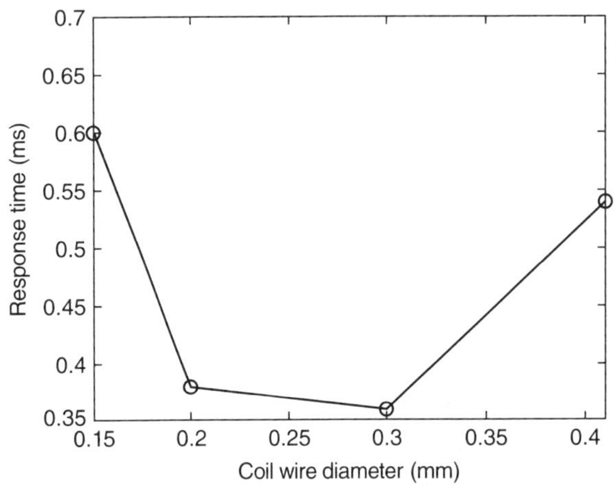

Figure 3.30 The influence of the wire diameter on the response of the solenoid valve.

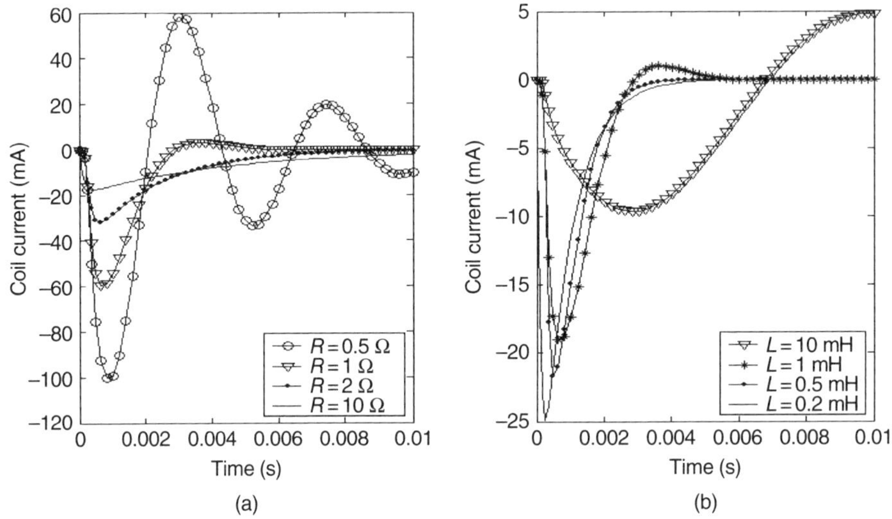

Figure 3.31 The influence of solenoid valve coil resistance and inductance on the changes of the coil current.

unable to work at all. Figure 3.31 shows, respectively, the influences of solenoid valve coil resistance and inductance of the coil on changes of coil current. Because the actual solenoid valve coil resistance and inductance use different winding structures and methods, the influence of both should be considered.

3.1.3.7 The Influence of Working Air Gap (Electromagnetic Valve Lift) of the Solenoid Valve

The size of the solenoid valve working air gap (solenoid valve lift) has a great influence on the solenoid valve magnetic permanence and electromagnetic attraction. According to the formula above, the electromagnetic attraction is inversely proportional to the square of the air gap value. That is, when the air gap is reduced, the corresponding electromagnetic attraction will increase significantly. Figure 3.32 shows the response time of

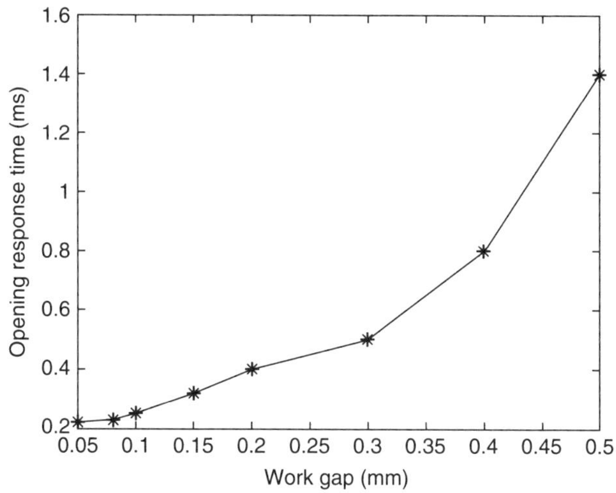

Figure 3.32 The influence of the working air gap on the opening response of the solenoid valve.

the solenoid valve opening in different operating air gaps. As shown by the diagram, the smaller working air gap has a better opening response of the solenoid valve. However, the working gap size also affects the control chamber fuel after turning on the solenoid valve. That is, when the solenoid valve is fully turned on, the minimum flow section should not be less than the cross-sectional area of the B orifice. Considering both the response time of the solenoid valve and the hydraulic response of the injector, the size of the working air gap is determined. The smaller operating air gap not only increases the electromagnetic force but also makes the response of the solenoid valve faster because of the shortening of the valve lift.

3.1.3.8 Material Selection of the Electromagnetic Valve

The core material of the solenoid valve should be using a soft magnetic material, which has the characteristics of high permeability, low coercive force, and low iron loss. However, soft magnetic materials have different magnetization curves, and the magnetic induction B and magnetic field strength H and $B = \mu H$ greatly influence the response characteristics of the solenoid valve. Therefore, soft magnetic materials with the solenoid valve according to the design requirements must be chosen.

The requirements for iron of the high-speed response characteristics of the solenoid valve of the electronically controlled injector are: higher saturation magnetic induction, high permeability, high resistivity, low coercivity, and little hysteresis loss. Table 3.5 gives the main performance parameters of several soft magnetic materials used in this chapter. Figure 3.33 shows the B–H curves of soft magnetic materials. Figure 3.33 and Table 3.5 show that the Fe–Ni alloy is sensitive to a weak magnetic field and is suitable for a high-sensitivity system. When used as a common rail injector solenoid valve material, under certain driving conditions (current changes in the slope in a certain range) it can reach the time of the opening response of the solenoid valve in a very short time, with T_{o1} turns on the solenoid valve as shown in Figure 3.34a. In the solenoid valve lifting process, due to its saturation magnetic induction intensity of B_s, which is relatively small, it soon reaches magnetic saturation, and the electromagnetic attraction is

Table 3.5 The main performance parameters of some soft magnetic materials.

Materials	Bs(T)	Br(T)	Hc (A/m)	μr × 1000	μm × 1000	ρ (μΩ cm)	proportion (g/cm³)
Electrical pure iron	2.15	1–1.4	48–88	0.2	3.5–6	10	7.85
Iron nickel alloy							
1j50	1.5	0.8	0.4	2.8	25–45	45–50	8.2
1j79	0.75	0.5–0.6	4	22	100–180	55–60	8.6
Fe Al alloy 1j6	1.35–1.5	0.5	2	0.5	3–6	70	7.2
Plain carbon steel	2.1	1.2	200–224	0.15	3	12	7.9

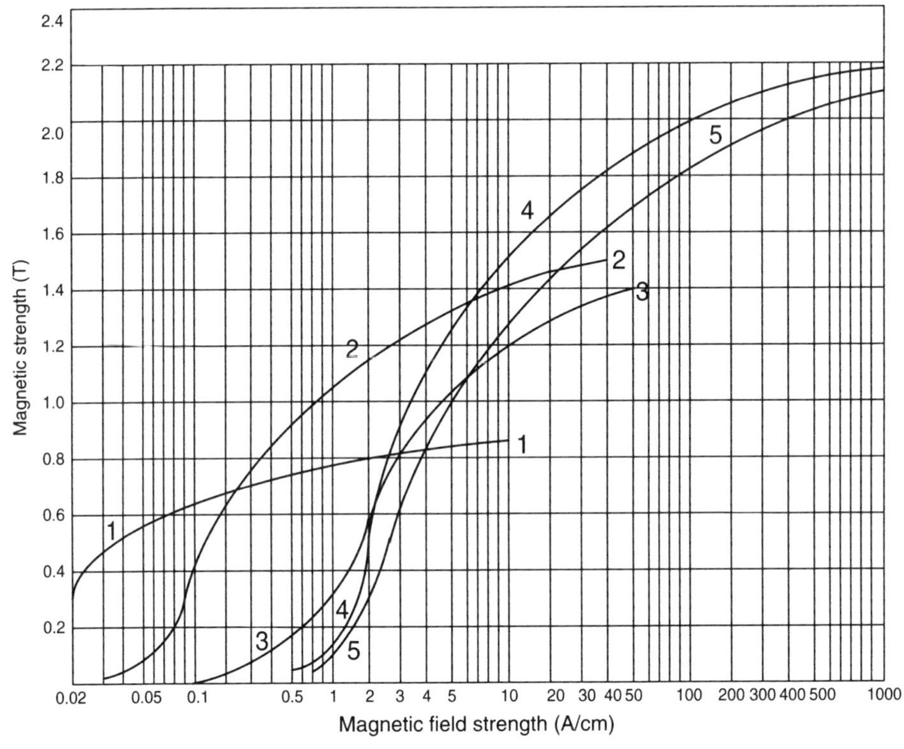

Figure 3.33 B-H curves of partially soft magnetic materials.

relatively small, so that the solenoid valve opening response time is larger, as shown in Figure 3.34b. The iron aluminum alloy 1j6 is characterized by a relatively large resistivity, and its sensitivity to the weak magnetic field is worse than that of the iron nickel alloy but better than that of the industrial pure iron; in addition, its saturation magnetic induction strength B_s is almost the same as that of iron nickel alloy 1j50. The advantages of iron are a relatively high saturation magnetic induction intensity B_s and its magnetization curve has a high permeability in a wide range, so the material is often used as the

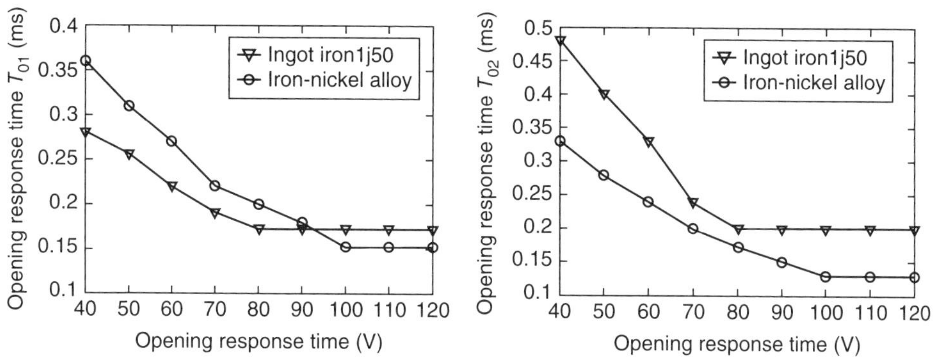

Figure 3.34 Influence of electromagnetic valve material on the opening response time T_{o1} and T_{o2}.

common rail injector solenoid valve material. When the driving voltage and the electromagnetic valve coil current changes in a certain range, it can obtain better response characteristics, as shown in Figure 3.34b.

3.1.4 What Should Be of Concern When Designing the Solenoid Valve

Using the magnetic circuit and basic law as a basis, establishing the mathematical analyzing model of dynamic characteristics of the solenoid valve, and using electromagnetic field finite element analysis software ANYSYS as the platform to study the structure, materials, and loading situations of solenoid valve, the following main conclusions can be made:

(1) The smaller the pre-load of the solenoid valve, the smaller the opening response time T_o, but the closing response time T_c increases. Under the premise of the guaranteed value of T_o, try to use a larger spring pre-load to make the solenoid valve comprehensive response characteristics better. Furthermore, the spring stiffness has a great influence on the closing response time T_{c1}, so a relatively large stiffness spring should be selected.
(2) The drive circuit has a great influence on the response of the solenoid valve. When the coil of the solenoid valve is determined, choosing the appropriate capacitance can obtain a greater growth rate of solenoid valve coil current. Increasing the drive voltage can improve the response characteristics of the solenoid valve, but for some solenoid valves that adopted a certain material and drive circuit, there is a voltage saturation valve, and when exceeding that value, there will be no obvious effect on improving the response characteristics.
(3) The size of structure, coil diameter, and number of turns have certain effects on a solenoid valve's response characteristics. A certain size and position of the window of the coil can obtain optimal response characteristics; a larger diameter leads to a good response characteristic.
(4) In order to reduce remanence, reduce magnetic resistance, and increase the peak value of electromagnetic attraction, both the armature and the iron core should be made of materials with higher permeability and higher saturation magnetic induction. The specific choice of materials can be determined by experiment.

3.2 Nozzle Design Technology

3.2.1 Mathematical Model and Spray Model Analysis of the Nozzle Internal Flow Field

Computational fluid dynamics (CFD) through computer numerical analysis and image display analyzes the system including fluid flow and heat conduction and other related physical phenomena. The basic idea of CFD can be concluded to be as follows: originality in the time domain and space domain continuous physical quantity field, such as the velocity field and the pressure field, together with a series of discrete points on the value of the variable substitution; establishing the algebraic equations about the relationship between these variables by discrete points by making sure the principles and methods are correct; and then solving the algebraic equations to obtain the approximate value of the field variables.

CFD can be regarded as the numerical simulation of flow under the control of the flow basic equations (mass conservation equation, momentum conservation equation, energy conservation equation). We can get the basic physical quantities of complicated flow problems from the position (such as velocity, temperature, pressure, concentration, distribution, etc.) and the physical quantity varying with time. Combined with CAD, the structure can be optimized.

Figure 3.35 shows the flow chart of a CFD numerical simulation for the three-dimensional flow of an injector nozzle, which is mainly composed of nozzle CFD modeling, mesh generation, numerical solution, and post-processing. The detailed description is given in the following.

FIRE software has been developed by Austria AVL Company for flow simulation of internal combustion engines. The spray and combustion process of the CFD software for

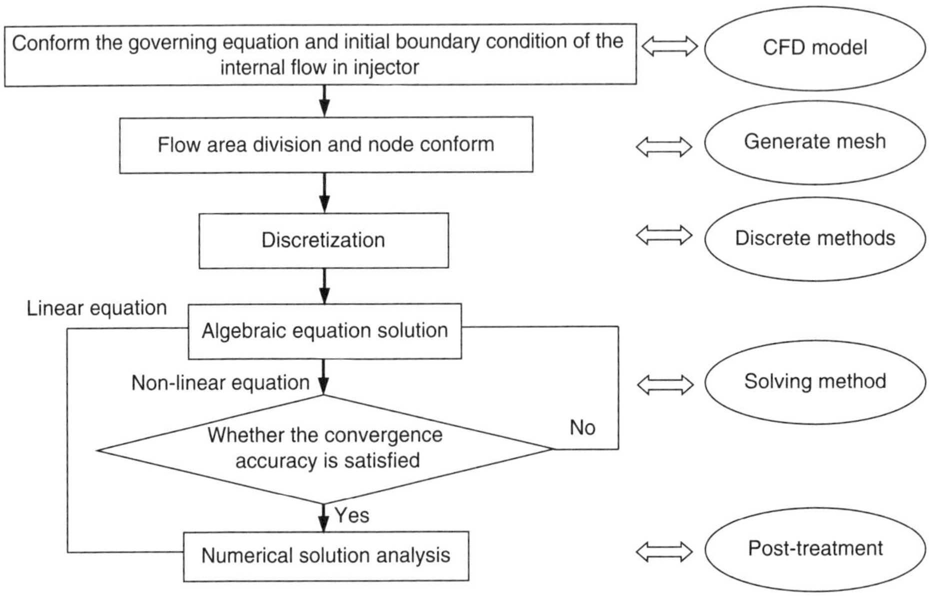

Figure 3.35 Flow chart for 3D flow in the nozzle of the injector.

FIRE, like most of the CFD software, consists of three parts – pre-processing, solver, and post-processing. In the CFD process, pre-processing is a very important part, because the grid type and quality generated by it can usually determine the reliability of the whole simulation calculation, the accuracy of the discrete calculation of the flow field, and the calculation time. FIRE has the ability to generate grid and moving mesh, including fully automatic, semi-automatic, and manual grid generation functions, and can make the high-quality mesh division of complex structures of the intake port in a very short time. At the same time, FIRE contains abundant turbulence, combustion, and emission prediction models. The turbulence models include the $k - \varepsilon$ turbulence model, RSM model, and AVL-HTM model. The combustion models include the EBU model, Flamelet model, PDF model, coherent model, and a characteristic time scale model. The emission model includes the Zeldovich NO prediction model, Kennedy–Hiroyasu–Magnussen soot model, Kennedy–Hiroyasu–Magnussen–Rad soot model, and the advanced soot model. At the same time, FIRE has a custom model for users. Users can easily choose different calculation models according to different objects and environments, in order to achieve the consistency between simulation and the actual situation.

3.2.1.1 CFD Simulation of the Nozzle Flow Field

3.2.1.1.1 Description of the Computational Model

The fluid flow is governed by the law of conservation of physics. The basic laws of conservation include the law of conservation of mass, the law of conservation of momentum, and the law of conservation of energy. If the flow contains a mixture or interaction of different components, the system also follows the law of conservation of composition. If the flow is in a turbulent state, the system must obey the additional turbulent transport equation.

The governing equation is the mathematical description of these conservation laws. The governing equations corresponding to these basic conservation laws and the governing equations for cavitation and turbulence are described below.

Mass Conservation Equation Any flow problem must satisfy the law of conservation of mass. The law can be expressed as the increased mass in the fluid micro unit in unit time equal to the net mass flowing into the micro body at the same time interval. According to this law, the equation of conservation of mass can be obtained:

$$\frac{\partial \rho}{\partial t} + \frac{\partial(\rho u)}{\partial x} + \frac{\partial(\rho v)}{\partial y} + \frac{\partial(\rho w)}{\partial z} = 0 \tag{3.25}$$

Introducing the vector symbol: $div(a) = \partial a_x/\partial x + \partial a_y/\partial y + \partial a_x/\partial z$, Eq. (3.25) can be written as

$$\frac{\partial \rho}{\partial t} + div(\rho u) = 0 \tag{3.26}$$

Some documents use the notation ∇ to represent divergence. That is:

$$\nabla a = div(a) = \partial a_x/\partial x + \partial a_y/\partial y + \partial a_x/\partial z$$

Then Eq. (5) can also be written as

$$\frac{\partial \rho}{\partial t} + \nabla(\rho u) = 0 \tag{3.27}$$

From Eqs. (3.25) to (3.27), ρ is the density, t is the time, $\vec{u}$ is the velocity vector, and u, v, w are the components of the velocity vector $\vec{u}$ in the direction of x, y, z.

According to the mass flow equation of a transient three-dimensional compressible fluid shown above, if the fluid is incompressible, the density ρ is constant and the formula (3.25) becomes

$$\frac{\partial(u)}{\partial x} + \frac{\partial(v)}{\partial y} + \frac{\partial(w)}{\partial z} = 0 \tag{3.28}$$

If the flow is in the steady state, the density ρ does not change with time:

$$\frac{\partial(\rho u)}{\partial x} + \frac{\partial(\rho v)}{\partial y} + \frac{\partial(\rho w)}{\partial z} = 0 \tag{3.29}$$

The equation of mass conservation (8) or (9) is often called the continuous equation.

Momentum Conservation Equation The law of conservation of momentum is also the fundamental law that any flow system must obey. The law can be expressed as the change rate of the momentum of fluid in the infinitesimal body that is equal to the sum of the forces acting on the element. The law is actually Newton's second law. According to this law, the momentum conservation equations can be derived in three directions x, y, z:

$$\frac{\partial(\rho u)}{\partial t} + div(\rho u \mathrm{u}) = -\frac{\partial p}{\partial x} + \frac{\partial \tau_{xx}}{\partial x} + \frac{\partial \tau_{yx}}{\partial y} + \frac{\partial \tau_{zx}}{\partial z} + F_x \tag{3.30a}$$

$$\frac{\partial(\rho v)}{\partial t} + div(\rho v \mathrm{u}) = -\frac{\partial p}{\partial y} + \frac{\partial \tau_{xy}}{\partial x} + \frac{\partial \tau_{yy}}{\partial y} + \frac{\partial \tau_{zy}}{\partial z} + F_y \tag{3.30b}$$

$$\frac{\partial(\rho w)}{\partial t} + div(\rho w \mathrm{u}) = -\frac{\partial p}{\partial z} + \frac{\partial \tau_{xz}}{\partial x} + \frac{\partial \tau_{yz}}{\partial y} + \frac{\partial \tau_{zz}}{\partial z} + F_z \tag{3.30c}$$

In the formula, p is the fluid pressure on the infinitesimal element: τ_{xx}, τ_{yx}, τ_{zx} are the components of the viscous stress acting on the surface of a micro element due to molecular viscosity τ. F_x, F_y, F_z are the forces on the infinitesimal element; if this force is gravity and the Z axis goes up vertically, then $F_x = 0, F_y = 0, F_z = -\rho g$.

Formulas (3.30a) to (3.30c) show a momentum conservation equation for any type of fluid (including a non-Newtonian fluid). For the Newton fluid, the viscous stress τ is proportional to the deformation rate of the fluid.

$$\begin{aligned}
\tau_{xx} &= 2\mu\frac{\partial u}{\partial x} + \lambda div(u) \\
\tau_{yy} &= 2\mu\frac{\partial v}{\partial y} + \lambda div(u) \\
\tau_{zz} &= 2\mu\frac{\partial w}{\partial x} + \lambda div(u) \\
\tau_{xy} &= \tau_{yx} = \mu\left(\frac{\partial u}{\partial y} + \frac{\partial v}{\partial x}\right) \\
\tau_{xz} &= \tau_{zx} = \mu\left(\frac{\partial u}{\partial z} + \frac{\partial w}{\partial x}\right) \\
\tau_{yz} &= \tau_{zy} = \mu\left(\frac{\partial v}{\partial z} + \frac{\partial w}{\partial y}\right)
\end{aligned} \tag{3.31}$$

where μ is the dynamic viscosity and λ is the second viscosity, where generally $\lambda = -2/3$. The formula (3.31) is replaced by Eqs. (3.30a) to (3.30c), giving

$$\frac{\partial(\rho u)}{\partial t} + div(\rho uu) = div(\mu grad\, u) - \frac{\partial p}{\partial x} + S_u \tag{3.32a}$$

$$\frac{\partial(\rho v)}{\partial t} + div(\rho vu) = div(\mu grad\, v) - \frac{\partial p}{\partial y} + S_v \tag{3.32b}$$

$$\frac{\partial(\rho w)}{\partial t} + div(\rho wu) = div(\mu grad\, w) - \frac{\partial p}{\partial z} + S_w \tag{3.32c}$$

In the equations, $grad(\,) = \partial(\,)/\partial x + \partial(\,)/\partial y + \partial(\,)/\partial z$, S_u, S_v, S_w are the generalized source terms of the momentum conservation equation. $S_u = F_x + s_x$, $S_v = F_y + s_y$, $S_w = F_z + s_z$, expressions of s_x, s_y, s_z, are as follows:

$$s_x = \frac{\partial}{\partial x}\left(\mu\frac{\partial u}{\partial x}\right) + \frac{\partial}{\partial y}\left(\mu\frac{\partial v}{\partial x}\right) + \frac{\partial}{\partial z}\left(\mu\frac{\partial w}{\partial x}\right) + \frac{\partial}{\partial x}(\lambda\, div\, u) \tag{3.33a}$$

$$s_y = \frac{\partial}{\partial x}\left(\mu\frac{\partial u}{\partial y}\right) + \frac{\partial}{\partial y}\left(\mu\frac{\partial v}{\partial y}\right) + \frac{\partial}{\partial z}\left(\mu\frac{\partial w}{\partial y}\right) + \frac{\partial}{\partial y}(\lambda\, div\, u) \tag{3.33b}$$

$$s_z = \frac{\partial}{\partial x}\left(\mu\frac{\partial u}{\partial z}\right) + \frac{\partial}{\partial y}\left(\mu\frac{\partial v}{\partial z}\right) + \frac{\partial}{\partial z}\left(\mu\frac{\partial w}{\partial z}\right) + \frac{\partial}{\partial z}(\lambda\, div\, u) \tag{3.33c}$$

where s_x, s_y, s_z are small amounts. For incompressible fluids with constant viscosity, $s_x = s_y = s_z = 0$.

Equations (3.33a) to (3.33c) give the momentum conservation equation, referred to as the momentum equation, also known as the equation of motion or the Navier–Stokes equation.

Equation of Conservation of Energy The law of conservation of energy is the fundamental law that must be satisfied for a flow system with heat exchange. The law can be expressed as the increase in the rate of energy in the infinitesimal body equal to the net heat flow into the infinitesimal body, plus the work done by the physical force and the surface force to the infinitesimal body. The law is the first law of thermodynamics.

The energy E of fluid is usually the sum of internal energy i, kinetic energy $K = \frac{1}{2}(u^2 + v^2 + w^2)$ and potential energy P. We can establish the equation of conservation of energy for total energy E, but the equation of conservation of energy E obtained in this way is not very easy to use. It is generally deduced from the change of kinetic energy, giving the conservation equation of internal energy i. However, we know that there is a relationship between the internal energy i and temperature T; that is, $i = c_pT$, where c_p is the specific heat capacity. In this way, we can obtain the energy conservation equation with the temperature T as the variable:

$$\frac{\partial(\rho T)}{\partial t} + div(\rho uT) = div\left(\frac{k}{c_p} grad\, T\right) + S_T \tag{3.34}$$

The expansion form of this formula is

$$\frac{\partial(\rho T)}{\partial t} + \frac{\partial(\rho uT)}{\partial x} + \frac{\partial(\rho vT)}{\partial y} + \frac{\partial(\rho wT)}{\partial z} = \frac{\partial}{\partial x}\left(\frac{k}{c_p}\frac{\partial T}{\partial x}\right) + \frac{\partial}{\partial y}\left(\frac{k}{c_p}\frac{\partial T}{\partial y}\right) + \frac{\partial}{\partial z}\left(\frac{k}{c_p}\frac{\partial T}{\partial z}\right) + S_T \tag{3.35}$$

where c_p is heat capacity, T is temperature, and k is the heat transfer coefficient of the fluid. S_T is the internal heat source of the fluid and is the part of the fluid mechanical energy converted into heat due to viscous action, and is sometimes referred to in viscous dissipative terms. The formula (3.34) or (3.35) is often referred to in abbreviated form as the energy equation.

Mass Conservation Equation of Components In a particular system, there may be a qualitative exchange, or there may be a variety of components, each component needing to obey the law of conservation of mass. For a given system, the component mass conservation law can be expressed as being in some chemical components equal to the time rate of change, through the system interface and the net flow and diffusion components of productivity generated by reaction.

According to the law of mass conservation law of components, the mass conservation equation of components s can be written as

$$\frac{\partial(\rho c_s)}{\partial t} + div(\rho u c_s) = div(D_s grad(\rho c_s)) + S_s \tag{3.36}$$

In the equation, c_s is the component of the volume concentration s, ρc_s is the mass concentration of the component, and D_s is the diffusion coefficient of the component. S_s is the mass of the component produced by a chemical reaction per unit volume within the unit time of the system, which is productivity. The first and second items on the left side and the first and second items on the right side are, respectively, called the time change rate, the convection term, the diffusion term, and the reaction term. The sum of the mass conservation equations of components is the continuous equation, because $\sum S_s = 0$. Therefore, if there are a total of z components, then there is only $z - 1$ independent mass conservation equations.

Each component conservation equation is expanded and the formula (3.35) can be rewritten as

$$\begin{aligned} &\frac{\partial(\rho c_s)}{\partial t} + \frac{\partial(\rho c_s u)}{\partial x} + \frac{\partial(\rho c_s v)}{\partial y} + \frac{\partial(\rho c_s w)}{\partial z} \\ &= \frac{\partial}{\partial x}\left(D_s \frac{\partial(\rho c_s)}{\partial x}\right) \frac{\partial}{\partial y}\left(D_s \frac{\partial(\rho c_s)}{\partial y}\right) + \frac{\partial}{\partial z}\left(D_s \frac{\partial(\rho c_s)}{\partial z}\right) + S_s \end{aligned} \tag{3.37}$$

The equation of mass conservation of components is often referred to as the component equation.

Linear Cavitation Model The linear cavitation model is used to describe the mass exchange between phases:

$$\Gamma_l = \rho_k N''' 4R^2 \dot{R} = -\Gamma_k \tag{3.38}$$

In the equation, N''' represents the bubble density, R represents the bubble radius, and its growth rate can be calculated by the Rayleigh equation:

$$R\ddot{R} + \frac{3}{2}\dot{R}^2 = \frac{\Delta p}{\rho_l} \tag{3.39}$$

where Δp is the effective pressure difference at the two sides of the gas–liquid interface. Linearizing the above equation and ignoring the inertia item, we could obtain that

$$\Gamma_l = \frac{1}{C_{CR}} sign(\Delta p) 3.85 \frac{\rho_k}{\sqrt{\rho_l}} N'''^{1/3} \alpha_k{}^{2/3} |\Delta p|^{1/2} = -\Gamma_k \tag{3.40}$$

In the cavitation process, the evaporation of the liquid phase counteracts the condensation of some vapors, resulting in a decrease in the actual condensation rate. The empirical coefficient C_{CR} represents the condensation factor, which is used to reduce the condensation rate and make it close to the true value. The number density of bubbles is expressed as

$$N''' = \begin{cases} N_0''' & \alpha_k \le 0.5 \\ 2(N_0''' - 1)(1 - \alpha_k) + 1 & \alpha_k > 0.5 \end{cases} \tag{3.41}$$

The initial value N_0''' is determined by the nature of the liquid phase. One of the most frequently used values is 10^{12}. The bubble number density value should not be less than the maximum bubble diameter contained in the bubble number density value:

$$N''' = \max\left(N''', \frac{6\alpha_k}{\pi D^3{}_{b,\max}}\right) \tag{3.42}$$

If the volume α_k and density of the fluid N''' are fixed, the diameter of the bubble can be determined:

$$D_b = \left(\frac{6\alpha_k}{\pi N'''}\right)^{1/3} \tag{3.43}$$

The momentum model between the gas–liquid exchange contains a turbulent flow discrete force and a resistance force:

$$M_l = C_D \frac{1}{8} \rho_l A_i''' \mid v_r \mid v_r + C_{TD} \rho_l k_l \nabla \alpha_k = -M_k \tag{3.44}$$

The interfacial area density equation of the bubble flow equation is

$$A_i''' = \pi D_b{}^2 N''' = (36\pi N''')^{1/3} \alpha_k{}^{2/3} \tag{3.45}$$

The resistance coefficient C_D is a function of the bubble Reynolds number Re_b. The expression for the Reynolds number is

$$\mathrm{Re}_b = \frac{v_r D_b}{v_l} \tag{3.46}$$

And the coefficient of bubble resistance is

$$C_D = \begin{cases} \frac{192}{\mathrm{Re}_b}(1 + 0.10\mathrm{Re}_b{}^{0.75}) & \mathrm{Re}_b \le 1000 \\ 0.438 & \mathrm{Re}_b > 1000 \end{cases} \tag{3.47}$$

The empirical coefficient C_{TD} takes into account the bubble dispersion caused by the turbulent mixing process. Its reasonable range of value is

$$C_{TD} = 0.05 - 0.5$$

Turbulent Diffusion Model Turbulence is a highly complex three-dimensional unsteady and rotating irregular flow. In turbulent flow, various physical parameters, such as velocity, pressure, temperature, and so on, change randomly with time and space. From the physical structure, the turbulence is a flow composed of vortices of various scales, and the size of these vortices and the direction distribution of the rotating axis are random.

The $k - \varepsilon$ model is adopted in this paper. The basic idea of the model is to reflect the effects of convection and diffusion in the flow field, and the parameters that determine the main characteristics of the turbulent viscosity coefficient must be taken to be dependent variables of the differential equation. In the two-equation model, the most widely used model is the $k - \varepsilon$ model. When the turbulent flow is calculated, two additional transport equations need to be solved: the turbulent kinetic energy equation k and its dissipation rate equation ε. According to k and ε, the turbulent transport coefficient, which is the turbulent viscosity μ_t, of momentum can be obtained by dimensional analysis. The FIRE program uses a standard $k - \varepsilon$ model for the simulation of turbulent flows. Table 3.6 gives the governing equations of the standard $k - \varepsilon$ model.

3.2.1.2 Determination of the Calculation Area and Establishment of the Calculation Model

To determine the computational region of the CFD analysis the definite range of the analytical problem needs to be found by setting the boundary of the problem where the

Table 3.6 Governing equations of the standard $k - \varepsilon$ model.

Governing equation	$\frac{\partial(\rho\phi)}{\partial t} + \frac{\partial(\rho u\phi)}{\partial x} + \frac{\partial(\rho v\phi)}{\partial y} + \frac{\partial(\rho w\phi)}{\partial z} = \frac{\partial}{\partial x}\left(\Gamma\frac{\partial\phi}{\partial x}\right) + \frac{\partial}{\partial y}\left(\Gamma\frac{\partial\phi}{\partial y}\right) + \frac{\partial}{\partial z}\left(\Gamma\frac{\partial\phi}{\partial z}\right) + S$
	For u, v, w, k, ε, T the generalized diffusion coefficient Γ is
	$u, k, \varepsilon \ : \ \Gamma = \eta_{eff} = \eta + \eta_t$
	$k : \Gamma = \eta + \frac{\eta_t}{\sigma_k}$
	$\varepsilon : \Gamma = \eta + \frac{\eta_t}{\sigma_\varepsilon}\Gamma$
	$T : \Gamma = \frac{\eta}{\Pr} + \frac{\eta_t}{\sigma_T}$
Source term	$u : S = -\frac{\partial p}{\partial x} + \frac{\partial}{\partial x}\left(\eta_{eff}\frac{\partial u}{\partial x}\right) + \frac{\partial}{\partial y}\left(\eta_{eff}\frac{\partial v}{\partial x}\right) + \frac{\partial}{\partial z}\left(\eta_{eff}\frac{\partial w}{\partial x}\right)$
	$v : S = -\frac{\partial p}{\partial y} + \frac{\partial}{\partial x}\left(\eta_{eff}\frac{\partial u}{\partial y}\right) + \frac{\partial}{\partial y}\left(\eta_{eff}\frac{\partial v}{\partial y}\right) + \frac{\partial}{\partial z}\left(\eta_{eff}\frac{\partial w}{\partial y}\right)$
	$w : S = -\frac{\partial p}{\partial z} + \frac{\partial}{\partial x}\left(\eta_{eff}\frac{\partial u}{\partial z}\right) + \frac{\partial}{\partial y}\left(\eta_{eff}\frac{\partial v}{\partial z}\right) + \frac{\partial}{\partial z}\left(\eta_{eff}\frac{\partial w}{\partial z}\right)$
	$k : S = \rho G_k - \rho\varepsilon$
	$\varepsilon : S = \frac{\varepsilon}{k}(c_1\rho G_k - c_2\rho\varepsilon)$
	$G_k = \frac{\eta_t}{\rho}\left\{2\left[\left(\frac{\partial u}{\partial x}\right)^2 + \left(\frac{\partial v}{\partial y}\right)^2 + \left(\frac{\partial w}{\partial z}\right)^2\right] + \left(\frac{\partial u}{\partial y} + \frac{\partial v}{\partial x}\right)^2 + \left(\frac{\partial u}{\partial z} + \frac{\partial w}{\partial x}\right)^2 + \left(\frac{\partial v}{\partial z} + \frac{\partial w}{\partial y}\right)^2\right\}$
	$T \ : \ S$ depends on the actual problem

condition is known. If you do not know the exact boundary conditions, do not set the boundary of the analysis close to the region of interest and do not put the boundaries in the gradient where the variables are changing. When you do not know where the gradient is largest, you need to make a tentative analysis and then modify the analysis area according to the result.

When the injector is in the fuel injection state, the flow path of the fuel in the nozzle is found as follows. The high-pressure fuel from the high-pressure oil pipe flows into the valve seat through the annular gap between the needle valve and the needle seat in the injector, then into the pressure chamber at the end of the nozzle, and finally is injected into the combustion chamber at a high speed from six orifices at a certain angle, as shown in Figure 3.36.

The needle lift motion can be divided into three stages:

Needle lift process: $h = 0 \rightarrow h = \max$
The needle valve is in the maximum lift range: $h = \max$
Needle closing process: $h = \max \rightarrow h = 0$

This section analyzes the whole movement process of the injector needle valve, where the duration from opening to closing is 2 ms and the maximum lift of needle valve is 0.25 mm.

The geometry of the nozzle flow region is very irregular and the modeling is complicated by general CAD software. Pro/Engineer Wildfire, a popular 3D modeling tool, adopts a new idea of parametric design and the entity model is generated by the size parameter of the entity. After setting the size parameters of the entity, the model can be obtained through the regeneration process, which is intuitive and fast. When analyzing the influence factors, such as the injection angle and the number of holes, the whole head of the nozzle is taken as the model because of the different arrangements of the spray holes. Other geometric models only capture the 30° rotation of the whole space in order to save computation time, as shown in Figure 3.37. The injector installed in position in the cylinder head is shown in Figure 3.38. In the generation of a FIRE file with an .stl file extension, set the Pro/E height to 0 and then this value will be automatically set to the acceptable minimum by the system. Then set Angle Control to 1. Finally, fill after the initial mesh is generated.

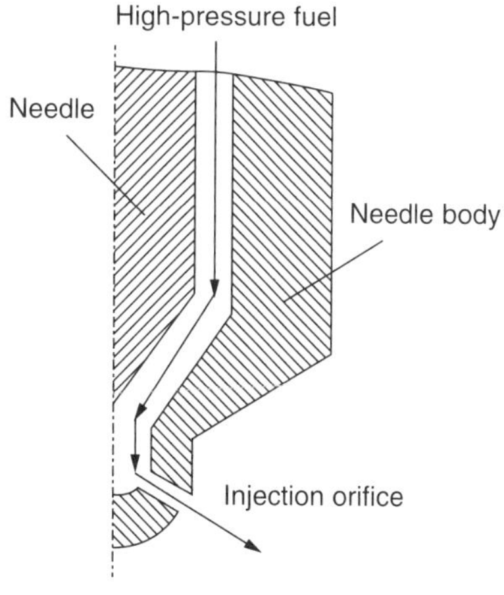

Figure 3.36 Schematic diagram of the fuel flow path.

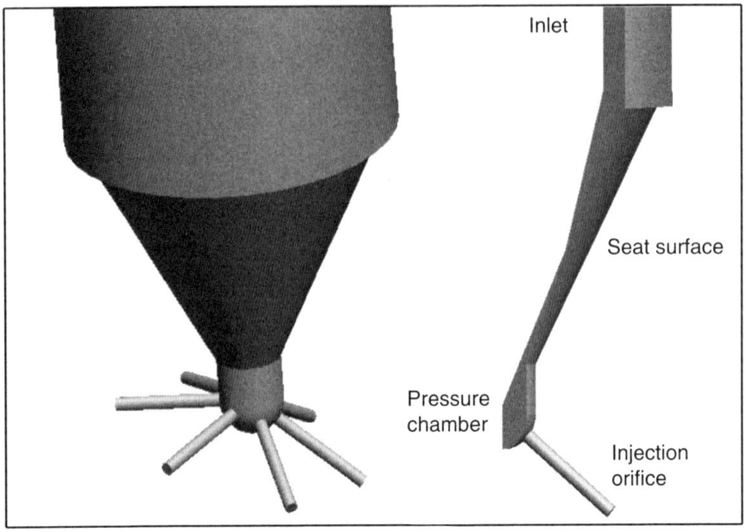

Figure 3.37 CAD model of the computational region.

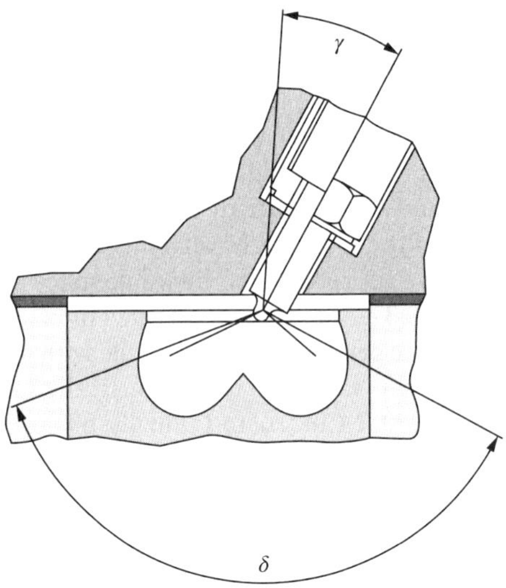

Figure 3.38 Nozzle installation position.

According to the structure limitation of the injector in the diesel engine cylinder head, a number of nozzle models with different sized parameters are established for numerical calculation. The calculation model of the original nozzle used for comparison needs the following nozzle size parameters:

Orifice number: $n = 6$
Orifice diameter: $D = 0.2$ mm
Orifice hole length: $L = 1.6$ mm
Orifice inclination angle: $\gamma = 20°$
Orifice cone angle: $\delta = 140°$
Orifice inlet fillet radius: $\gamma = 0$ mm
Needle lift: $h = 0.25$ mm

3.2.1.3 Discrete Computational Model of the Finite Volume Method

3.2.1.3.1 Computational Mesh Generation

For the numerical analysis of flow, the mesh must first be divided; that is, the space region of the whole fluid flow is divided into many polyhedral subregions, where the nodes of each region are determined. The grid distribution is the foundation of numerical flow control equations of the discrete units. Therefore, grid technology is one of the key technologies to realize numerical simulation of turbulent flow, as the grid quality directly affects the analysis of convergence and the accuracy of the results. The difficulty and cost of grid generation occupies a large proportion in the whole simulation analysis process. Most of the flow fields in the fluid machinery are complex irregular regions, and it takes a lot of time to complete the grid division of these regions. "From some point of view, it is even more difficult to automatically generate ideal meshes around complex bodies than to program a three-dimensional flow solver. Even in highly developed CFD countries, grid generation still accounts for about 60% of all manual labor and time for resolution tasks." We therefore need to pay attention to the following basic principles in the process of meshing:

Grid number. The number of grids will affect the accuracy of calculation results and the size of the calculation. Generally speaking, as the number of mesh increases, the calculation accuracy will be improved, but at the same time the scale of calculation will increase. Therefore, in determining the number of grids, we should weigh the two factors to consider and find the best combination of the two points.

Mesh density. Mesh density refers to the size of the grid in different parts of the structure, which is to adapt to the distribution characteristics of the calculation data. In order to better reflect the changing rules of data, more intensive grids are needed to calculate the gradient of data variation (such as narrow flow). In order to reduce the calculation scale, the relatively sparse grid should be divided in the region where the change gradient of the data is small. Thus, the whole structure shows different mesh forms with different densities.

Body fitted boundary. The mesh is orthogonal or nearly orthogonal to the boundary, so that the boundary conditions can be discretized, the discretization error can be reduced, and the accuracy of the calculation results can be improved. The grid line should be consistent with the flow direction, which is helpful in reducing the false diffusion error. In the case of an unknown flow, the grid should be updated according to the actual flow during the calculation to meet the requirements.

In order to make the grid meet, as much as possible, the above principles and requirements, using the semi-automatic mesh generation method in FIRE, first, its nodes are divided again and then automatically generate the appropriate grid using FAME. They are then processed manually on the grid to make the grid division uniform and reasonable. The method of block coupling is used to do local encryption between the import and export boundaries.

Hexahedral mesh is used for mesh generation in this section. The number of units in the calculation area of the single hole nozzle is about 45 000, while that of the overall nozzle is about 250 000. Considering the continuity requirement of the fluid calculation, the minimum volume of the grid is determined when the needle lift $h = 0.02$ mm is the initial time of solution, as shown in Figure 3.39a. When the needle valve lift is $h = 0.25$ mm, the grid volume is the largest, as shown in Figure 3.39b.

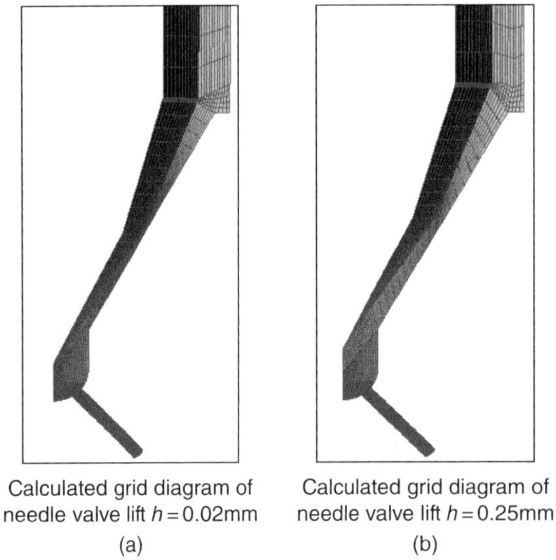

Figure 3.39 Meshing of computational models.

3.2.1.3.2 Definition of Boundary and Initial Conditions

Flow Inlet Boundary The flow inlet boundary refers to the development of flow references at the import boundary. The commonly used flow inlet boundaries include the velocity inlet boundary, the pressure inlet boundary, and the mass inlet boundary. For example, the velocity inlet boundary represents the velocity value of a node on a given inlet boundary. The mass inlet boundary is mainly used for compressible flow. Some flow parameters, such as absolute pressure, turbulent kinetic energy, and the dissipation rate, should be involved in the use of the inlet boundary. To this end, the boundary conditions are described as follows.

When examining the reference pressure, in the flow calculation program, the pressure is always expressed by the relative value, and the actual solution pressure is not an absolute value but is relative to the inlet pressure (i.e. the reference pressure field). Therefore, in some cases, the pressure of other points can be determined by setting the inlet pressure to 0.

On the estimated values of k and ε at the inlet boundary, when using various $k - \varepsilon$ models to calculate turbulence, the estimated values of k and ε on the inlet boundary should be given. At present, there is no theoretical accurate formula to calculate these two parameters, which can only be obtained through experiments. However, it is impossible to do experiments on a variety of flows, so we have to use the approximate formulas already available in the literature to estimate them. In the absence of any known conditions, according to the turbulent kinetic energy intensity T_i and characteristic length L, the values of k and ε are roughly estimated from the following equation:

$$k = \frac{3}{2}(\overline{u}_{ref} T_i)^2; \quad \varepsilon = C_\mu^{3/4} \frac{k^{3/2}}{l}; l = 0.07L \tag{3.48}$$

where $\overline{u}_{ref}$ is the average speed at the inlet and the characteristic length L can be calculated using the equivalent pipe diameter.

Flow Outlet Boundary The flow outlet boundary condition refers to the given flow parameters at a specified position (geometric outlet), including speed, pressure, etc. The flow outlet boundary conditions are combined with the boundary conditions of the flow import.

Setting Pressure Boundary When the detailed information of the flow distribution is unknown, but the pressure value of the boundary is known, the constant pressure boundary condition can be used. The typical problems of this condition include external disturbance, free surface flow, natural ventilation and combustion, and internal flows with multiple outlets.

Wall Boundary The wall boundary is the most commonly used boundary in the flow problem. For the conditions of the wall boundary, except for the pressure correction, the source terms of the discrete equations need to be treated separately. Especially for turbulent computations, the turbulence evolves into the laminar flow near the wall region. Therefore, we need to use the wall function method to apply the known values on the wall to the source terms of the discrete equation of the inner node for the near wall region.

Periodic Boundary The periodic boundary condition is also called the cyclic boundary condition, which is often proposed for the symmetric problem. For example, in axial flow turbines and pumps, the flow of the impeller can be divided into subdomains that have an equal number of blades, and the periodic boundary is on the initial boundary and ending boundary of the subdomains. The flows on these two borders are exactly the same.

To use these periodic boundary conditions, we must make the flux of all the flow variables flowing out of the boundary of the circulation outlet equal to the flux related to corresponding variables entering the circulation boundary. This can be achieved by taking the node variables on the left and right sides of the inlet surface equal to the node variables on the respectively left and right sides of the outlet surface.

Initial Condition In the transient problem, in addition to the given boundary conditions, the initial value of the flow variable of each calculation point in the flow area, that is the initial condition, is also required. To sum up, other special processing is needed in addition to initializing the relevant data before the calculation. Thus the initial conditions are relatively simple.

In the given initial condition, two points need to be emphasized:

(1) The initial conditions for all the units in the whole area should be given to all the calculated variables.
(2) The initial condition must be physically reasonable, otherwise it will lead to unreasonable results, even if the result is not convergent.

3.2.1.3.3 Numerical Solution

Before computing, relevant parameters and control items need to be set, including activating the momentum equation and the continuity equation. The $k - \varepsilon$ model should be chosen for the turbulence model equation, and the flow in the nozzle is set as an incompressible fluid.

The setting of convergent factors is very critical, because they are the factors that most directly affect the convergence or divergence of calculation and are also related to the number of iterations of calculation convergence. It is easy to converge if the factor is small but the number of iterations is long and the time of calculation is long. It is easy to diverge if the factor is large but the number of iterations is short and the time of calculation is short. Therefore, these convergent factors have to be tested many times in order to find the right value.

In order to achieve a tradeoff between computational accuracy and convergence, the blending factor is used in the softness factor in order to knead in the upwind format in the higher-order format. In order to ensure the conservation of mass, the continuous equation needs to select the center difference in every application and the flexible factor is defined as 1, that is, the pure two-order difference.

The total number of iterations should be sufficient, but if the iteration does not converge, the total number of iterations will be forced to end the cycle. The setting of convergent conditions should be appropriate: too large or divergent or able to converge but the result is not accurate enough; if it is too small though the result is accurate but the number of iterations is too much and leads to a longer time to calculate.

The amount of data can be output during the calculation, which is a backup of the calculation process. When there is an unexpected stop in the calculation process (stop or crash), we can continue computing from the nearest backup point instead of restarting it from the beginning. If the change is stable and the change trend is in the direction of convergence, it can continue to proceed.

3.2.1.4 Spray Model of the Nozzle

In the last 20 years, with the extensive application of various advanced testing methods and the rapid development of numerical simulation technology based on CFD and HPC (high-performance computing), people's understanding of liquid splitting and the atomization process is gradually increasing. On this basis, a series of mathematical models for the formation of liquid injection splitting and spray formation have emerged. It can be said that the study of the spray model is the most fruitful field in the whole combustion science of an internal combustion engine in recent years. Although these models have not yet been developed to the level of perfection and ideal, they have begun to be applied in scientific research and engineering development, and play a very important role.

Fuel atomization in an internal combustion engine can be divided into two processes: primary breakup and secondary breakup. The former process occurs in the high Weber number nozzle area. It is not only formed by the interaction between the two phases of the gas and liquid but is also influenced by the flow phenomena in the nozzle, such as turbulence and cavitation. The classic broken models, such as TAB, the Reitz–Diwakar model, and the WAVE model, do not distinguish between these two processes. The development of the spray model has paid more and more attention to the primary breakup process. For example, the Huh–Gosman model attaches great importance to the effect of turbulence in the nozzle on the subsequent spray fragmentation. However, the primary breakup model of the diesel engine describes the effect of the cavitation and turbulence on the spray breakage at the same time. Models such as FIPA and KH-RT provide the possibility of treatment for primary breakup and secondary breakup, respectively. Because of the lack of experimental data in the primary breakup area, these models need to adjust the additional model parameters in order to match the experimental data of the secondary breakup area. These parameters are not only related to the shape of the nozzle but also to the numerical solution. The adjustment process is sometimes tedious. At the same time, there are more and more models considering the effect of hole flow on spray breakage in the nozzle, and the hole nozzle flow model is one of them. Applications of the hole nozzle flow model, the WAVE model, the KH-RT model, and the primary breakup model of the diesel engine are discussed in the following sections.

3.2.1.4.1 Hole Type Flow Nozzle Model

Most CFD software regards fuel spray as a representative statistical sample of a number of discrete droplet group (particles). The initial position, size, speed, and temperature of each representative droplet should be given at the outlet of the nozzle. Due to the lack of experimental data near the nozzle spray breakup, we have to use the traditional application of experimental data as the inlet conditions downstream of the spray nozzle. For the hole nozzle spray, the most commonly used theoretical calculation method is the discrete blob droplet model. It is assumed that the radius and velocity of the initial droplet are, respectively, the geometric radius and the average injection speed of the outlet. In fact, the blob model neglects the influence of the flow of the nozzle inside the nozzle on the flow state at the outlet of the nozzle. The possible flow in the nozzle can be a single-phase flow, cavitation flow, and medium reflux, including laminar and turbulent flow. The high-pressure injection in a modern diesel engine makes the most likely hole flow in the orifices, as shown in Figure 3.40. In the figure, R is the inlet radius of the nozzle, D is the outlet diameter of the orifices, and L is the length of the orifices. On the basis of the blob droplet model, the nozzle flow model proposed the concept of an effective injection velocity and an effective injection hole area. To determine whether the flow is in a cavity state, the static pressure is calculated of the contraction flow area (C point) in the inlet of the nozzle:

Figure 3.40 Hole type flow model.

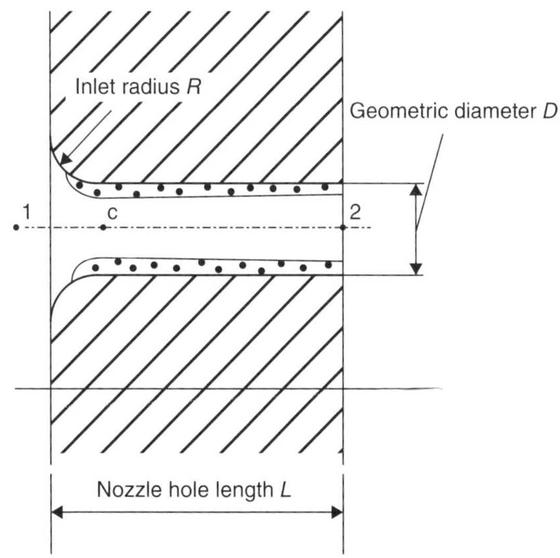

$$p_c = p_1 - \frac{\rho_l}{2}U_c^2, \qquad U_c = \frac{U}{C_c}$$

$$C_c = \frac{1}{\sqrt{\frac{1}{C_{ct}^2} - \frac{11.4R}{D}}} \tag{3.49}$$

$$p_1 = p_2 + \frac{\rho_l}{2}\left(\frac{U}{C_d}\right)^2 \tag{3.50}$$

In the equation, $C_d = \frac{1}{\sqrt{K_{inlet} + fL/D + 1}}$, ρ_l = fuel density, C_d = discharge coefficient, p_1, p_2 = pressure at the inlet (first point) and inlet (second point); U_c = flow velocity at the contraction (C point), C_c = contraction coefficient, C_{ct} – theoretical constant ≈ 0.611, f = wall friction factor, and K_{inlet} = loss coefficient at the inlet (first point).

When p_c is greater than the saturated vapor pressure of liquid p_{vapor}, the nozzle flow is in the flow state; we then set the droplet velocity equal to U and the initial droplet Sauter mean diameter (SMD) D_{32} equals the diameter of the nozzle outlet. When p_c is smaller than the saturated vapor pressure of the liquid p_{vapor}, we then set the flow in a full cavitation state. A new pressure p_1 and flow coefficient C_d at the inlet should be recalculated:

$$p_1 = p_{vapor} + \frac{\rho_l}{2}U_c^2, \quad C_d = C_c\sqrt{\frac{p_1 - p_{vapor}}{p_1 - p_2}} \tag{3.51}$$

The effective injection speed and effective injection area are obtained by solving the mass continuity and momentum conservation equations between the constricting orifice and the outlet:

$$U_{eff} = U_c - \frac{p_2 - p_{vapor}}{\rho_l U}, \quad A_{eff} = A\frac{U}{U_{eff}} \tag{3.52}$$

In the equation, A is cutting area of the injection orifices. Therefore, the spray SMD D_{32} is

$$D_{32} = \sqrt{4A_{eff}/\pi} \tag{3.53}$$

The geometry of the nozzle affects the formation of holes. The lower the R/D ratio, the sharper the inlet of the nozzle, and the earlier the hole will take place. The faster the generation speed of the sharpened nozzle, the smaller the effective diameter will be. The L/D ratio only affects the time of the cavitation and does not affect the extent of the cavitation. KIVA, FIRE, and FLUENT provide a hole nozzle flow model, which is used to describe the inlet conditions of high-pressure spray droplets in diesel engines.

3.2.1.4.2 WAVE Model

The process of forming solid cone spray is by the pressure chamber nozzle or the porous nozzle under the action of high pressure. Figure 3.41 shows the breakup process of the high-speed liquid circular injection. The mechanism of high-speed injection breaking is complicated. The WAVE model is a kind of model of the injection breakup mechanism using surface wave instability theory. The WAVE model considers that the breakup of

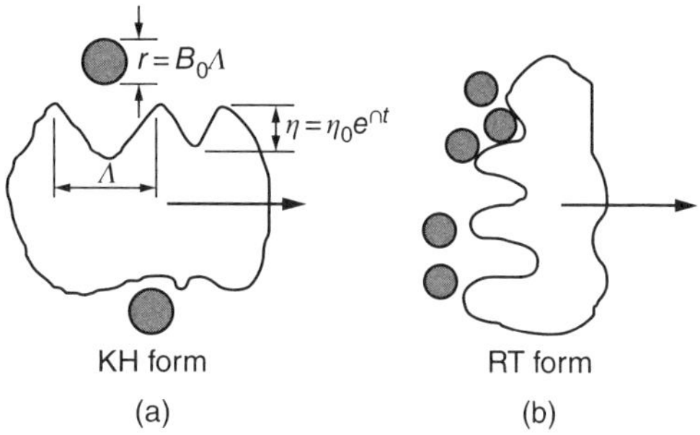

Figure 3.41 A schematic diagram of the breakup mechanism of KH and RT droplets.

the liquid injection is due to the relative velocity between the two gas–liquid phases. The growth of the Kelvin–Helmholtz unstable wave (KH wave) on the injection surface causes the droplets to be cut from the liquid surface, as shown in Figure 3.41.

Analysis of the Stability of the Injection The model considers that a cylindrical injection with a density of ρ_l, a kinematic viscosity of v_l, and a radius of a is injected into a steady, incompressible and sticky steady gas environment with density ρ_g and a relative velocity of U.

In the cylinder coordinate system with injection motion, an arbitrary infinitesimal axis symmetric surface disturbance acts on the initial state in the form of Eq. (3.54) and generates unstable surface waves on the injection surface:

$$\eta = \eta_0 e^{ikz+\omega} \tag{3.54}$$

In the equation, η_0 is the amplitude of initial disturbance, k is the wave number, and ω is the rate of wave growth. Among them, the fastest growing is the most unstable wave (or the most unstable disturbance) with the largest wave growth rate $R_e(\omega)$ as Ω. It will lead to the breakup of the injection directly. By solving the linear equation of motion of the liquid, the dispersion equation, which is associated with the wave growth rate ω and wave number k, is

$$\begin{aligned}\omega^2 + 2vlk^2\omega\left[\frac{I_1'(ka)}{I_0(ka)} - \frac{2kL}{k^2+L^2}\frac{I_1(ka)}{I_0(ka)}\frac{I_1'(La)}{I_0(La)}\right] \\ = \frac{\sigma k}{\rho_l a^2}(1-k^2a^2)\left[\frac{L^2-a^2}{L^2+a^2}\right]\frac{I_1(ka)}{I_0(ka)} \\ + \frac{\rho_g}{\rho_l}\left[U - i\frac{\omega}{k}\right]^2\left[\frac{L^2-a^2}{L^2+a^2}\right]\frac{I_1(ka)}{I_0(ka)}\frac{k_0(ka)}{k_1(ka)}\end{aligned} \tag{3.55}$$

where

$$L^2 = k^2 + \omega/vl \tag{3.56}$$

In the equation, I_1, I_2 are the modified Bessel functions of the first kind, k_0, k_1 are the modified Bessel functions of the second kind, and σ is the surface tension of droplets.

Reitz gives the numerical solution of the equation to the curve fitting, and gives the maximum wave growth rate Ω and the corresponding wavelength Λ:

$$\frac{\Lambda}{a} = 9.02\frac{(1 + 0.45Oh^{0.5})(1 + 0.4Ta^{0.7})}{(1 + 0.87We_g^{1.67})^{0.6}} \tag{3.57}$$

$$\Omega\left[\frac{\rho_l a^3}{\sigma}\right]^{0.5} = \frac{(0.34 + 0.38We_g^{1.5})}{(1 + Oh)(1 + 1.4Ta^{0.6})} \tag{3.58}$$

In the equations, Oh is the Ohnesorge number of the liquid, $Oh = We^{0.5}/Re$; We is the liquid Weber number, $We = \rho_l U^2 a/\sigma$; Re is the Reynolds number, $Re = Ua/vl$; Ta is Taylor number, $Ta = Oh\sqrt{We_g}$; and We_g is gas Weber number, $We_g = \rho_g U^2 a/\sigma$.

Droplet Breakup The WAVE model assumes that the KH unstable wave with the maximum growth rate causes the droplet diameter r_{stable} and breaking time of the injection breakup τ as follows:

$$r_{stable} = B_0\Lambda \tag{3.59}$$

$$\tau = \frac{3.726B_1 a}{\Lambda\Omega} \tag{3.60}$$

In the equation, the Reitz model constant is set to 0.61. B_1 is a model constant used to modify the breakup time, which is related to the initial disturbance of the liquid injection, and the different applicable values need to be adjusted for different nozzles.

When $B_0\Lambda > a$, the parent droplets with a radius of a will break into a new droplet with a radius of r_{stable}. When $B_0\Lambda \leq a$, the change rate of the radius of the parent droplet is

$$\frac{\mathrm{d}a}{\mathrm{d}t} = \frac{-(a - r_{stable})}{\tau} \tag{3.61}$$

The WAVE breakage model is suitable for the high-speed injection of a high Weber number ($We > 100$), such as the simulation of a high-pressure diesel engine spray. Only in this condition is it reasonable for the injection surface wave to be controlled by the KH unstable wave. The WAVE breakup model is provided in KIVA, FLUENT, and FIRE. The effect of the hole and flow turbulence on the injection hole is considered by adjusting the model constant. If we use the WAVE model in the "blob" discrete droplet model, we often find that there is hardly any fuel vapor near the nozzle. The reason is that the droplet near the nozzle is still large, so it is difficult to evaporate. FIRE provides a WAVE Child fragmentation model. A bimodal spectrum is given at the outlet of the nozzle and 90% mass droplets have the same diameter as the injection hole while the diameter of a 10% mass droplet is very small. Breakup will cause a large increase in the calculated droplets, while appropriate control is required for the number of initial droplets.

3.2.1.4.3 KH-RT Model

The HK-RT model developed on the basis of the WAVE model considers that the KH surface wave and the RT disturbance have been in a competitive relationship in the process of spray fragmentation. The KH mechanism is suitable for spray breakup at a high relative velocity and high environmental density. The RT mechanism is suitable for describing the rapid deceleration of droplets, which results in the rapid growth of surface waves on the leeward side of the droplet, causing deformation to break into small droplets, as shown in Figures 3.2–3.7. The KH-RT model is used to simulate the KH breakup by using the WAVE model formula. The frequency Ω of the surface wave with the maximum growth rate and the corresponding wave number k of the RT disturbance is

$$\Omega = \sqrt{\frac{2}{3\sqrt{3\sigma}} \frac{g_t|\rho_l - \rho_g|^{1.5}}{\rho_l + \rho_g}} \tag{3.62}$$

$$k = \sqrt{\frac{g_t \mid \rho_l - \rho_g \mid}{3\sigma}} \tag{3.63}$$

The RT breakup time is

$$\tau = \frac{B_2}{\Omega} \tag{3.64}$$

The drop diameter generated by the RT perturbation is

$$r_{stable} = B_3 \frac{\pi}{k} \tag{3.65}$$

In the equation, g_t is the velocity reduction in the direction of droplet motion and B_2, B_3 are the model constants.

The KH-RT model is provided in KIVA and FIRE. The combination of the nozzle flow model and the KH-RT model can have a more accurate prediction effect on the spray of a high-pressure diesel engine.

3.2.1.4.4 Primary Breakup Model of Diesel Engine

Based on the measurement of the pulsation of a diesel engine's liquid core under different nozzle configurations and working conditions, FIRE considers that the intensity and frequency of liquid core pulsation are related to the turbulent flow of nozzles, the cavitation, and the aerodynamic performance of nozzles. In the primary breakup of a diesel engine established by FIRE, the influence of nozzle turbulence on the breakup is considered by solving an additional equation about turbulent kinetic energy and the turbulent energy dissipation rate in the liquid core area. The effect of hole blasting on the breakup is reflected in the additional source term of the turbulence model. The influence of turbulence and cavitation on breakup is in competition with aerodynamic forces until they reach the secondary breakup area, which is dominated by the aerodynamic force located downstream at a certain distance from the nozzle expansion.

Assuming that the effect of diffusion is ignored, the turbulence equation near the liquid core can be written as

$$\frac{\mathrm{d}k}{\mathrm{d}t} = -\varepsilon + S_k \tag{3.66}$$

$$\frac{\mathrm{d}\varepsilon}{\mathrm{d}t} = -C_1 \frac{\varepsilon}{k}(\varepsilon - S_k) \tag{3.67}$$

$$S_k = C_2 \frac{\mathrm{d}E_k}{\mathrm{d}t} \frac{N_L}{m_L} \tag{3.68}$$

$$E_k = 2\pi\rho_l R^3 \left[\frac{\mathrm{d}R}{\mathrm{d}t}\right]^2 \tag{3.69}$$

In the equation, S_k is the Hole source term; E_k is the liquid kinetic energy around a hole; m_L is the mass of the annular liquid; N_L is the number of vacuoles in a circular liquid; and C_1, C_2 are the model constants.

Based on the concept of banded oil droplets in the concentrated spray area near the outlet of the nozzle, the Euler Lagrange two-phase flow method is used to solve the main crushing model to simulate the characteristics of the oil droplets and their interactions. In the downstream area away from the nozzle, the discrete droplet method is still used. The conversion between the two is determined by the drop in critical number and diameter.

The spray atomization mechanism of the internal combustion engine is complex. The spray model in CFD software simplifies and assumes a spray process. Theoretically speaking, the spray breakup model established by the orifice flow model and the surface wave instability theory has greatly improved the accuracy of numerical simulation of spray fragmentation near the nozzle. It is not an easy task to choose the appropriate crushing model for the simulation analysis of an internal combustion engine, but by choosing the correct parameters of the model, we can get better simulation results and match the experiment with the calculation results.

3.2.2 Analysis of the Influence of Injection on the Electronically Controlled Injector

Under conditions of high pressure, the shape and geometry size of a fuel injector have a more significant influence on the fuel injection form. Injector orifices must be designed and studied in detail. Therefore, using mature commercial software to study a TBD234 diesel injector involves finding the diameter, length diameter ratio, inlet radius, needle tip shape, and how the injection angle and number of orifices of different structures affect the internal flow field of the nozzle and the fuel injection law. The grid built in the single orifice model is basically the same as that shown in Figure 3.39, the only difference being about the geometric feature that is emphasized. In addition, we need to establish a complete calculation model for analyzing the injection angle and number of orifices, so the number of grids needed is larger and the computation time is longer.

The calculation is based on the time (crankshaft angle) for the calculation step, the engine speed is 1500 r/min, and the duration of injection is 2 ms. In order to simulate fully the injection process of the TBD234 diesel fuel injector, this section simulates the motion curve of the needle valve, as shown in Figure 3.42. The $k - \varepsilon$ standard equation is chosen for the turbulence model and the near wall surface function method is used

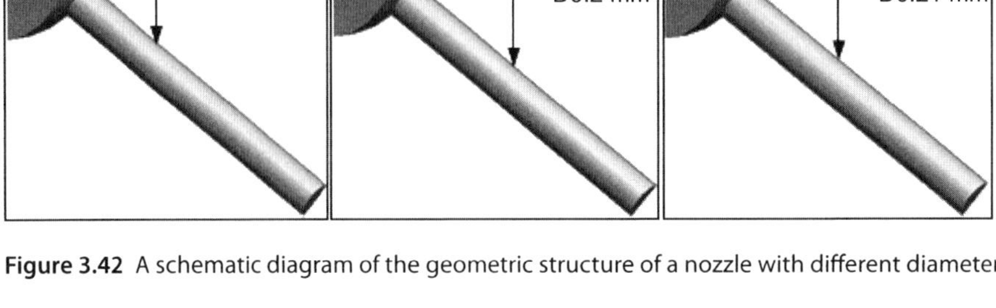

Figure 3.42 A schematic diagram of the geometric structure of a nozzle with different diameters.

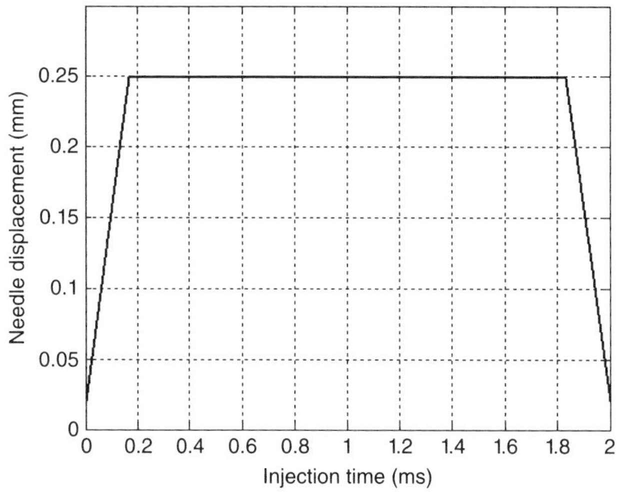

Figure 3.43 Needle lift.

to solve the near wall surface. Diesel oil is regarded as an incompressible medium and so the energy conversion and loss in the flow process are not considered. To solve the equation of momentum conservation and continuity, a simple algorithm is used for the coupling of pressure and velocity and the Euler algorithm is used to calculate the gas–liquid two-phase flow in the nozzle.

3.2.2.1 The Effect of Injector Orifices

Figure 3.42 shows the three-dimensional modeling of the geometric structure of the nozzle shape with the diameter of the nozzle. Spray nozzles are the same except for their diameters. The diameters of the orifices are 0.19, 0.2, and 0.21 mm, and the length is 1.6 mm.

In the calculation, all parameters are consistent. Each calculation result is selected to compare the cloud map at the same time in the early, middle, and late stage of the injection, and the flow field inside the nozzle can be observed intuitively. The calculated results are as follows.

Figure 3.43 shows the calculated injection law curve by three different apertures of orifices. The mass flow of the nozzle with the large aperture is the largest, and the smaller the aperture, the smaller is the mass flow. However, as the aperture increases, the mass flow rate increases slowly. The reason is that the cavitation area inside the large diameter orifice is large, which decreases the discharge coefficient and reduces the mass flow rate (Figure 3.44).

Figure 3.45 shows the pressure cloud of the three nozzles at the center section of the nozzle at the same time and before, during, and after the injection. It can be seen that with the increase of the aperture size, the pressure field is set up slowly, the pressure changes in the pressure chamber and the injection orifices are more obvious, and the pressure distribution gradient varies greatly. This can easily form a swirl, causing an increase in the loss of momentum and affecting the flow velocity of the fuel in the nozzle. The pressure gradient in the pressure chamber and the orifices in the nozzle with the smaller aperture is small and evenly distributed. From the opening of the needle valve

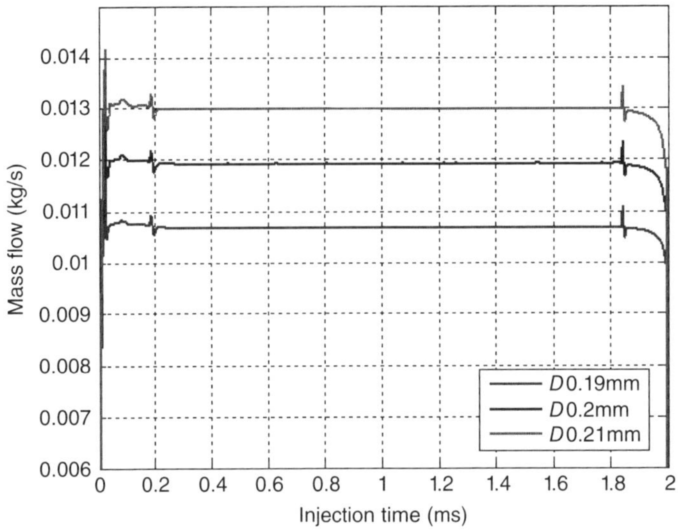

Figure 3.44 Mass flow of nozzles with different apertures.

to the closing, the pressure difference of the injector to the orifice is very large. This is because when the fuel flows from the pressure chamber to the orifices at high speed, the cross-sectional area decreases sharply, the direction of flow changes greatly, and the pressure gradient changes violently. It is easy to form a low-pressure area or even negative pressure on the corner of the orifice inlet, causing cavitation. In the numerical value, the absolute value of the negative pressure of the large aperture is larger, which will reduce the flow coefficient of the orifice and make the cavitation form more easily. Then the pressure at the outlet of the orifice is close to the pressure in the cylinder at the top of the compression stroke and so the cavitation will collapse. The pressure drop in the nozzle is different, which will cause the internal velocity of the orifice to be distinctly different.

Figure 3.46 is the gas volume fraction of the three nozzles at the center section of the orifice and the wall of the nozzle at the same time at early, middle, and late stages of the injection. The blue region represents the liquid phase, the green region represents the gas phase, and the dark red part indicates the area with the largest cavitation intensity. From the following three pictures, with the opening of the needle valve, the occurrence of the cavitation becomes more and more obvious. In the center section of the orifice, the size of the nozzle in the injection process is small, the volume of the gas phase in the nozzle is relatively small, and the distribution area is close to the upper edge of the wall. The cavitation in the aperture of the larger nozzle is formed earlier, the proportion of the gas phase is larger, and the strength of the cavitation is thicker. At the same time, it can be observed that the cavitation along the orifice extends to the outlet of the nozzle and the distribution of the gas phase area also varies greatly due to the different opening of the needle valve. The difference between the cavitation of the three nozzles on the wall is very small. However, the difference in the size of the cavitation area inside the nozzle will affect the subsequent spray characteristics. The larger air hole area is good for atomization of the fuel and the cone angle of the spray is greatly reduced.

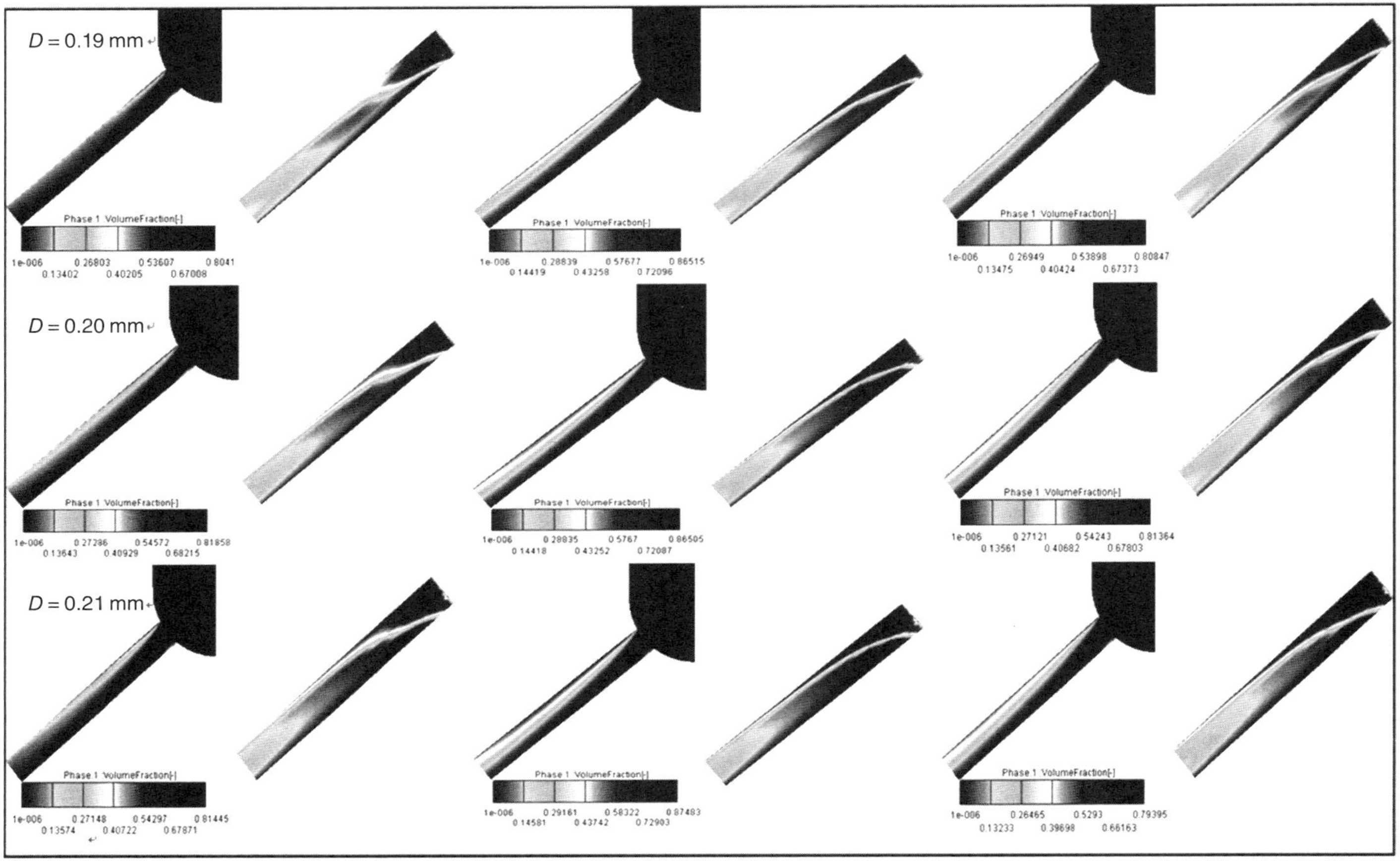

Figure 3.45 Gas phase volume distribution on the center section of the orifice and the wall surface.

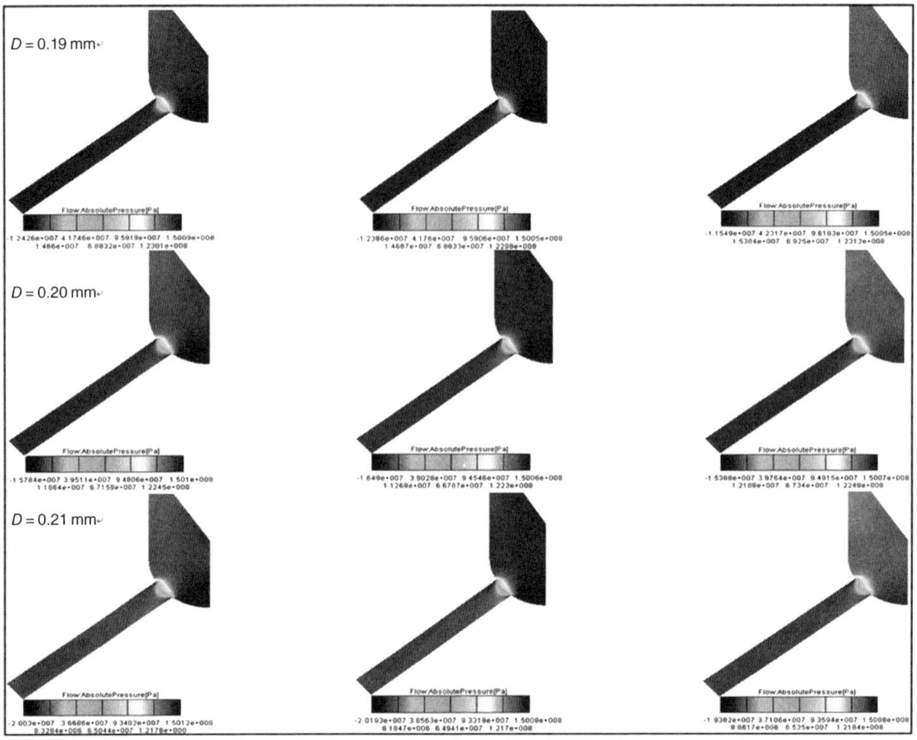

Figure 3.46 Pressure distribution on the central section of the nozzle.

Figure 3.47 shows the velocity cloud of the three nozzles at the same time for early, middle, and late stages of the injection. The three kinds of nozzle have a smaller orifice speed at the outlet of the nozzle when the needle lift is small, so the spray characteristics will be relatively poor at the initial stage of the injection. When the needle valve lift is larger, the speed of the outlet of the orifice is relatively high and the atomization and mixing will be better. In addition, it can be seen that the inner velocity distribution of the orifice of the nozzle with a small aperture is more uniform and the velocity is higher. The main reason is that the cross-section area of fuel injection from the pressure chamber into the orifice is relatively small, so the flow rate is increased. From Figure 3.5, we can see that the small aperture has a smaller cavitation intensity, so the velocity distribution is uniform.

The density of the curves shown in Figure 3.48 represents the intensity of the gradient change, while the depth of the color represents the flow rate and the turbulence kinetic energy. The turbulence near the outlet of the orifice plays an important role in the primary atomization process of the fuel injection, and the increase of turbulence will increase the initial disturbance of the primary atomization. The higher the average velocity of the nozzle outlet, the better is the atomization quality.

By looking at the figure, we can see that the larger the aperture, the more obviously the change is of the gradient of flow velocity and turbulence kinetic energy. This indicates that the cavitation intensity affects the degree of turbulent flow inside the orifice

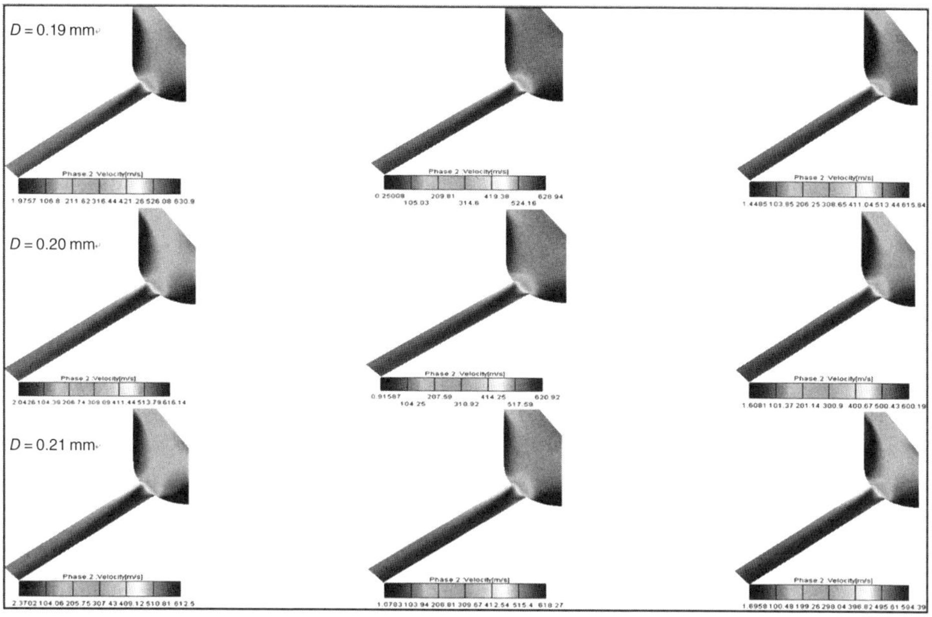

Figure 3.47 Velocity distribution on the center section of the nozzle.

and transfers the effect to the outlet of the orifice. The turbulent kinetic energy of the 0.21 mm aperture is larger, the intensity of the turbulence increases, and the energy loss is relatively large. The range of flow velocity (538–591 m/s) in the middle section of the fuel injection is greater than that of the 0.19 mm orifice (550–590 m/s). The velocity gradient of the fluid in the orifice is large, while the turbulent kinetic energy gradient of the 0.2 mm orifice is small. The outlet velocity is the smallest, but the high-speed area is larger than that of the 0.19 mm aperture.

In addition, it can be seen that with the increase of the needle valve lift, the average speed of the outlet of the orifice will increase. When the needle valve lift is smaller, the average speed of the outlet of the nozzle with the smaller aperture is larger. When the lift of the needle valve is large, the range of $d = 0.21$ mm is the largest of the outlet speeds than that of the former two. The high-speed area is larger and the average rate becomes larger. This indicates that the larger aperture of the orifice makes the effective outlet area smaller because of the larger cavitation area, which makes the outlet speed increase.

When the cavitation is formed, the greater the strength of the cavitation becomes, and the velocity gradient and the turbulent kinetic energy gradient on the outlet cross-section change fiercely.

3.2.2.2 The Influence of the Ratio of the Length to the Diameter of the Orifice

In view of the specific nozzle structure, the flow coefficient is at a maximum when the length and diameter ratio is at a certain value, and the flow coefficient will be reduced when the value is less or more than that. Some studies have shown that the best proportion of orifice length and orifice diameter is 4. At this time, the quality of fuel atomization and the penetration distance of the oil beam are all in the best state; otherwise, the injection pressure will be weakened and the spray characteristics will be affected.

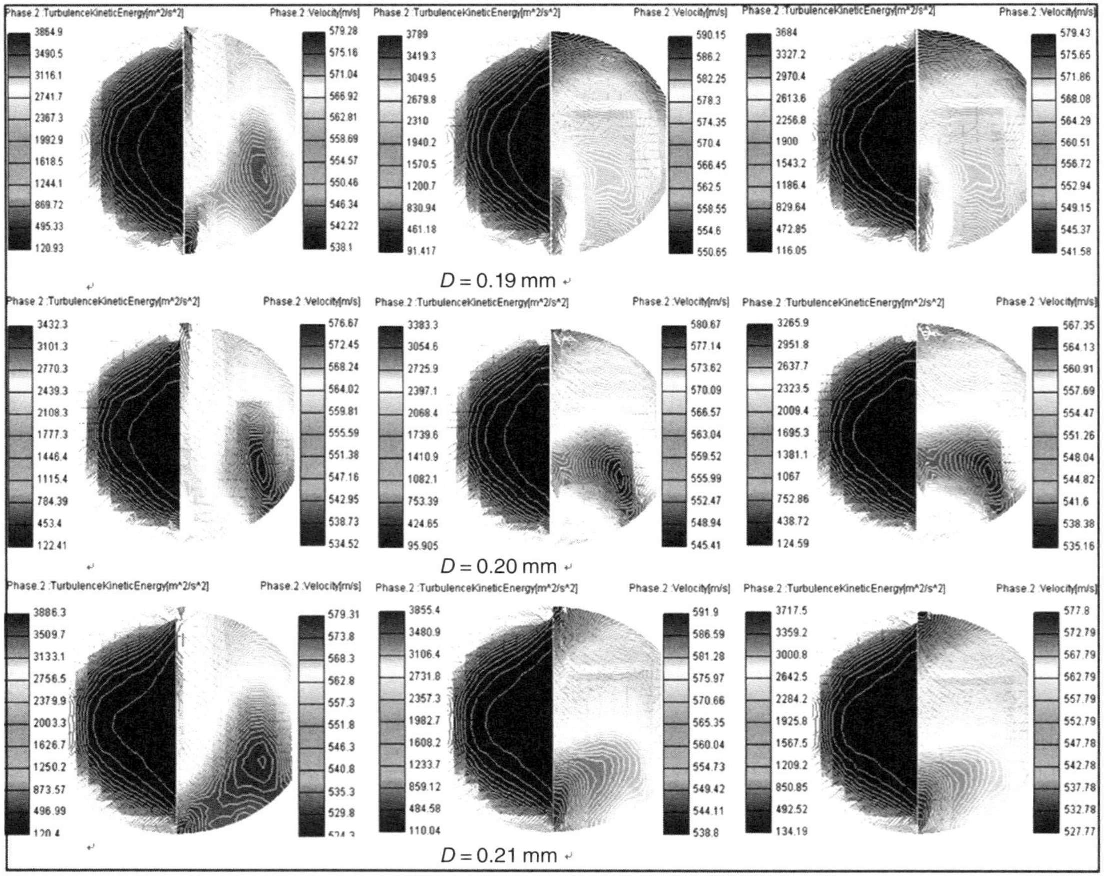

Figure 3.48 The orifice outlet turbulent kinetic energy distribution (left) and velocity (right).

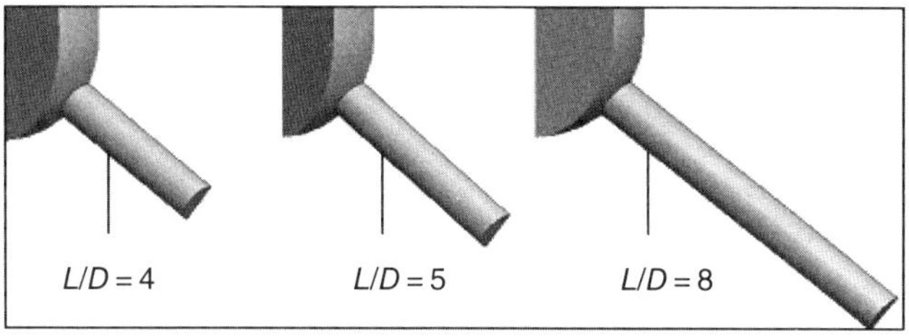

Figure 3.49 A schematic diagram of different ratio nozzle structures.

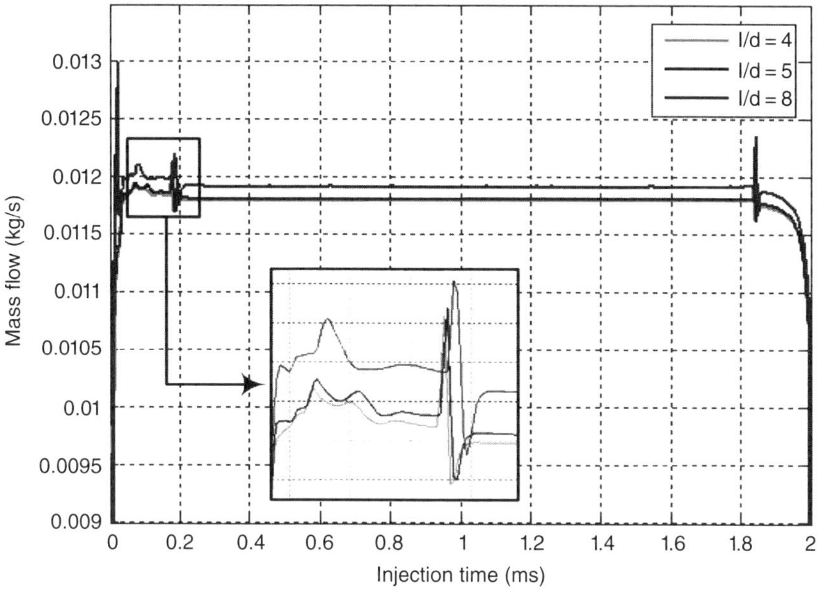

Figure 3.50 Mass flow of different ratio nozzles.

A schematic diagram of the nozzle structure is shown in the nozzle length to diameter ratio in Figure 3.49. The nozzle diameter of the model is D of 0.2 mm and the ratio of the length of L to the nozzle diameter D is 4, 5, and 8, respectively.

In the calculation, all parameters are consistent, and each calculation result is compared with the cloud map at the same time at the early, middle, and late stages of the injection, so that the flow field inside the nozzle can be observed intuitively. The results of the simulation are shown in Figure 3.50. From Figure 3.50, we can see that the influence of the orifice length to diameter ratio on the fuel injection rule is that in the early stage of fuel injection, the mass flow rate at the early stage of oil injection with the smaller ratio is larger and the injection time is longer. In addition, the mass flow of a longer orifice is larger, but the difference is not very great. The main reason is that the greater the ratio, the more the drag force increases and the flow coefficient decreases.

From Figure 3.51, it can be seen that the form of pressure field of the nozzle with a large ratio is relatively slow and the pressure gradient is larger. Therefore, under the same initial condition, there is a small lag of the nozzle with a large ratio compared to that of

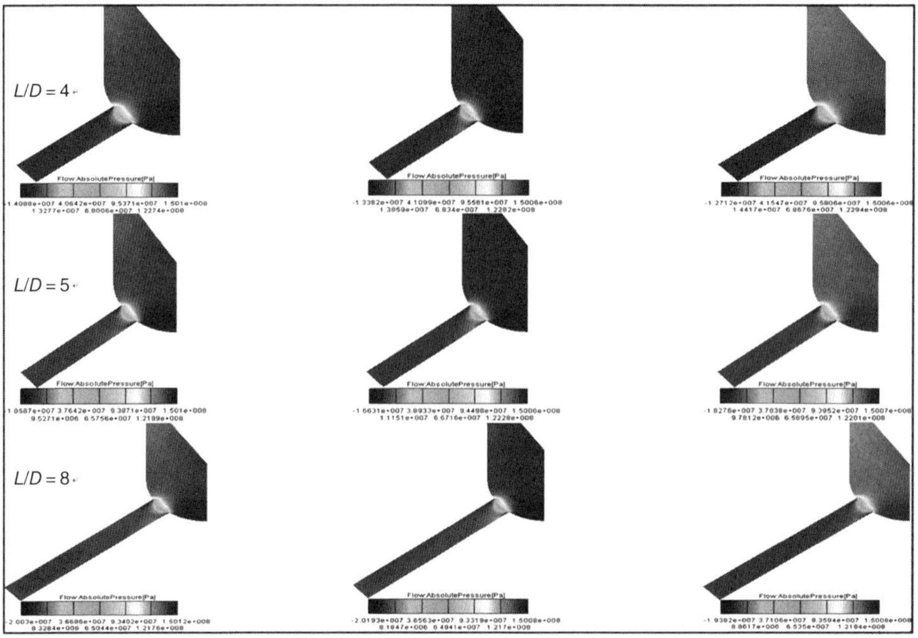

Figure 3.51 Pressure distribution on the center section of the orifice.

a small ratio nozzle. In addition, the nozzle that has a ratio of 5 has the largest negative pressure at the inlet of the orifice.

Figure 3.52 shows the calculated center section of the orifice and the gas volume fraction distribution of the wall. It can be seen that the cavitation of the three different ratio orifices has been produced under a high injection pressure. The cavitation is extended along the axis of the orifice to the outlet of the nozzle, that is, the so-called "super cavitation." With the increase in the ratio of length to diameter, the cavitation layer becomes thicker and thicker, and the cavitation is gradually spread to the center of the hole. Starting from the injection cavitation atomization theory, when cavitation bubbles provide maximum perturbation kinetic energy for jet atomization in the injection, the center collapses, so the longer the orifice diffusion to spray the hole center location continues the more likely it is that the injection atomization of the liquid is increased, which has been proved by Tamaki Nobushige. It is also found that, with the increase in the length to diameter ratio L/D, the hole strength on the orifice cross-section is not obvious. The area of the strong hole on the wall of the orifice decreases, and decreases gradually at the outlet of the nozzle.

From Figure 3.53, the turbulent kinetic energy distribution and gas volume distribution cloud distribution in the outlet of the orifice can be found, where the hole strength in the outlet section decreases as the length to diameter ratio L/D increases. However, the distribution surface of the section is more extensive and even, which can achieve an effect similar to those of the gas fuel injection and the flash fuel injection, but is more beneficial to fuel atomization. The larger the ratio of the turbulent kinetic energy gradient, the greater is the gradient, which indicates that the difference in the cavitation area causes the difference in the degree of disturbance of the fuel flow in the outlet section.

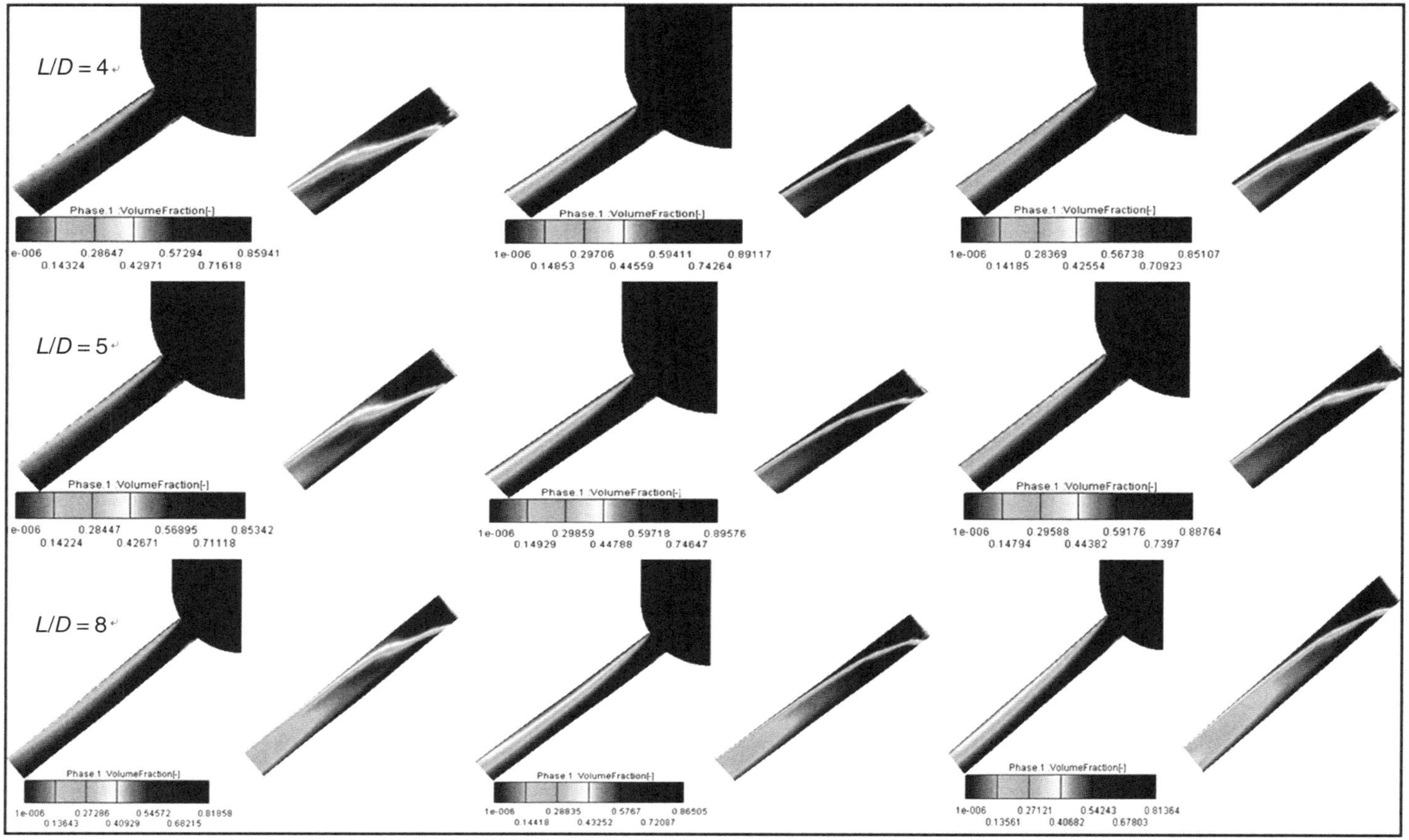

Figure 3.52 Gas phase volume distribution on the center section of the orifice and the wall surface.

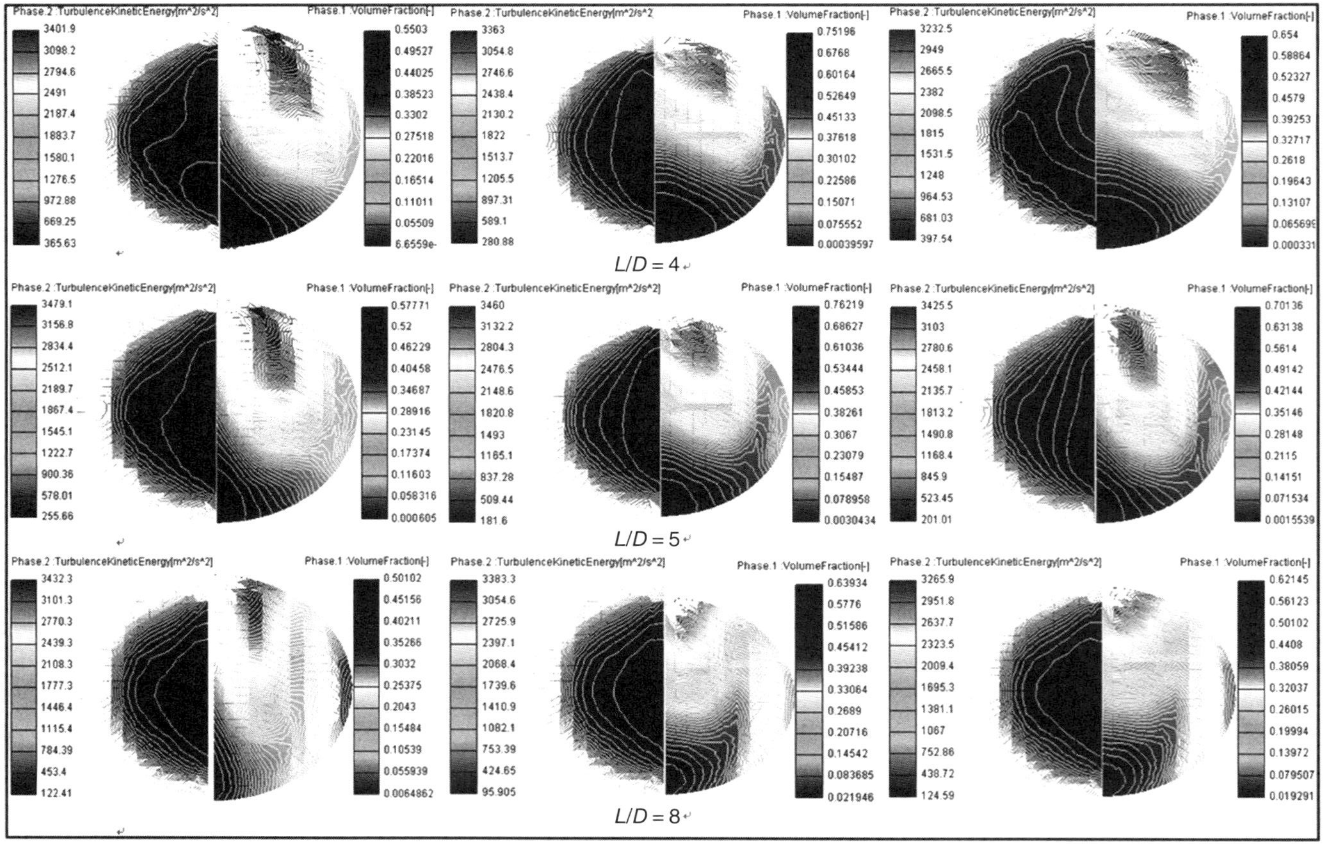

Figure 3.53 The orifice outlet turbulent kinetic energy distribution (left) and the gas phase volume distribution (right).

It can be speculated that the conical angle of the oil beam is distinctly different because of the difference between the ratio and the disturbance caused by the hole in the orifice to the liquid.

From the velocity cloud shown in Figure 3.54, it can be seen that with the increase in the length to diameter ratio, the overall flow velocity in the nozzle becomes larger, and the kinetic energy and the initial velocity are higher. Therefore, it can be seen that the average velocity of the length and diameter is slightly larger than the one with a larger ratio in the center section of the orifice. On the outlet section, the nozzle with a ratio of 5 is higher than that with a ratio of 4. Although the high-velocity value and the region range of nozzle with the ratio of 8 are smaller, the low velocity is much higher and the average velocity is larger. This is reflected in Figure 3.53, which shows that the strength of the cavitation hole is large but the whole area is smaller on the surface of the outlet, causing the outlet speed to be higher in a certain area, while the air hole strength of the orifice with the larger ratio is weaker. However, the whole area is large, which greatly reduces the effective flow area of the outlet and increases the average velocity. Therefore, this will have a great influence on the penetration distance of the oil beam.

3.2.2.3 The Influence of the Round Angle at the Inlet of the Orifice

According to the experience of Italy OMT Company, the flow coefficient of the orifice at the sharp edge is about 0.66. With the increase of the round angle of the inlet of the nozzle, the flow coefficient will increase gradually, until it is close to 0.95; then the round angle of the inlet of the nozzle is further enlarged, after which there is no obvious effect. According to the relevant data, the flow coefficient is constant when the dimensionless amount of R is greater than 0.3 (D is the diameter of the orifice and R is the mean radius of the round angle of the inlet of the nozzle). When R is greater than 0.3, the flow coefficient can be increased by 20%. On the basis of keeping the other structure of the original orifice unchanged, an attempt is made to reverse the fillet at the inlet of the nozzle to improve the flow of fuel at the inlet of the nozzle. In accordance with the same method, the three-dimensional model of the radius ratio of the round angle of a different entrance is set up, as shown in Figure 3.55.

All the parameters in the nozzle model after the fillet setting are consistent with the model of the round angled nozzle. Each calculation result can only be compared with the cloud map at the same time during the injection, while the flow field in the nozzle can be observed intuitively. The simulation results are as follows.

Figure 3.56 shows the injection law curve of six different nozzles. For simplicity, the six kinds of nozzle are numbered from 1 to 6. We can see that if the nozzle diameters are the same, the mass flow for a closed angle orifice is larger than that of a round angle nozzle. If the mass flow rate of the nozzle is equal, the inlet structure of the fillet orifice can reduce the diameter of the orifice, and the mass flow rate increases with the increase of fillet radius. It can be seen that the fillet can increase the injection volume and discharge coefficient simultaneously, as shown in Figure 3.57, but when the fillet radius increases to a certain degree, the flow coefficient obviously also increases. In the rising stage of the needle valve, the mass flow curves of nozzles 2 and 6 are smaller than those of the other four nozzles, and the curves of the nozzles change more evenly. It is obvious that the radius of the round angle of the entrance is larger, which is beneficial to the increase of the flow coefficient of the orifice and the decrease in the sensitivity to the change of the round angle, ensuring the stability of the jet flow and uniformity of the pore flow

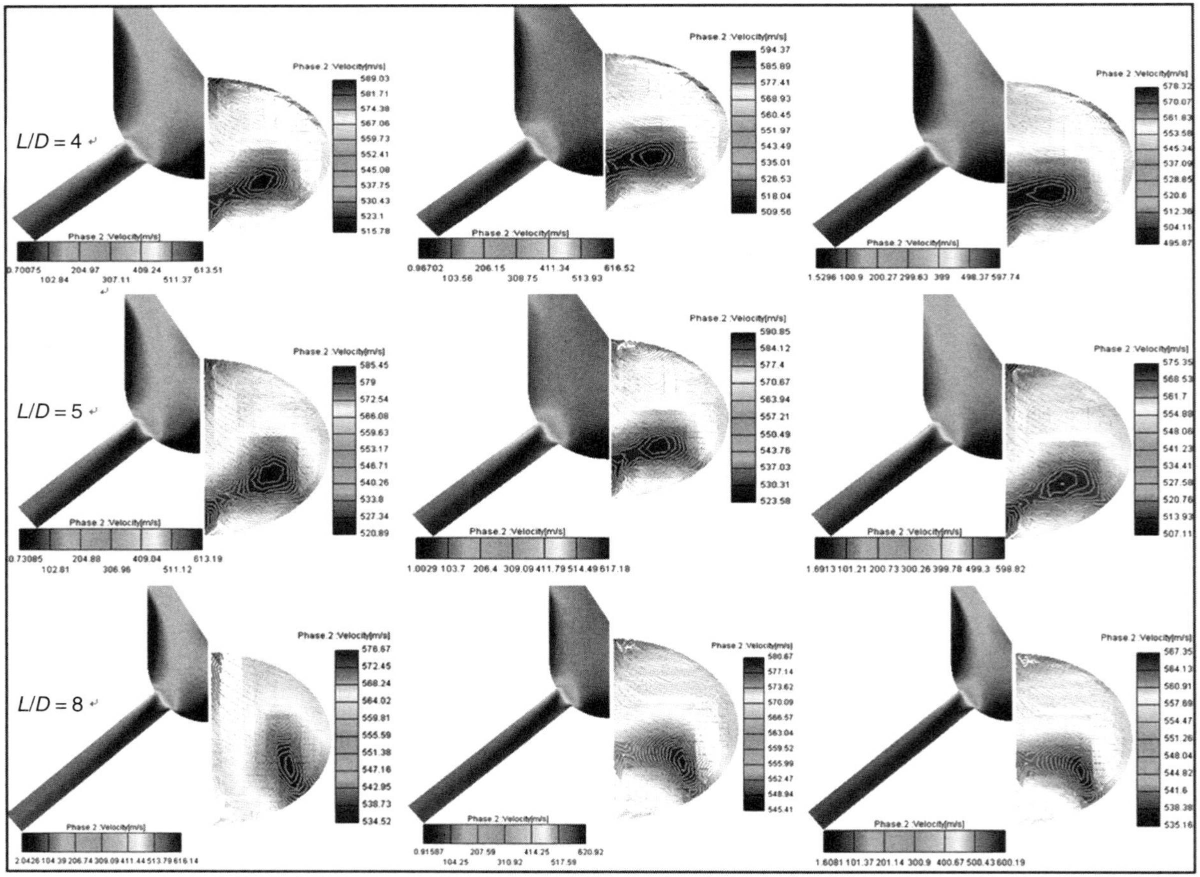

Figure 3.54 Velocity distribution on the center section of the orifice and the cross-section of the outlet.

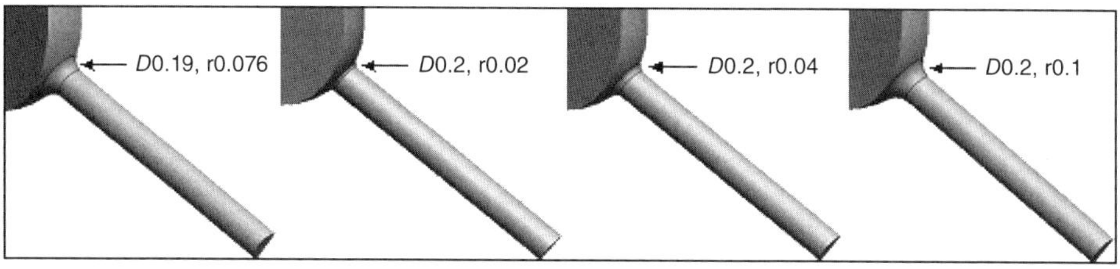

Figure 3.55 A schematic diagram of the geometric structure of different radius ratios.

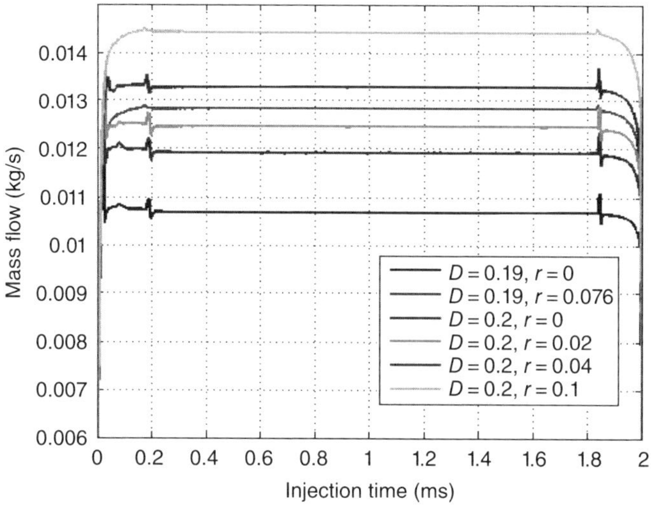

Figure 3.56 Mass flow curve of the nozzle with different radii of a round angle.

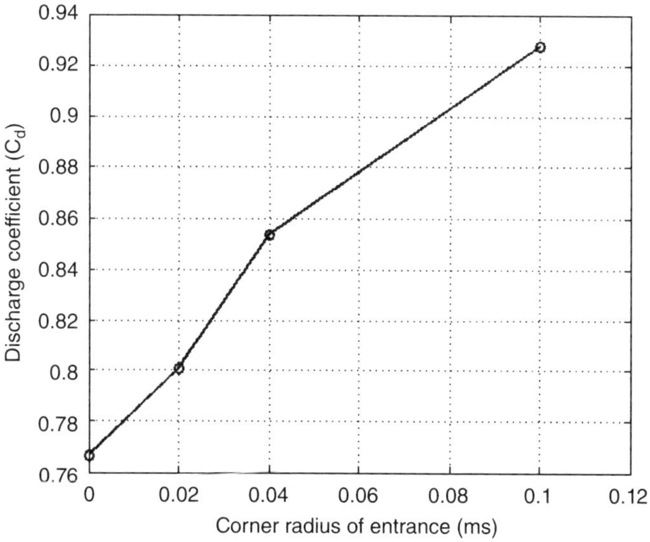

Figure 3.57 The influence of the radius of the entrance round angle on the flow coefficient.

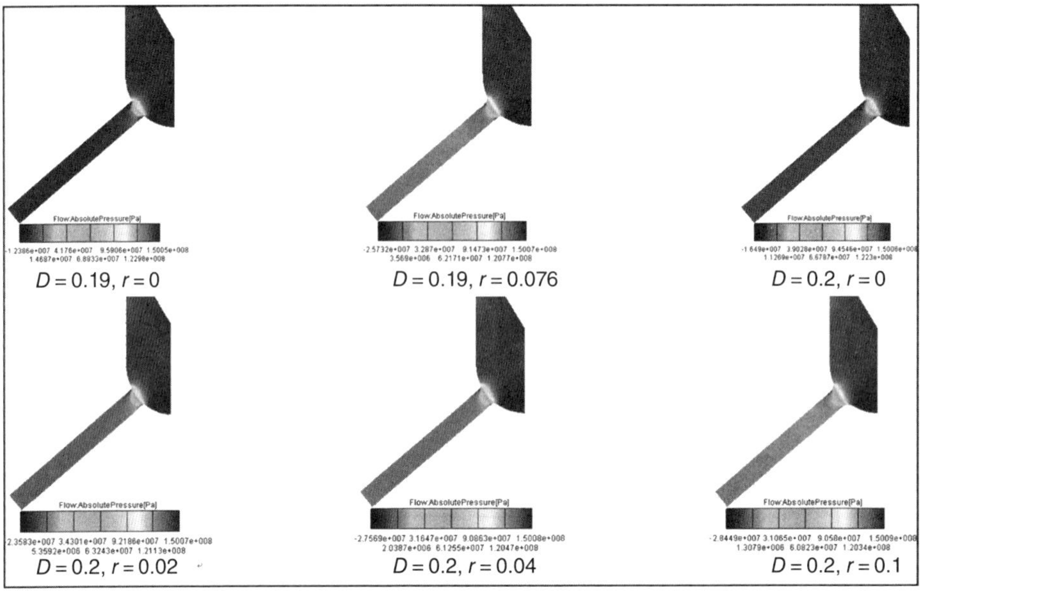

Figure 3.58 Pressure distribution on the center section of the orifice.

rate of the porous nozzle. Through comparison, it can be seen that the influence of the entrance round angle on the mass flow ratio compares to that of the aperture and the length diameter ratio.

Comparing the pressure cloud (Figure 3.58) of these nozzles and observing the pressure distribution at the entrance to the orifice, the green area of the entrance is deepened with the radius of the round angle in the axial direction of the orifice. This indicates that increasing the radius of the inlet round angle decreases the amount of fuel along the low-speed area near the wall of the pressure chamber, improves the direction of fuel flow, and thus makes the pressure loss decrease, which makes the pressure at the inlet of the orifice higher. The negative pressure area of the inlet round angle of the orifice is greatly reduced and the absolute value of the negative pressure obviously decreases, for the greater the radius of the round corner, the more obvious is the decrease. This is due to the change in the fuel flow direction at the inlet of the round angle and the smaller pressure drop than that of the sharp angle nozzle. In addition, the effective injection pressure is increased after the inlet of the orifice is rounded.

From the gas distribution map (Figure 3.59), it can be seen that the cavitation phenomenon of the sharp angle orifice is more obvious than that of the round angle orifice. The effective flow area of the orifice outlet is also greatly reduced. This is because the sharp angle orifice flows fiercely at the corner, and it is easy to form a low pressure or even a negative pressure area, making cavitation forms obvious. With the increase in the radius of the inlet round angle, the thickness of the cavitation and the distribution area on the wall obviously decrease, and the length of the extension also becomes shorter. The dimensionless R of nozzle 6 is the largest, and the cavitation at the lower edge of the nozzle completely disappears, with only a small area remaining on the upper wall. Although the radius of the corner increases, the influence of the cavitation factor is reduced and the uniformity of the flow rate is improved. However, the overall strength of the cavitation is weakened, especially the weakening of the cavitation strength at the

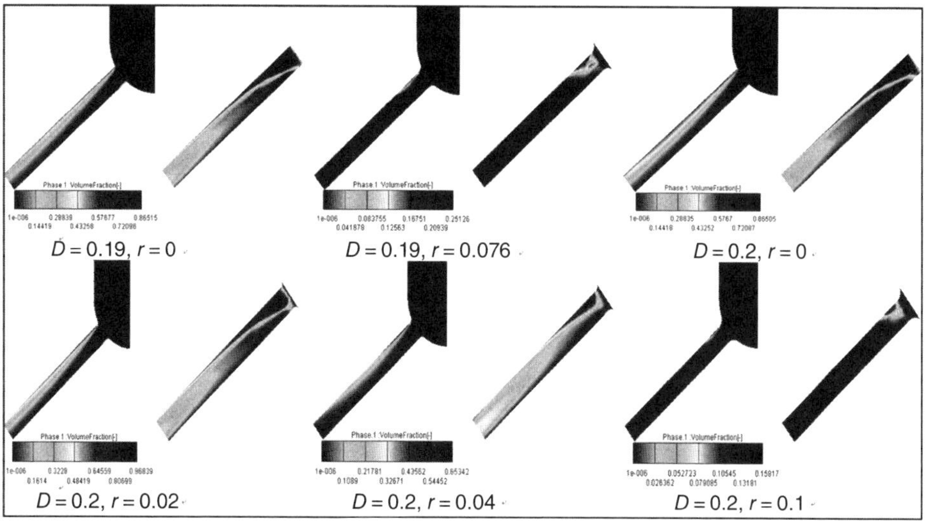

Figure 3.59 Gas phase volume distribution on the center section of the orifice and the wall surface.

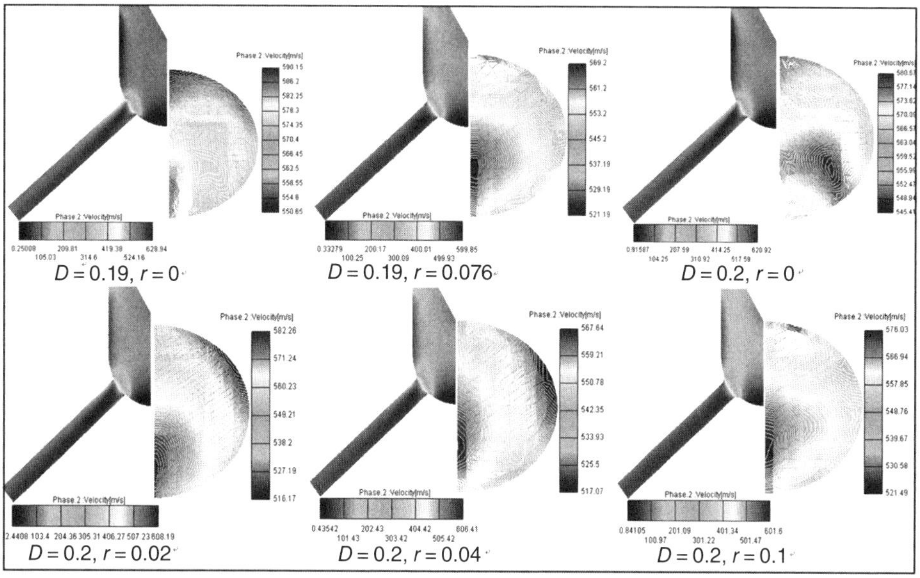

Figure 3.60 Velocity distribution on the center section of the orifice and the cross-section of the outlet.

center of the nozzle, which will not be beneficial for atomization of the jet flow. Therefore, it is not good for the radius of the circle to be too big. When processing the round angle of the nozzle, we need to think comprehensively about it in order to achieve better performance of the diesel engine, with the radius of the inlet round angle properly chosen for that size of nozzle.

It can be seen from the velocity distribution cloud (Figure 3.60) that the velocity range of oil nozzle 1 is 550–590 m/s. The speed of nozzle 2 with a radius of 0.076 mm round angle is reduced to 521–569 m/s. This is because the sharper the inlet of the orifice, the more serious the cavitation is, and the effective flow area of the outlet of the orifice is reduced while the speed is obviously increased. The total and exit speed of nozzles 3 and 6 is reduced as the radius of the round angle increases. The reduction of velocity of nozzle 4 is the smallest, while nozzle 5 is the slowest with a radius of 0.04 mm round angle, and thus the average speed is the minimum. However, the increase of the nozzle outlet velocity with the 0.1 mm round angle shows that the speed decrease is weaker. At the same time, it can be seen that the flow velocity inhomogeneity on the section of the orifice outlet is smaller and the upper flow velocity is increased.

According to the relevant literature, a variety of processing methods can be used to round off the angle of the nozzle at the inlet of the orifice, including the hydraulic extrusion grinding method, the thin liquid flow grinding method, the electric spark or electrolysis method, etc. The direct injection engine used in the car's hole nozzle has widely used EDM plus the abrasive grinding processing method to control the flow dispersion. At the same time, because of the method of abrasive extrusion processing, it has a positive impact on forming the round angle on the inner edge of the nozzle, improving the flow coefficient of the nozzle, improving the service life of the nozzle, improving the surface roughness of orifice, and improving the spray shape. The DYNETICS Company of the United States uses a thin liquid flow grinding method to grind small holes and to remove burrs and round holes. The Japanese DENSO Corporation also has the ability to use the thin liquid flow grinding method to round off the angle of the nozzle at the inlet of the orifice, and the processing accuracy has reached 0.001 mm. With the increase of the flow coefficient of the orifice, the diameter of the orifice will gradually decrease, and the application trends of the porous nozzle in the direct injection diesel engine is due to small hole technology. At present, some famous foreign nozzle manufacturers, such as BOSCH and DENSO, have applied EDM technology to process small orifices.

3.2.2.4 The Influence of the Shape of the Needle Valve Head

In order to understand the influence of the head shape of different needle valves on the flow field, a nozzle model of a single cone needle valve and a double cone needle valve has been established. Aiming at the phenomenon that the sealing head seat of the flat head needle nozzle with the head cutting off is not closed and the spray is poor during the test, a model was established where and the internal flow field is analyzed and the reasoning to support it is analyzed. A schematic diagram of the shape of a different needle valve head is shown in Figure 3.61. A double cone side valve front chamfer is 100°.

In the calculations, all parameters are set together. Each calculation result is selected to compare the cloud data at the same time in the early, middle, and late stages of the injection, and the flow field inside the nozzle can be observed intuitively. The simulation results are as follows.

The mass flow comparison of the nozzles of three different needle valve head shapes are shown in Figure 3.62. You can clearly see that the double cone needle valve nozzle outlet mass flow is large, with a smaller mass flow for a single side valve cone nozzle. This is because the flow section is smaller and the double cone needle valve nozzle internal flow and outlet velocity is larger than that of the single cone valve. In the pre-injection period, the mass flow of the flat head needle valve is much smaller than the previous two. In the middle of the injection, although the mass flow is slightly larger than that of single cone side valve, the mass flow is very unstable, showing a lot of fluctuation.

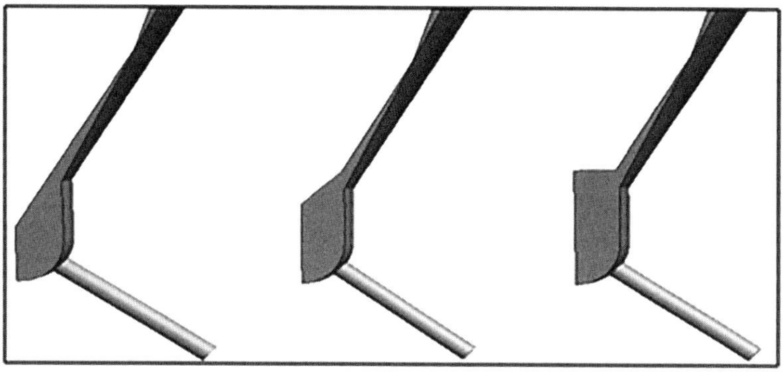

Figure 3.61 A schematic diagram of three kinds of needle valve structures.

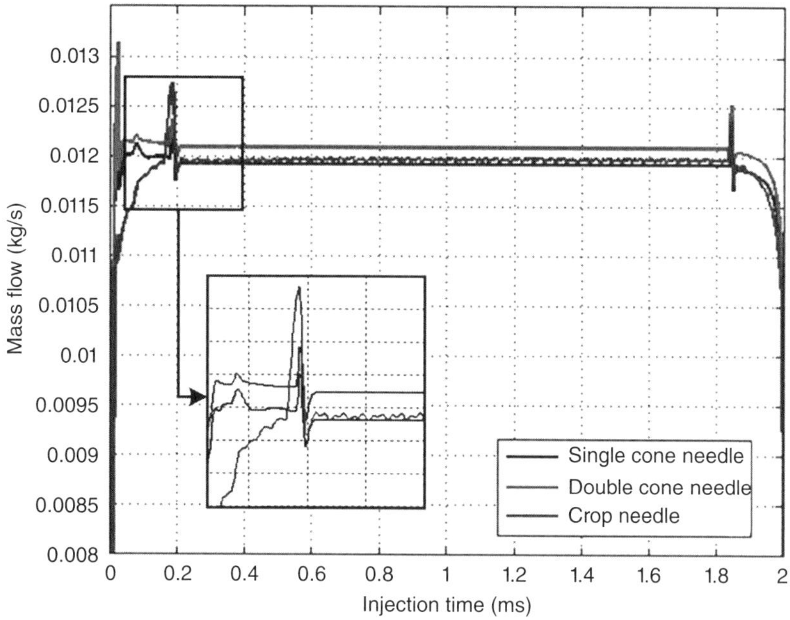

Figure 3.62 Mass flow of three needle valve shaped nozzles.

Figure 3.63 shows the pressure cloud map distributions of three kinds of nozzle. The head structures of the three kinds of needle are different, causing the shape and volume of the pressure chamber to be different, so the resistance of the fuel oil entering the pressure chamber is different. It can be seen from the diagram that the pressure gradient distribution in the pressure chamber of the three kinds of nozzle is very different. The pressure gradient in the pressure chamber of the single cone nozzle is smaller than that of the flat head needle valve. The pressure gradient of the nozzle of the double cone is the smallest and the distribution is uniform. The flow area at the inlet of the pressure chamber is large, and it can be seen that the pressure gradient of the pressure chamber that occurred between the needle valve and the seat surface becomes smaller. This shows that the throttle loss between the needle valve and the seat surface is increased under

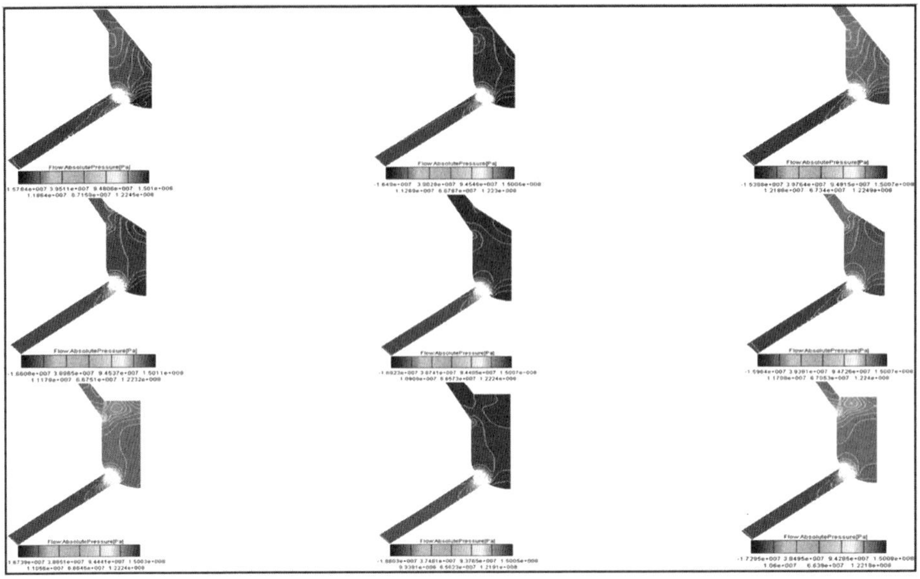

Figure 3.63 Pressure distribution on the center section of the orifice.

the pressure of second conical surfaces, and the pressure loss is controlled. Under the same initial conditions, the formation of the pressure field of the flat head needle valve is very slow. The pressure gradient under the head at the early and late stages of injection is very large, which is probably the cause of the closure of the needle valve, which leads to the formation of swirl here, causing momentum loss and affecting the rate of flow and the velocity of flow. It can be seen that the volume of the pressure chamber of the flat head needle valve is the largest, which explains why there is leakage of fuel oil and an increase in deposits of carbon on the head of the needle valve.

Figure 3.64 shows the velocity distribution cloud map. It is known by comparison that the difference in velocity distribution is caused by the different circulation area at the inlet of the nozzle pressure chamber. The velocity distribution of the nozzle pressure chamber of the double cone side valve is relatively uniform, and at the inlet and the outlet of the nozzle the velocity is high. This shows that by improving the flow area at the inlet of the pressure chamber, we can reduce the loss of throttle and increase the flow rate of fuel. In the early and late injections, the flow velocity in the pressure chamber of a single cone flow side valve nozzle is relatively high, which is because the pressure chamber inlet flow section of a single cone side valve nozzle is small, due to the increased fuel flow caused by the throttling effect. The cross-section area of the fuel flow of the pressure field at the inlet of the double cone needle valve increases suddenly due to the chamfering of the head of the needle valve, and the fuel flow rate will naturally decrease but the pressure will also expand here. In addition, it can be seen that the velocity distribution in the inner part of the nozzle pressure of the flat head needle valve is very uneven and the vortex is formed under the head of the needle valve, resulting in momentum loss. In the stage of small needle valve lift, the flow rate of the fuel is much lower than that of the first two kinds of needle valve, which makes the mass flow smaller and has a great influence on the spray.

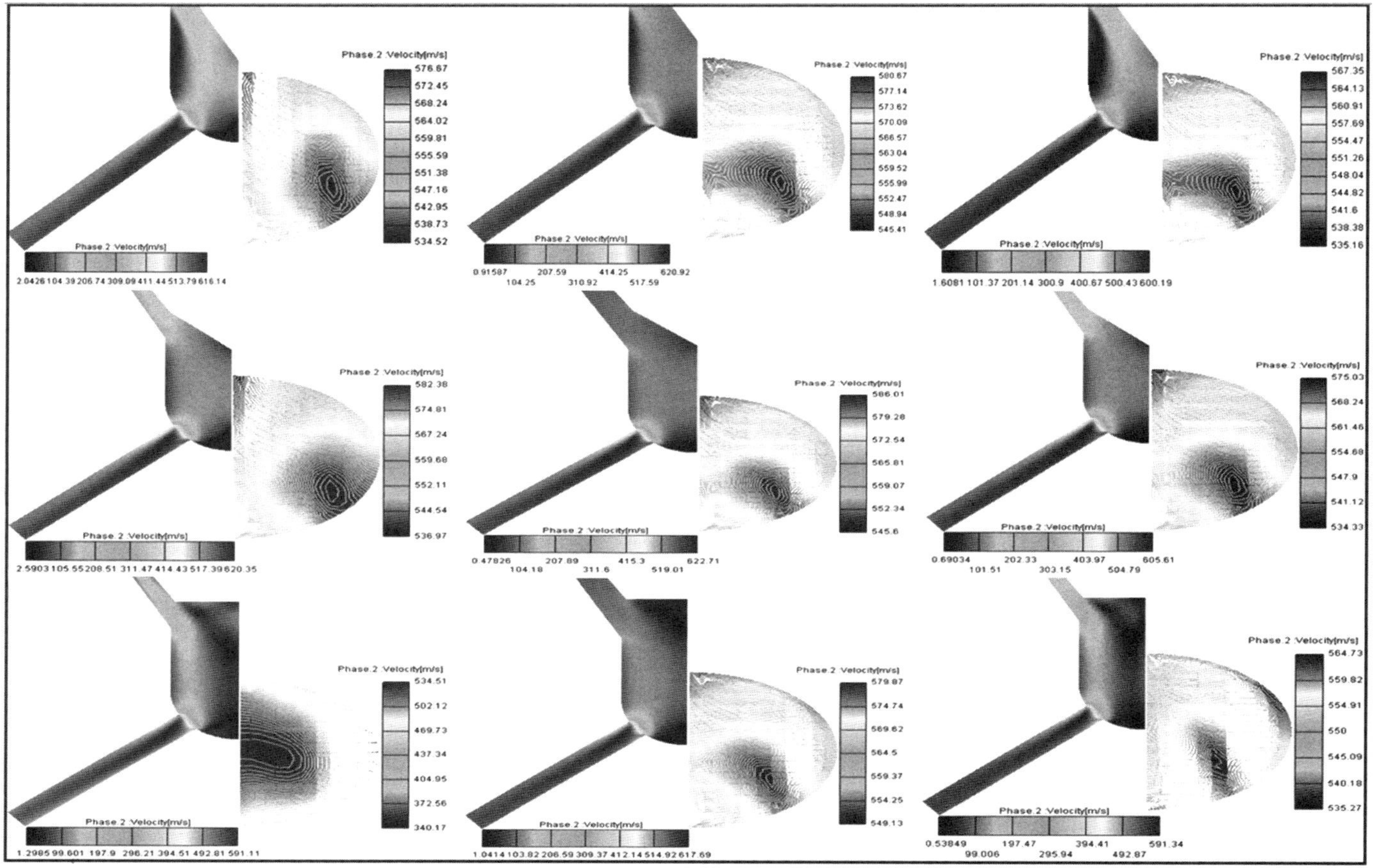

Figure 3.64 Velocity distribution on the center section of the orifice and the cross-section of the outlet.

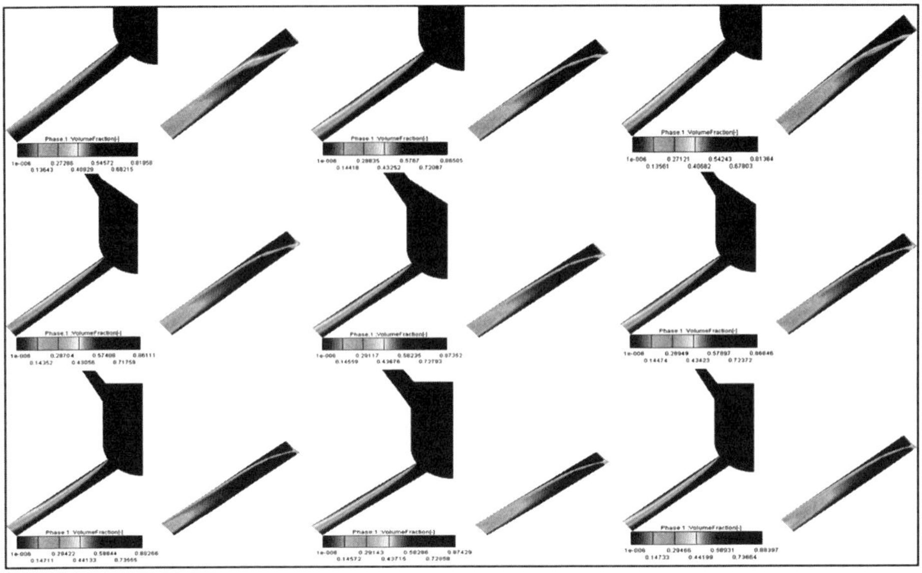

Figure 3.65 Cavitation distribution on the center section of the orifice and the cross-section of the outlet.

From the diagram in Figure 3.65, it can be seen that the cavitation area in the orifice of the single cone nozzle is smaller and the thickness of the cavitation is thinner during the small needle valve lift. The volume of the gas phase in the nozzle with the flat head valve is the largest, which greatly reduces the mass flow. Although cavitation of the double cone side valve is thicker than that of the single cone, the size distribution of the gas phase region is more concentrated and stable. The difference in the size of the cavitation area between the three kinds of nozzle leads to the differences in the velocity of the nozzle and the subsequent spray characteristics.

The simulation results show that the difference in the shape of the needle valve head is one of the main factors that cause the inhomogeneous flow of each orifice in the injector. By increasing the chamfering of the head of the needle valve, it will be beneficial to reduce the pressure loss at the inlet of the pressure chamber and increase the speed of the outlet of the orifice and increase the mass flow. There is a need to pay attention to the fact that the increase in the outlet velocity of the orifice can also increase the penetration distance of the oil beam and increase the possibility of the oil wall impinging. Therefore, it is necessary to further understand the influence of the cone shape and the pressure chamber size on the convection field.

3.2.2.5 Effect of the Injection Angle

The size of each injection orifice of the injector has an important influence on the performance indexes of the diesel engine, such as power, torque, oil consumption, and emission. Because the arrangement of the injector on the cylinder head of the diesel engine is restricted by structure, there is a certain amount of offset and angle between the center line and the central line of the cylinder and the central line of the combustion chamber. To enable the fuel bundles to spray out of the fuel injector and distribute reasonably in the combustion chamber, the angles between the orifices and the axes of the

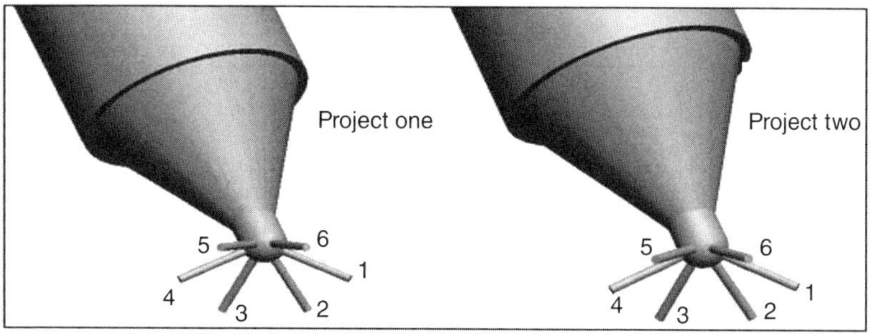

Figure 3.66 A schematic diagram of two kinds of scheme nozzle structures.

injector must be different, which will inevitably cause a difference in the flow volume of the different orifices. It is very important to obtain the jet flow and its mass in order to improve the combustion process of the internal combustion engine.

Establish two nozzle models as shown in Figure 3.66. The first one is the orifice with a cone angle of 140°, the second is the orifice with a cone angle of 120°, and the other structure dimensions remain the same. The calculated grid number is about 250 000 units and the total calculation time is 90 hours. The simulation results show that the mass flow of the two nozzles in each hole is inconsistent.

According to the configuration features of the injector on the cylinder head of the diesel engine, it is necessary to ensure that the height of the wall of the oil beam is consistent, with the axis of the needle valve body as the reference. The angle size of the six orifices in the two schemes are in turn: (orifice number) $4 > 3 = 5 > 2 = 6 > 1$. The inconsistency between the orifices and the angle of the nozzle axis will affect the consistency of the mass flow of orifices. Even if the two angles of orifices are the same, the mass flow difference can also be caused by the interaction of the flow between the two orifices. You can see the size of the mass flow of each hole in Figure 3.67. The mass flow of orifices 5 and 3 is relatively large, because the angle is the same, so the mass flow is basically the same. Although the angle of orifices 6 and 2 are the same, the mass flow is different, which is related to the size of the distribution of the cavitation area and the speed of the outlet. Orifices 1 and 4 are relatively small, with orifice 4 being the smallest. This is because their angle is the largest, the direction of fuel flow changes greatly, and the flow loss and fluid resistance are also large, resulting in a decrease in the discharge coefficient. The angle distribution of each orifice is different and the influence of interaction between orifices is different, so the flow loss and fluid resistance are different. At the same time, the distribution of turbulence in the pressure chamber is different, which leads to differences in the flow of each orifice. The uniformity of the mass flow of each hole in Figures 3.68 and 3.69 is better than that in plan 1, while the uniformity of the mass flow of the same angle nozzle is better. Due to the smaller injection angle of orifice 4, the flow loss is reduced and the mass flow is improved. However, it can be seen that the mass flow of orifices 1, 2, and 6 is reduced after changing degrees, causing the total mass flow to be smaller than that of plan 1, as shown in Figure 3.70. The main reason is that the fuel rate decreases as the angle becomes smaller and the distance between the fuel and the seat to the orifice is a long way away, causing the pressure drop to become larger and the fuel rate to reduce.

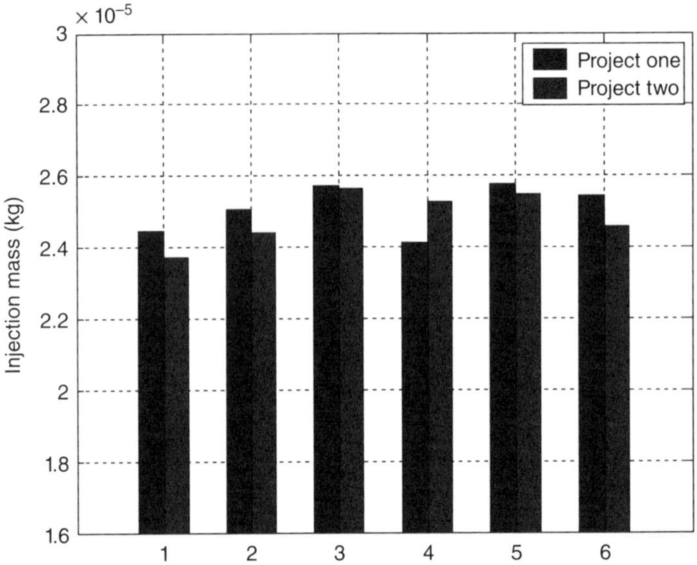

Figure 3.67 Flow distribution of each hole in two schemes.

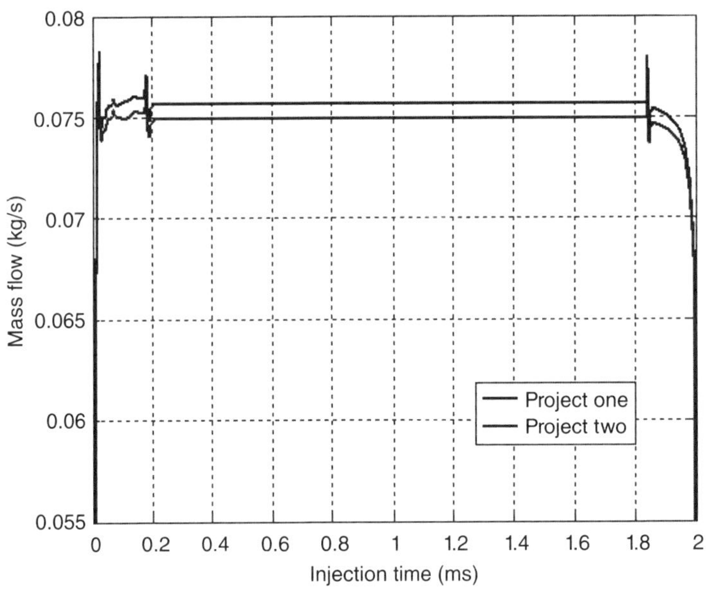

Figure 3.68 Total mass flow comparison of two schemes.

Figures 3.71 and 3.72 are two plans showing the distribution of pressure, cavitation, and velocity of the nozzle models in the early, middle, and late stages of injection. Because of symmetry, the velocity cloud on the outlet section is only compared with the injection orifices between 1 and 4 in the middle of the fuel injection. As shown in Figure 3.70a, the gradient distributions of the pressure fields of each hole are not the

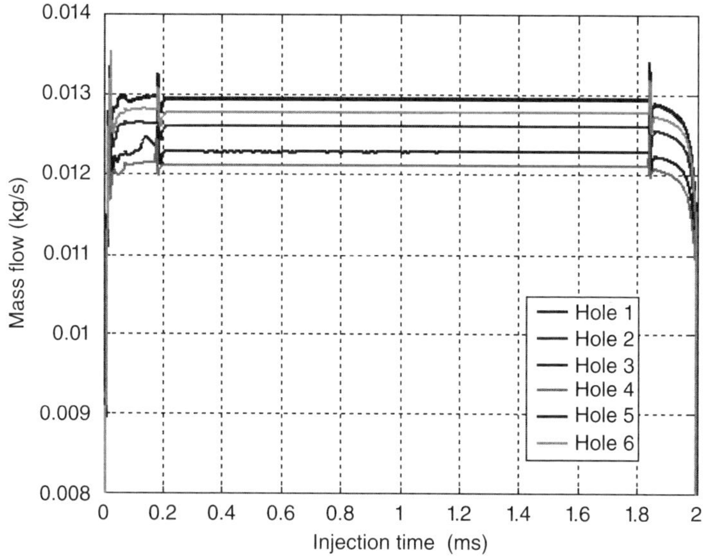

Figure 3.69 Mass flow of each orifice in a nozzle.

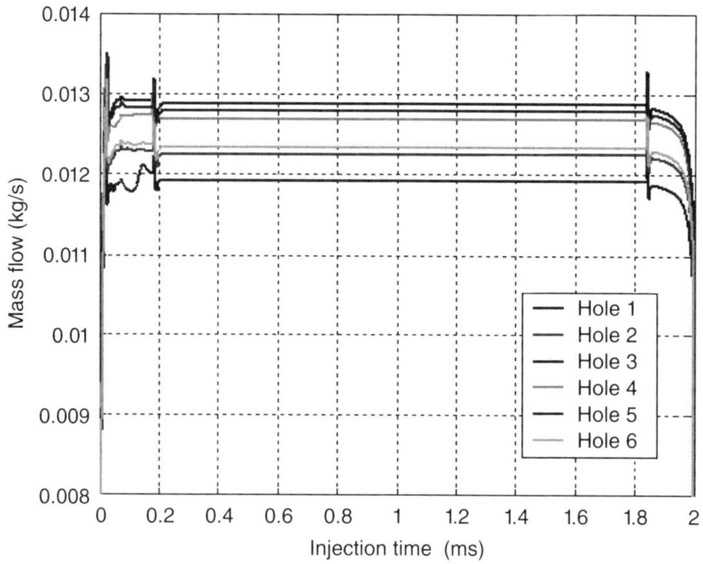

Figure 3.70 Mass flow of each hole in a scheme with two nozzles.

same because of the different angles. The fuel flows along the seat surface and flows in orifice 4 at the inlet of the orifice after largely changing direction. The cavitation is almost full of the upper half of the orifice due to the large bending degree of the corner. As a result, the effective flow area of the fuel is greatly reduced and the velocity under the orifice is very high, but the flow coefficient is reduced and the mass flow is also reduced. The distribution of the cavitation area is related to the angle of the

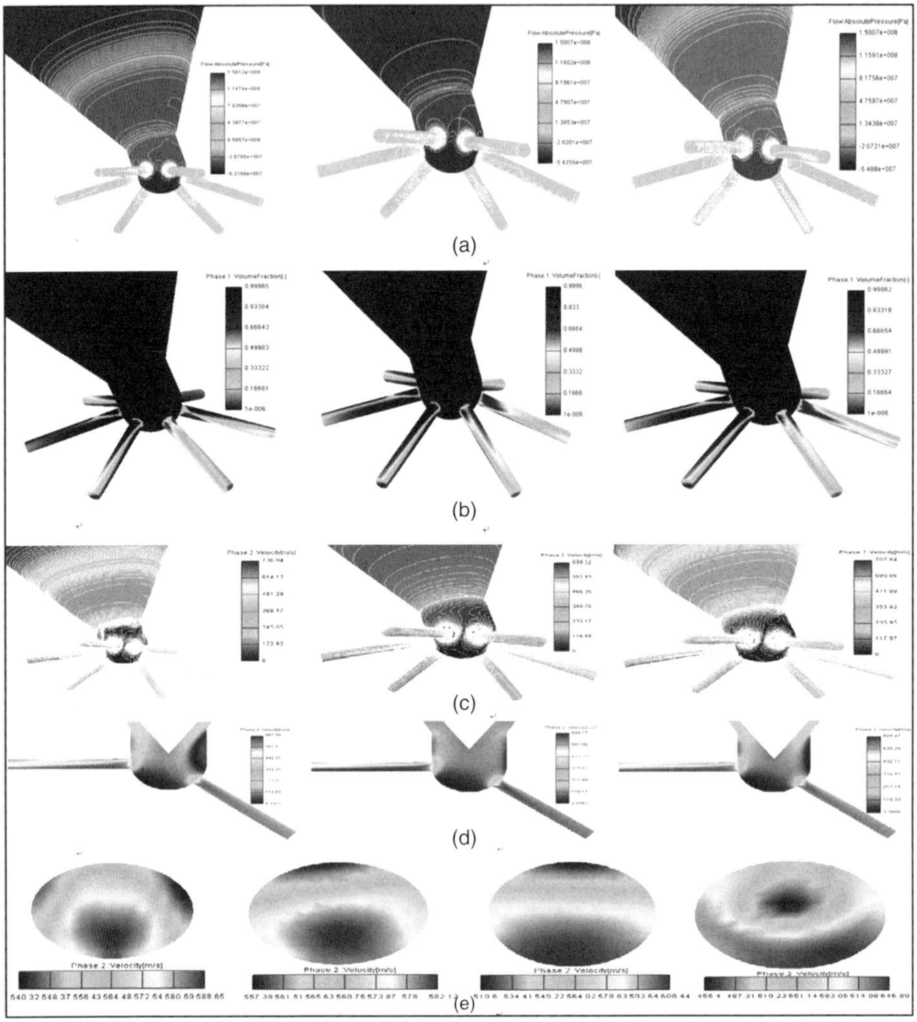

Figure 3.71 Plan 1 of nozzle pressure (a), gas phase volume (b), velocity (c, d, e) distributions.

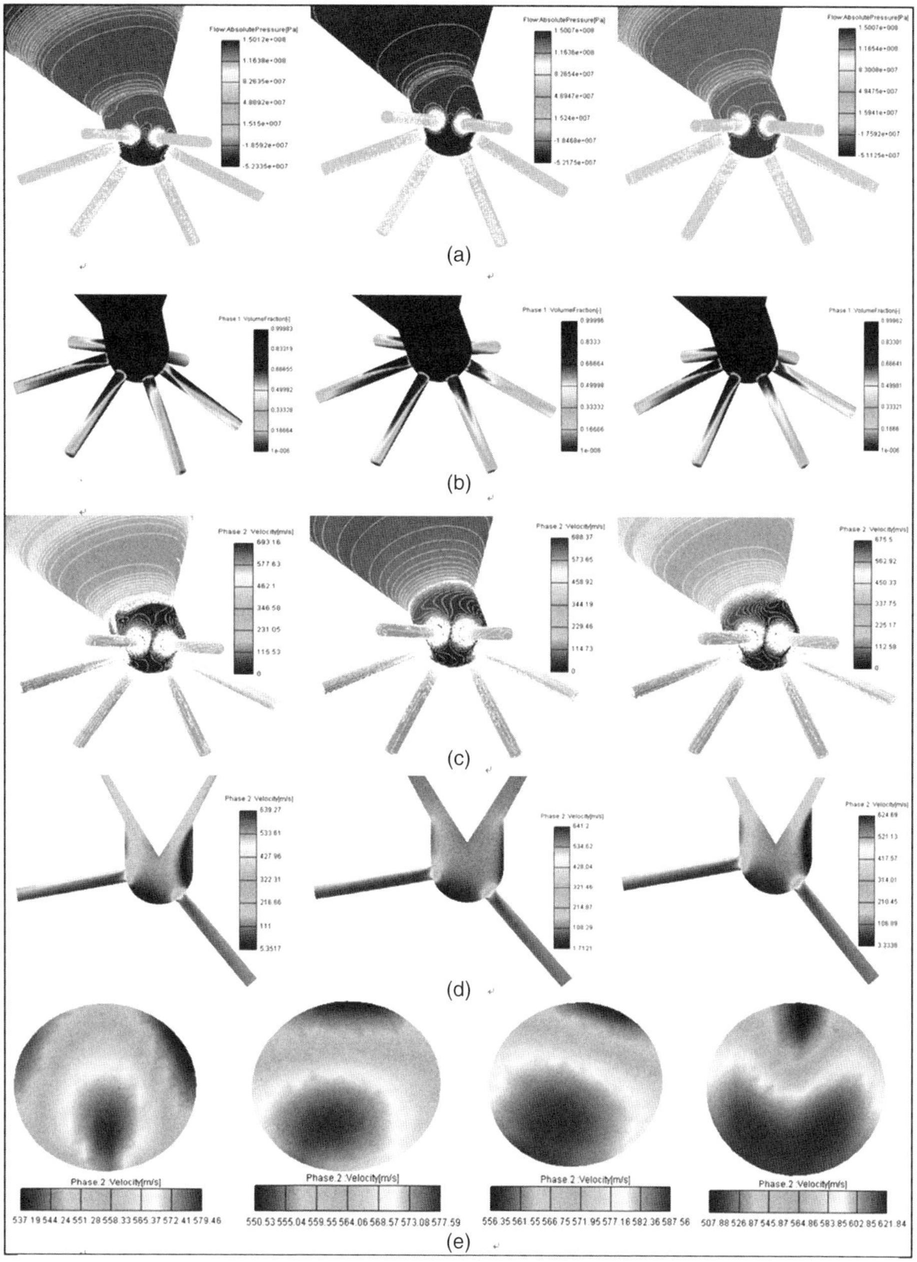

Figure 3.72 Plan 2 of nozzle pressure (a), gas phase volume (b), velocity (c, d, e) distributions.

orifice. We can see that the larger the orifice area, the smaller is the cavitation area, which means that the angle of the inlet of the orifice is also one of the factors that affect the cavitation distribution. The flow velocity in each orifice is different because of the pressure drop of each orifice and the size of the distribution of the cavitation area. It can be seen from Figure 3.71c to e that the average flow velocity inside orifice 3 is higher. Although the angle of the orifice is larger than that of orifice 1, the internal cavitation is evenly distributed and the area is small, so the mass flow is also larger. In addition, from Figure 3.72, it is found that the uniformity of the pressure field gradient and the gas hole distribution of each hole in plan 2 is improved after changing the orifice angle. On the flow rate, the speeds of orifices 1 and 2 decrease after changing the intersection angle, and the velocity of orifice 4 increases greatly, mainly because the intersection angle increases, thus improving the direction of the fuel flow, and the pressure loss is reduced, the cavitation intensity is weakened, and the distribution area is uniform.

The flow rate of each orifice in the nozzle is normally distributed and the fuel can be distributed in the whole combustion chamber in a relatively uniform way, which is beneficial to the formation of a more uniform mixture. If due to various factors the flow distribution is abnormal, on the one hand the spray pattern will be distorted, but, on the other hand, the spray range of the orifice changes, resulting in the orifice of the nozzle spraying unevenly, which affects the uniformity of mixing oil and gas, resulting in deterioration of the performance of the diesel engine.

3.2.2.6 The Influence of the Number of Orifices

A nozzle model with eight orifices is set up in Figure 3.73.

Figure 3.74 shows the mass flow characteristic curve and Figure 3.75 shows the flow distribution of each orifice in the nozzle. It can be seen that the mass flow of each hole deviates, which is caused by the difference between the angles of each orifice relative to the axis of the needle valve body. When the number of orifices is increased, the consistency of the flow of symmetric orifices is increased and the deviation is small. The mass flow of orifices 3 and 7 is relatively large, and the mass flow of orifices 1, 2, 4, 6, and 8 is the same. Orifice 5 with the maximum angle has the smallest mass flow.

Figure 3.76 is the distribution cloud map of pressure, cavitation, and velocity in the early, middle, and late stages of the injection of nozzle with eight orifices. Because of

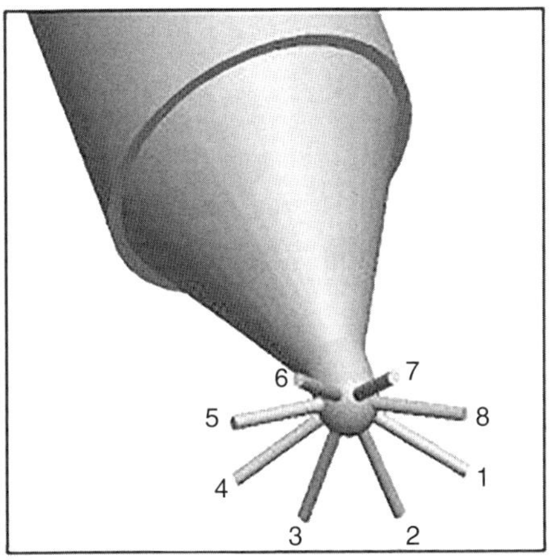

Figure 3.73 Schematic diagram of the nozzle structure with eight orifices.

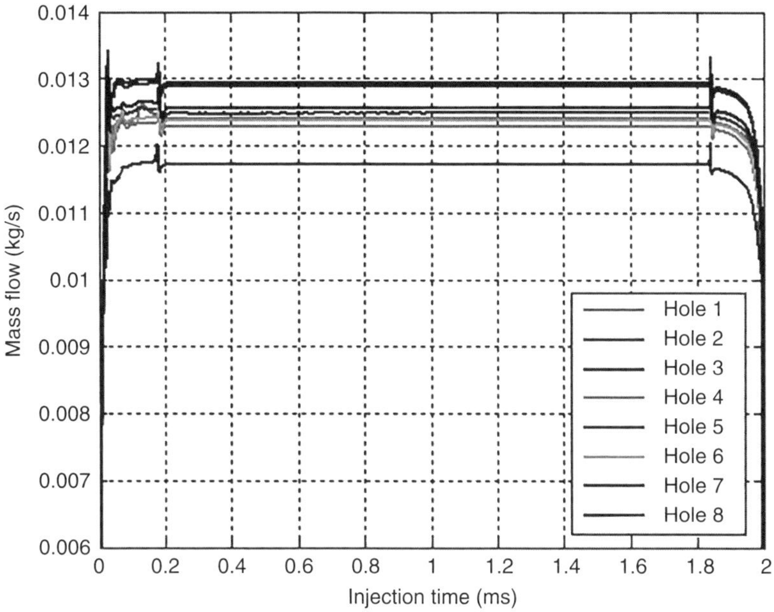

Figure 3.74 Mass flow of each orifice in a hole nozzle.

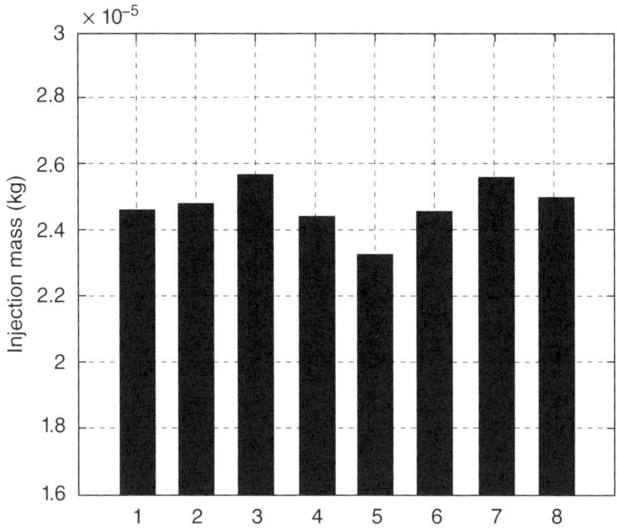

Figure 3.75 Flow distribution in each orifice of the nozzle.

symmetry, the velocity cloud on the outlet section is only compared with the injection orifices between orifices 1 and 5 in the middle stage of the injection. Comparing Figures 3.76a and 3.71a, it can be seen that the pressure field in the nozzle is set up slowly and the gradient is larger after the number of orifices is increased. In the middle stage of the injection, the pressure gradient at the inlet of the pressure chamber is

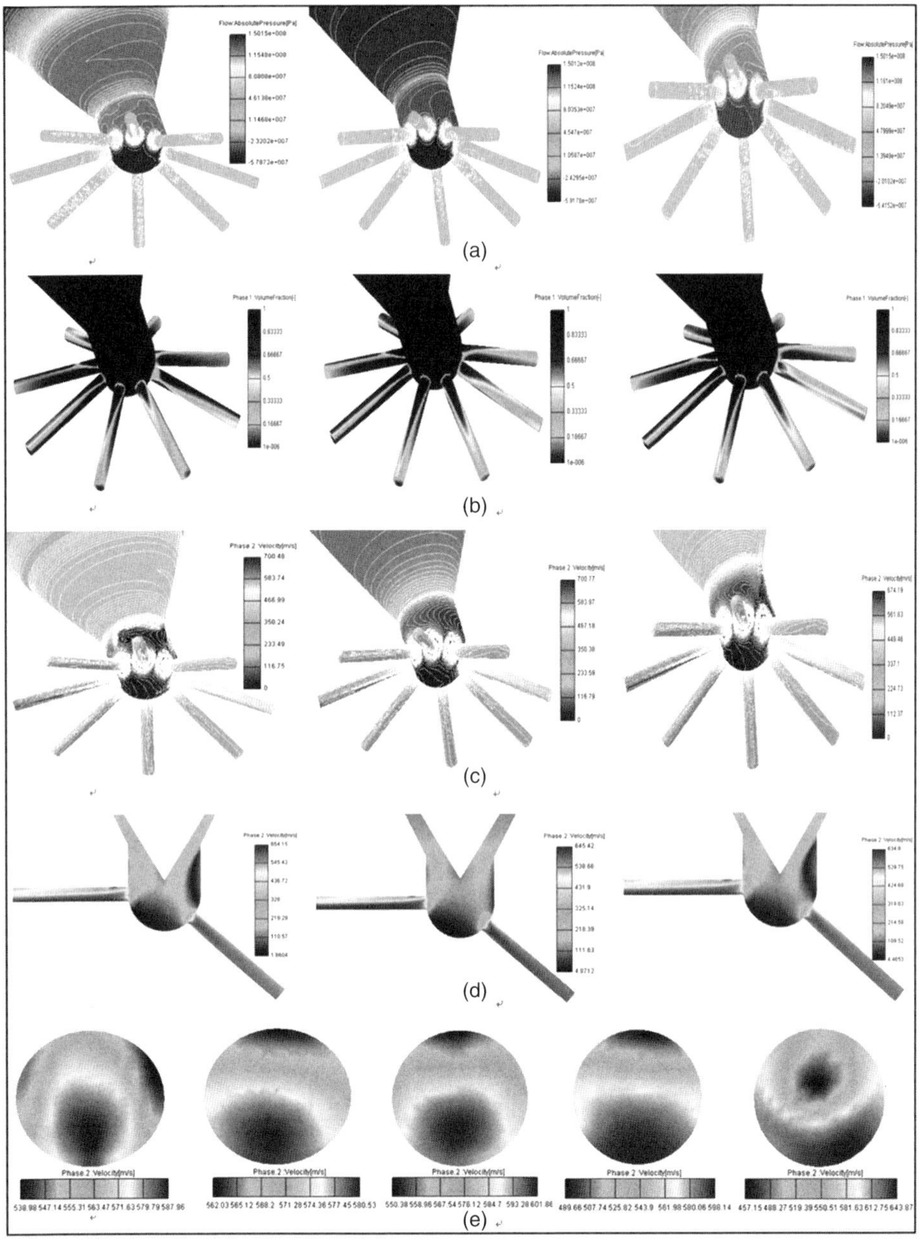

Figure 3.76 Nozzle pressure (a), gas phase volume (b), velocity (c, d, e) distribution (nozzle with eight orifices).

larger and the pressure difference resistance increases, which will inevitably affect the flow velocity of the fuel. From the velocity cloud of the two nozzles, the fuel flow rate of the nozzle with eight orifices is lower than that of the nozzle with six orifices.

Through the calculation result shown in Figure 3.76, it can be seen that the nozzle with eight orifices is better than the nozzle with six orifices in the symmetry of the symmetrical orifices. Though the outlet speed is smaller, the total mass flow rate is greatly improved. Therefore, we should increase the number of orifices as much as possible to ensure uniform mixing of the fuel and air under the premise of ensuring the duration of the injection.

3.2.3 Simulation and Experimental Study of Spray

The study of fuel atomization is the most critical part of combustion technology. In the engine combustion chamber, the fuel atomization directly affects the formation quality of the mixture gas and affects the engine's economic, dynamic, and emission levels. Especially for the diesel engine, because of poor volatilization of diesel, the mixture of diesel and air is achieved through atomization of the nozzle under the effect of high pressure. Therefore, the atomization quality depends on the characteristics of the fluid dynamic field inside and outside the orifices. The shape of the oil spray nozzle is an important index to use when evaluating the fuel injection process, including the penetration distance, the cone angle, and the atomization quality of the oil beam. Through the long distance, a large amount of fuel can be ejected to the lower wall of the combustion chamber, reducing the mixing effect of oil and gas and the evaporation rate of the fuel, thus prolonging the post-combustion period and affecting the combustion process. If the penetration distance is too short or the oil beam of each hole is not uniform, the space of the combustor cannot be fully utilized and the air is wasted.

So far, many important conclusions about the characteristics of diesel engine spray are obtained by using the spray simulation test system. Compared with the real machine test, the simulation test is greatly simplified and can produce very valuable results under the conditions of time and cost savings. These results provide a solid foundation for real and theoretical research and are also partly applied to the design and improvement of diesel engines. Although the test will gradually develop to the real machine test, the simulation test is still a powerful tool to use for the diesel spray test.

3.2.3.1 Test Scheme

The test bench mentioned above can be used to analyze the spray characteristics of high-pressure fuel injection. The pressure in the high-pressure vessel can reach 3 MPa, which can basically simulate the pressure in the cylinder during injection in the diesel engine, and the injection pressure of the common rail system can reach 150 MPa. The test scheme is shown in Table 3.7.

3.2.3.2 Simulation Calculation of the Nozzle Flow Field

Due to the uniform distribution and the same anglc of the four nozzles in the electronically controlled injector, only a single orifice model is established for calculation. Figure 3.77 shows the nozzle head grid sketch. The standard equation $k - \varepsilon$ is used to set up the turbulence model in the calculation parameters, and the near wall surface function is used to solve the near wall surface. Diesel oil is regarded as an incompressible

Table 3.7 Test scheme for spray characteristics.

	Test scheme
Type of injector	4 mm × 0.2 mm, $L/D = 5$
Container size	Diameter 135 mm, length 140 mm
Injection pressure (MPa)	130
Pressure for high-pressure vessel (MPa)	2
Temperature of high-pressure vessel (K)	298
Air density (kg/m^3)	23.54
Injection duration (ms)	2
Fuel	0$^{\#}$ light diesel oil
Fuel density (kg/m^3)	845

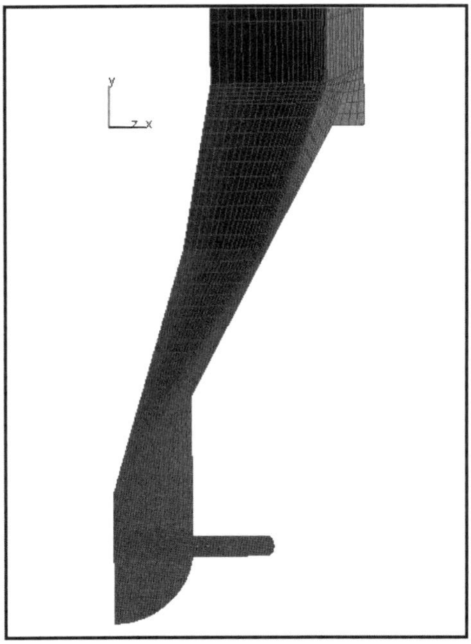

Figure 3.77 Mesh diagram of the nozzle head.

medium, and the energy conversion and loss in the flow process are not considered. In solving the equation of momentum, conservation, and continuity, the coupling of pressure and velocity is adopted by a simple algorithm. The Euler algorithm is used in the calculation of gas–liquid two-phase flow in the nozzle. The inlet and outlet boundary are set as the pressure boundaries, the inlet pressure is 130 MPa, and the outlet pressure is 2 MPa. The maximum lift of the needle valve is 0.25 mm and the duration of the needle valve from opening to closing is 2 ms. The calculated results are analyzed in three stages of early (0.05 ms), middle (0.83 ms), and late (1.95 ms). The results are as follows.

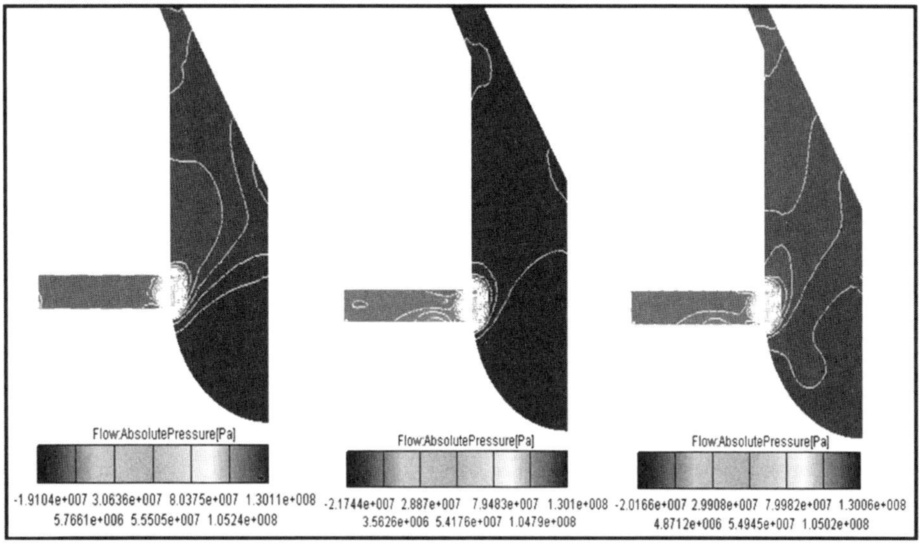

Figure 3.78 Pressure distribution on the central section of the nozzle.

Figure 3.78 is the pressure cloud on the center section of the orifice. It can be seen that the pressure gradient from the needle valve begins to rise to the highest level, while the pressure gradient in the pressure chamber becomes smaller and more stable. The pressure difference between the injector body and the orifices varies greatly because the area decreases sharply and the flow direction changes greatly when the fuel flows from the pressure chamber to the orifice. When the needle valve is the highest, the pressure gradient inside the spray hole varies greatly, so it is easier to form a low-pressure area or even negative pressure on the corner of the nozzle entrance, causing cavitation.

Figure 3.79 is a cloud map of the volume distribution of the gas phase on the wall, the center section, and the outlet of the orifice. It can be seen that the cavitation is formed near the corner of the orifice entrance and develops downstream along the wall toward the orifice. Then the diffusion reaches the outlet and the intensity also decreases gradually from the center of the orifice. The cavitation distribution in the upper and lower flow areas of the orifice is asymmetrical, with a thick upper side area and a thin lower side area, and decreases with the increase of the needle valve. The main reason is that the curvature of the upper fuel flow is larger and the distribution of the cavitation can make the cone angle of the lower side of the fog head different. Figure 3.80 is a cloud map of the center section of the orifice and the velocity distribution on the outlet. When the needle valve has just opened, the flow rate of fuel in the inner part of the nozzle is higher. As the needle valve rises to the maximum, the internal velocity is gradually stable and the overall speed reaches the maximum. As a result, the speed of the outlet of the nozzle is relatively high, and the atomization and mixing will be better. In addition, it can be seen that the high-velocity distribution area in the orifice is mainly concentrated on the lower side, and the effective flow area is very small because the influence of the cavitation is transferred to the outlet of the orifice.

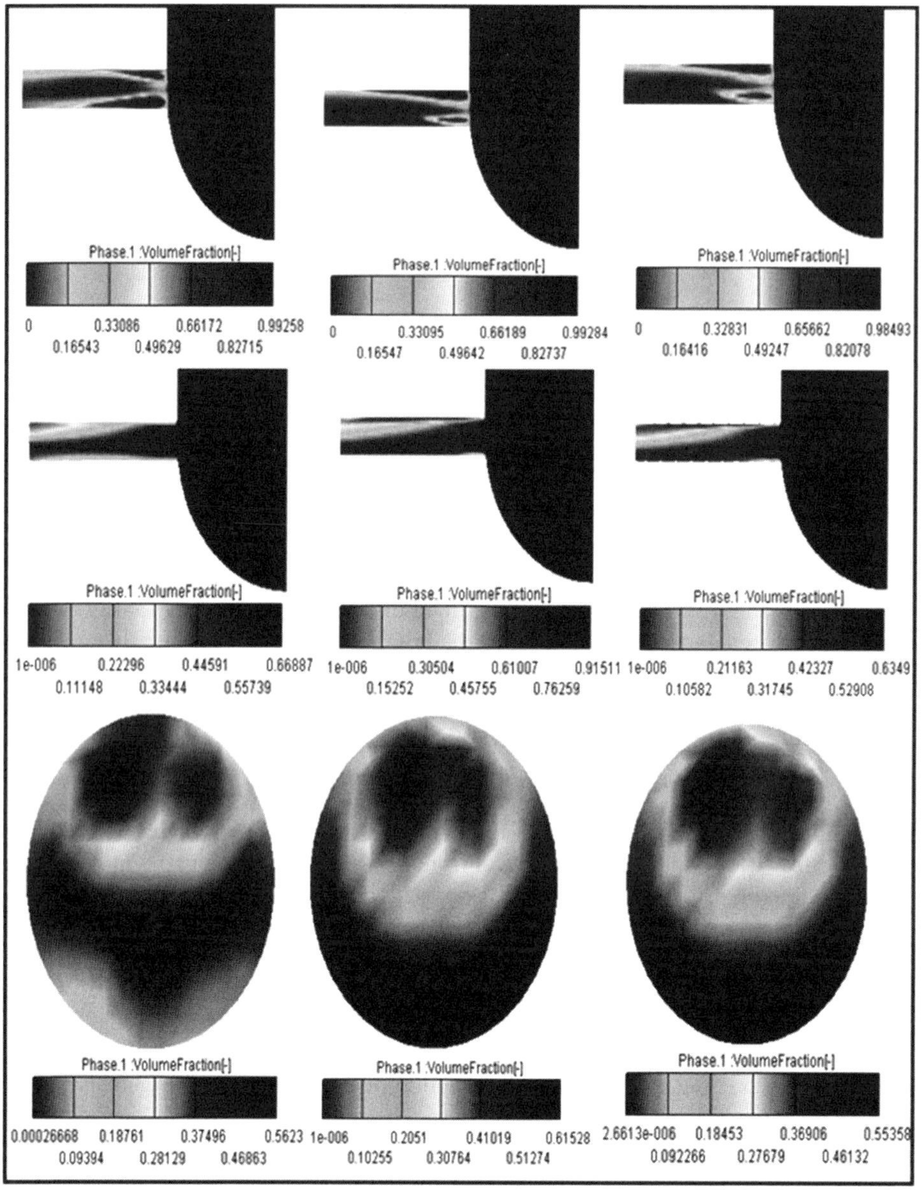

Figure 3.79 Gas volume distribution on the outlet of the wall surface and central section.

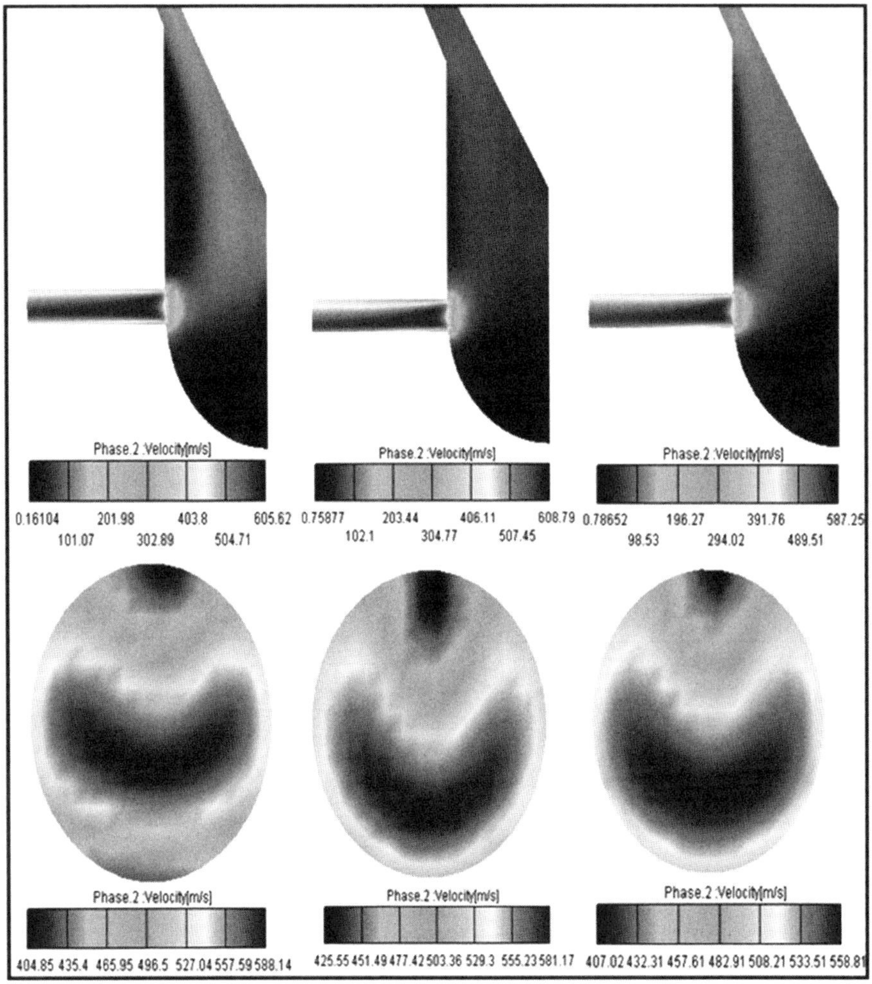

Figure 3.80 Velocity distribution on the center section of the orifice and the cross-section of the outlet.

3.2.3.3 Simulation and Test Verification of Spray

Because the image is not from the same cycle, the average value of pixels is taken by averaging three pixels in the same time interval in different cycles, thus reducing the influence of uncertain factors such as sloshing. At the same time, the calculation grid of the high-pressure vessel is set up to simulate the atomization development of fuel in it. There are 86 400 grids, as shown in Figure 3.81. In addition, the breakup of oil particles in the area near the nozzle is mainly influenced by turbulence and cavitation. With the increase of the fog beam, the breakup of oil particles is mainly influenced by the aerodynamic force. As a result, the previous flow calculation of the nozzle is used as a boundary condition to simulate the spray, including the calculated mass flow rate of the nozzle, the location vector of the orifice, and the outlet velocity, cavitation, and turbulent kinetic energy. In this way the simulated spray results will be closer to the real spray form.

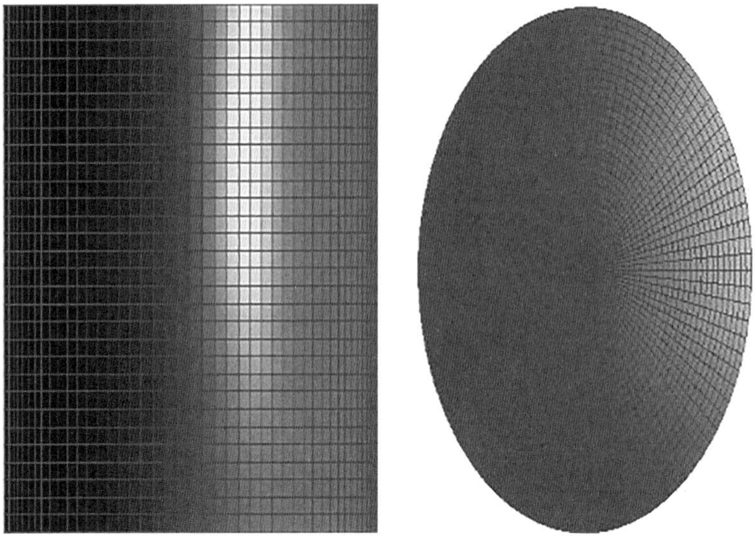

Figure 3.81 High-pressure vessel grid.

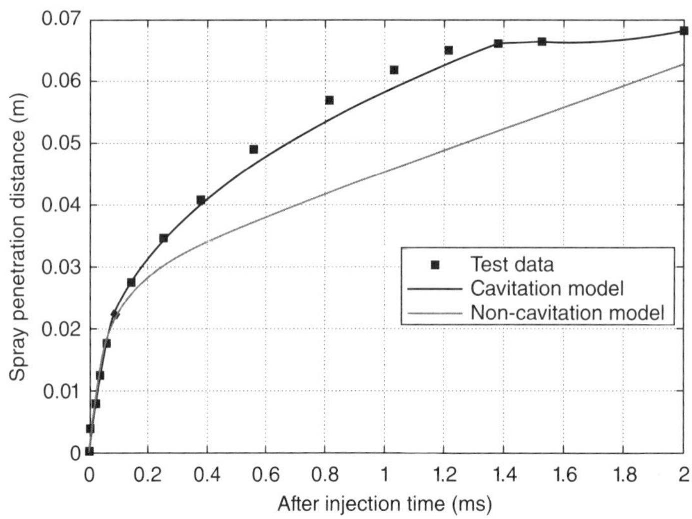

Figure 3.82 Contrast of spray penetration.

Figure 3.82 is a comparison of the spray penetration distance and the test value calculated under the injection pressure of 130 MPa and the back pressure of 2 MPa. It can be seen from the graph that the simulation results of the relative mass flow as boundary conditions and the increase of cavitation flow can be well matched with the experimental results. When the injection time tASOI < 0.38 ms, the calculated value is slightly larger than the experimental value, and are relatively close to each other. When tASOI > 0.38 ms, the calculated value is a little shorter than the experimental value, which may be the reason for the excessive resistance value in the RT model.

Table 3.8 Photos of spray forms and simulation comparison.

t_{ASOI}	Experimental photos of Spray forms	Model with Cavitation	Model without Cavitation
	0 0.035 0.07 m		
0.05 ms			
0.14 ms			
0.26 ms			
0.38 ms	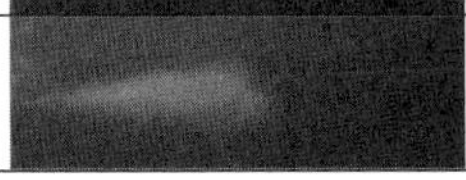		
0.56 ms			
0.83 ms			
1.04 ms	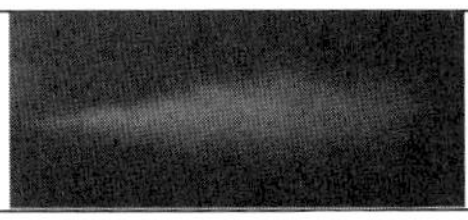		
1.22 ms	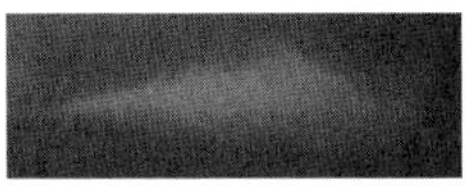		
1.39 ms	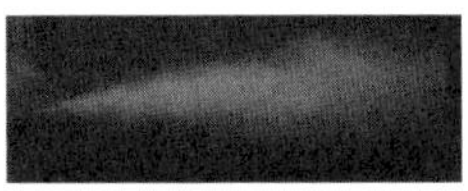		
1.53 ms	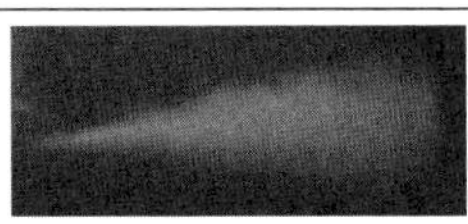		

Table 3.8 is a comparison of the photos of the calculation results of the development process of some spray bodies and simulated spray in this test. All the images are processed after taking an average. It can be seen that the calculated image is in accordance with the experimental image. The first atomization occurs mainly near the orifice, as shown in the image with 0.38 ms in the figure. With the continuous development of the oil beams, the second atomization started. The oil droplets in the front of the fog bundle are divided into many small droplets from those inside to those outside. Due to the reduction of the relative velocity between the small droplets and the high-pressure air, the air is fully sucked into the oil bundle, which makes the volume of the fog head larger. At the time of tASOI > 1.39 ms, the oil beam has already touched the inner wall of the inner cylinder, and is of no significance when estimating its range. At the time of tASOI > 1.39 ms, the oil beam has already touched the inner wall of the inner cylinder and there is no need to estimate its range. The spray cone angle can be defined as the angle between the outer contour and the orifice connection line at the maximum section of the spray cone, which is usually used as a parameter to reflect the quality of spray atomization. It can be seen from the results that there is a certain error in the prediction of the spray cone angle, which may be related to the insufficient strength of the cavitation flow in the calculation of the nozzle model. Comparing the spray penetration length and the actual measured results of the spray growth process from Table 3.8, both the simulated spray morphology and the specific value are in good agreement with the actual measurement results, and the accuracy of the nozzle flow simulation and the accuracy of the spray simulation are verified.

4

High-Pressure Fuel Pump Design Technology

Providing high-pressure fuel to the system, the high-pressure fuel pump is the key component to guarantee the high-pressure fuel. Its main function is to pressurize the low-pressure fuel into a high-pressure fuel and to store the fuel in the common rail, which waits for the electronic control unit (ECU) injection order. The design criteria of the fuel supply quantity of the high-pressure pump is to ensure the fuel quantity demand of the diesel engine injection and controlled fuel under any condition, as well as the fuel quantity change during start and acceleration. Also, it needs to reduce the driving power consumption and improve the efficiency of the fuel supply system.

The working pressure of the high-pressure fuel pump of the common rail system is very high. To reduce the volatility of pressure in the rail lumen and peak torque of the pump shaft, pumps with a multiple fuel supply such as in-line pumps, radial plunger pumps, and rotary plunger pumps are often used. To reduce the energy consumption of the high-pressure fuel pump, the common rail system high-pressure pump changes from a full stroke fuel supply in the early time to a dynamical fuel supply adjusted by the diesel engine load state. The main problems faced by the high-pressure fuel pump in a common rail system are the leakage caused by high pressure, the intensity problem of the high-pressure pump drive system caused by high pressure, and the control problem of fuel quantity caused by high pressure.

4.1 Leakage Control Technique for the Plunger and Barrel Assembly

With the constant increase in the required fuel supply pressure of the high-pressure fuel pump, the plunger and barrel assembly deforms under high pressure, which causes the gap between the assembly to become larger, which thus decreases the sealing performance and volumetric efficiency, and even cannot establish the high-pressure condition. If the gap is too small, the plunger will be stuck, affecting the reliability of the high-pressure fuel pump. In order to solve the contradiction between the stability and the sealing performance of the plunger and barrel assembly, it might be restricted by the current domestic processing condition to only rely on the material and machining process to ensure a proper gap between the plunger and barrel assembly. The use of "deformation compensation technology" can effectively solve this problem: appropriately increasing stiffness of the barrel, processing a suitably shaped hole at the top of the

Common Rail Fuel Injection Technology in Diesel Engines, First Edition.
Guangyao Ouyang, Shijie An, Zhenming Liu and Yuxue Li.

plunger, and using the radial deformation of the working plunger to keep the clearance between the plunger and barrel in an appropriate numerical range to compensate for the deformation of the barrel under high pressure.

To achieve the high pressure more than 150 MPa and ultra-high-pressure injection requires the processing gaps between plunger and barrel reach 2–3 μm, whose processing technic requirement is difficult to meet by the domestic machining standard and the material science standard. In addition, the deformation of the working plunger and barrel assembly is relatively large, so there is a difference between the gap of the processing plunger and barrel assembly and the gap of its working state. Therefore, the sealing requirement of the ultra-high-pressure working condition of the diesel engine high-pressure fuel pump cannot be satisfied. According to the finite element analysis, under working conditions, the barrel produces a relatively large radial extrusion deformation under the action of the high-pressure fuel. Under the combined action of the pressure of the side fuel and the driving force of the high-pressure fuel and camshaft, the plunger on the one hand produces radial extrusion deformation and on the other hand produces slight radial shrinkage at the assembly mating sealing strip. The deformation of the plunger and the barrel makes the working fit clearance increase, so it is difficult to meet the requirements of the good sealing performance of the fuel system when the plunger and barrel assembly works. Figure 4.1 gives the radial clearances of a plunger and barrel assembly calculated under different pressures.

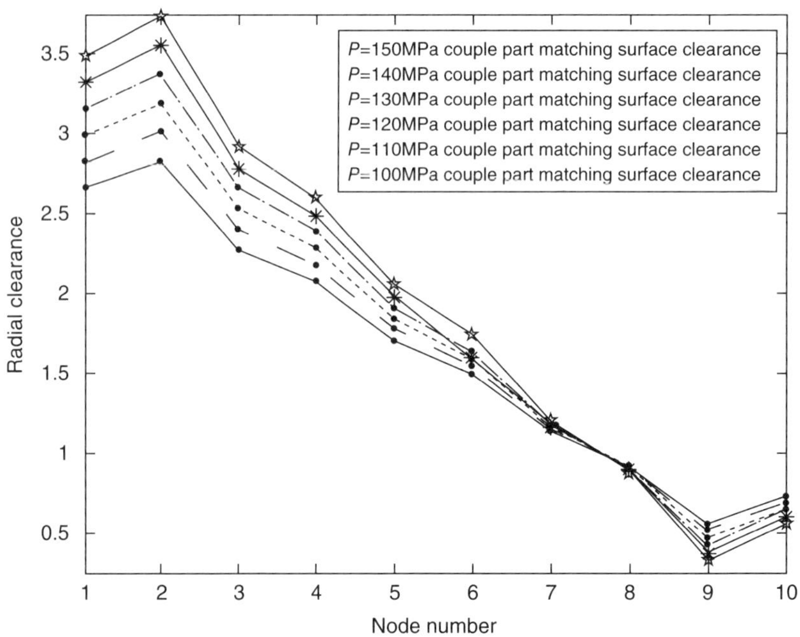

Figure 4.1 The axial distribution of the radial clearance of each node assembly part under different pressures.

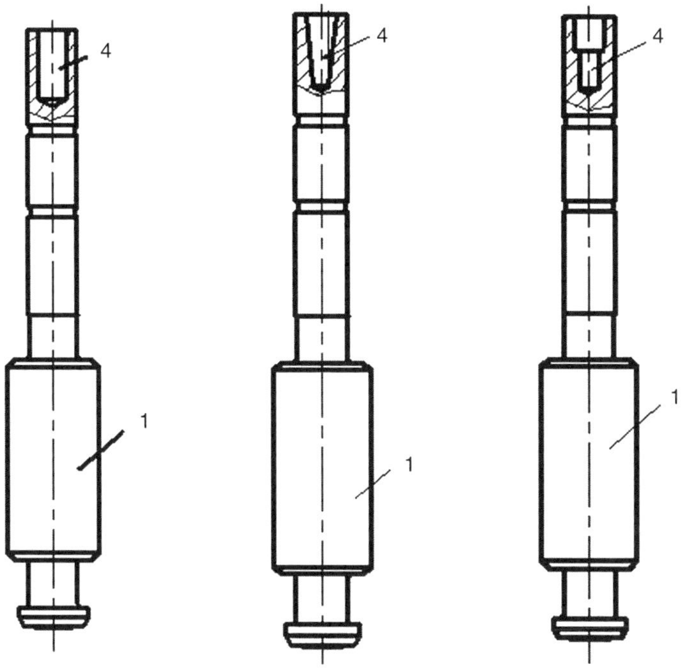

Figure 4.2 Sketch of the plunger structure processed by "upper compensating groove deformation compensation technology."

To solve the technical problem, a hole with a suitable shape, which is a compensation hole, can be processed in the top end of the plunger. By introducing the high-pressure fuel to fill the hole, a certain radial deformation is generated. Because this kind of radial deformation increases with the increase of fuel pressure and barrel deformation, the clearance between the assemblies always remains in a proper numerical range, and thus the dynamic seal of the plunger and barrel assembly in a non-stick condition is achieved. The structure of the plunger is shown in Figure 4.2: (a) is an open cylindrical groove, (b) is an open cone groove, and (c) is a ladder-shaped groove, which shows the deformation compensation technology. At the same time, the use of modern design theory and the calculation method can optimize the design of the technical scheme, realize a good sealing performance and a stable power performance of the plunger and barrel assembly within the full work pressure range, and meet the working requirements of an ultra-high-pressure fuel supply pump.

The use of this technology in the improvement of the high-pressure fuel pump plunger and barrel assembly design and structure, under the condition that it is appropriate to reduce the requirements for a machining gap in the assembly, can further reduce the gap in the working state. The reduction also increases the parts wear margin, greatly improves the service life, and improves the sealing performance of the assembly under the ultra-high-pressure working conditions.

4.1.1 Finite Element Analysis of the Fluid Physical Field in the Plunger and Barrel Assembly Gap

4.1.1.1 Similarity Principle

The similarity principle is the main theoretical basis of model experimental research. The similarity principle studies the relation between similar physical phenomena. The main content is the third law.

The similarity principle first theorem states that phenomena that resemble each other must have numerical similarity criterion.

The similarity principle second theorem states that in the same kind of phenomenon, if the monodromy conditions are similar and the similarity criterion of the single physical quantities with monodromy are numerically equal, these phenomena must be similar.

The similarity principle third theorem states that the relation between various physical quantities that describes a phenomenon can be expressed as a function of the similarity criterion: $\pi_1, \pi_2, \ldots, \pi_n$.

The necessary and sufficient conditions for the similarity of the two flows are geometric similarity, kinematic similarity, and dynamic similarity.

(1) Geometric similarity. The model is the same as the shape of the object, the angles of the corresponding parts are equal, and the length of the corresponding parts (including roughness) is proportional to each part.
    Length: $\delta_l = \frac{l_1}{l_2}$; area $\delta_A = \frac{A_1}{A_2} = \frac{l_1^2}{l_2^2}$; volume $\delta_V = \frac{V_1}{V_2} = \frac{l_1^3}{l_2^3}$; angle $\alpha_1 = \alpha_2$; $\beta_1 = \beta_2$; $\gamma_1 = \gamma_2$.
    Geometric similarity can be expressed by scale δ_l, and as long as δ_l remains unchanged, geometric similarity holds.
(2) Kinematic similarity. The time that the corresponding mass points flow through the corresponding distance is proportional. Because $\delta_t = \frac{t_1}{t_2}$, according to kinematic similarity and geometric similarity, the similarity relation between velocity and acceleration is derived:

$$\begin{cases} \delta_a = \dfrac{a_1}{a_2} = \dfrac{u_1/t_1}{u_2/t_2} \\ \delta_u = \dfrac{u_1}{u_2} = \dfrac{l_1/t_1}{l_2/t_2} = \dfrac{\delta_l}{\delta_t} \end{cases} \tag{4.1}$$

    As mentioned above, motion similarity depends on δ_u and δ_a, while velocity similarity and acceleration similarity depend on δ_l, δ_t, which means that under geometric similarity, if δ_t is constant, kinematic similarity holds.
(3) Dynamical similarity. The corresponding forces on the fluid of the object and on the fluid of the model are proportional.

$$\delta_F = \frac{F_1}{F_2} = \frac{M_1}{M_2}\frac{a_1}{a_2} = \frac{\rho_1 V_1}{\rho_1 V_2}\delta_a = \delta_\rho \delta_l^{\,3} \delta_a$$

$$\text{and} \quad \delta_a = \frac{\delta_u}{\delta_t} = \frac{\delta_l}{\delta_t^{\,2}};\ \delta_u = \frac{\delta_l}{\delta_t}$$

$$\text{so} \quad \delta_F = \delta_\rho \delta_l^3 \frac{\delta_l}{\delta_t^2} = \delta_\rho \delta_l^{\,2} \frac{\delta_l^2}{\delta_t^2} = \delta_\rho \delta_l^2 \delta_u^2$$

$$\text{or} \quad \frac{\delta_F}{\delta_\rho \delta_l^2 \delta_u^2} = 1 \tag{4.2}$$

Namely, the dynamical similarity conditions of two kinds of geometric similarity system are density similarity coefficients.

4.1.1.2 Similarity Criterion

(1) Newton similarity criterion:

Sort out formula $\dfrac{F_1}{F_2} = \dfrac{\rho_1 {l_1}^2 u_1^2}{\rho_2 l_2 u_2^2}$

get $\dfrac{F_1}{\rho_1 l_1^2 u_1^2} = \dfrac{F_2}{\rho_2 l_2^2 u_2^2}$

or $N_e = \dfrac{F}{\rho l^2 u^2}$ (4.3)

This is the Newton similarity criterion or the inertial force similarity criterion. That is, if the two fluid systems are similar in dynamics, the number is equal, and vice versa.

(2) Reynolds similarity criterion. When the viscous force and the inertia force play a major role in the flow phenomenon, the size of the viscous force $T = \mu A du/dy$ can be expressed by $T = \mu lu$. Expressed as Newton's similarity criterion this gives

$$\frac{l_1 u_1}{v_1} = \frac{l_2 u_2}{v_2} \text{ or } \frac{lu}{v} = Re \tag{4.4}$$

In the equation, $v = \mu/\rho$ and Re is the Reynolds similarity criterion.
According to the formula derivation, the Reynolds similarity criterion is the ratio of the viscous force to the inertial force.

(3) Froude similarity criterion. If the main forces are gravity and inertial forces, gravity can be measured by $\rho g l^3$. When put into the Newton similarity criterion:

$$\frac{u_1^2}{g_1 l_1} = \frac{u_2^2}{g_2 l_2} \text{ or } Fr = \frac{u^2}{gl} \tag{4.5}$$

In the equation, *Fr* is the Froude similarity number.

(4) Euler (similarity) criterion.
If the main function is the total pressure and inertial force, the total pressure can be measured by pl^2. When put into the Newton similarity criterion:

$$\frac{p_1}{\rho_1 u_1^2} = \frac{p_2}{\rho_2 u_2^2} \text{ or } Eu = \frac{p}{\rho u^2} \tag{4.6}$$

In the equation, *Eu* is the Euler similarity number.

(5) Mach (similarity) criterion. If the compressibility is dominant in the flow, the square root of the ratio of the inertia force to the elastic force is called the Maher similarity number (abbreviated to Maher number *M*):

$$M = \sqrt{\frac{\rho u^2 l^2}{k l^2}} = \frac{u}{\sqrt{\frac{k}{\rho}}} = \frac{u}{a} \tag{4.7}$$

In the equation, a is the velocity of sound in the fluid: $a = \sqrt{k/\rho}$.
The flow at $M < 1$ is called the subsonic flow; the flow at $M > 1$ is called the supersonic flow.

4.1.1.3 Dimensional Analysis and the Pion Theorem

(1) The harmony principle of physical dimension. In a complete physical equation, not only are the values on both sides equal, but each dimension of them must be the same, and the physical quantities of different dimensions cannot be added or reduced.

When a physical phenomenon is expressed, the function formula between the physical quantities is $y = f(x_1, x_2, \dots, x_n)$, and the dimension of the above formula must be the same according to the harmony principle of physical dimension, that is,

$$y = kx_1^{a_1}, x_2, \dots, x_n^{a_n} \tag{4.8}$$

In the equation,

$x_1, x_2, \dots, x_n$ are factors affecting the variable y,
$a_1, a, \dots, a_n$ are undetermined coefficients,
k is the dimensionless coefficient of proportionality.

(2) Buckingham theory (π theorem). This theorem provides a simple method for solving the undetermined coefficients of the multidimensional function formula by using the dimensional harmony principle. The physical quantity y is the function of other physical quantities $x_1, x_2, \dots, x_n$, which is $y = f(x_1, x_2, \dots, x_n)$. In hydromechanics, velocity v, fluid density ρ, and arbitrary characteristic lengths of the fluid are usually chosen to be the fundamental quantities. They are independent of each other, in line with the requirement of fundamental quantities, and cannot be derived from each other. The dimensions of the remaining y and $x_4, x_5, \dots, x_n$ in the function formula can be expressed by the product of some power of the dimension of the three fundamental quantities, which is

$$[x_4] = [v]^{a_4}[l]^{b_4}[\rho]^{c_4}$$
$$[x_5] = [v]^{a_5}[l]^{b_5}[\rho]^{c_5}$$
$$\dots$$
$$[x_n] = [v]^{a_n}[l]^{b_n}[\rho]^{c_n}$$

If each of the basic quantities is multiplied by a_1, a_2, a_3,

$$v' = \alpha_1 v$$
$$l' = \alpha_2 l$$
$$\rho' = \alpha_3 \rho$$

The rest of the physical quantities correspondingly change to

$$y' = \alpha_1^a \alpha_2^b \alpha_3^c y$$
$$x_4' = \alpha_1^{a_4} \alpha_2^{b_4} \alpha_3^{c_4} x_4$$
$$\dots$$
$$x_n' = \alpha_1^{a_n} \alpha_2^{b_n} \alpha_3^{c_n} x_n$$

Put the formula $y = f(x_1, x_2, \dots, x_n)$ into $y' = \alpha_1^a \alpha_2^b \alpha_3^c yy'$; then

$$y' = \alpha_1^a \alpha_2^b \alpha_3^c f(x_1, x_2, \dots, x_n)$$
$$y' = f(\alpha_1 v, \alpha_2 l, \alpha_3 \rho, \alpha_1^{a_4} \alpha_2^{b_4} \alpha_3^{c_4} x_4, \dots, \alpha_1^{a_n} \alpha_2^{b_n} \alpha_3^{c_n} x_n)$$

In the equation, $\alpha_1, \alpha_2, \alpha_3$ can be chosen arbitrarily. If $\alpha_1 = 1/v; \alpha_2 = 1/l; \alpha_3 = 1/\rho$ are put into the equation,

$$\frac{y}{v^a l^b \rho^c} = f\left(1, 1, 1, \frac{x_4}{v^{a_4} l^{b_4} \rho^{c_4}}, \dots, \frac{x_n}{v^{a_n} l^{b_n} \rho^{c_n}}\right)$$

$$\text{or } \pi = f(1, 1, 1, \pi_4, \dots, \pi_n) \tag{4.9}$$

As can be seen from the upper function, the functional relations of $n + 1$ dimensional quantities y, x_1, x_2, x_n can be simplified to functional relations of $n + 1 – 3$ dimensionless quantities. Then the physical equation can be solved by the principle of dimensional harmony. Buckingham first proposed this equation, where π is used to represent the dimensionless quantities, so the theorem is called the Buckingham theorem (π theorem).

4.1.1.4 Similarity Model and Finite Element Analysis of the Clearance Flow Field

(1) Determination of fluid flow state in the gap. Combined with the machining accuracy of the plunger barrel assembly, and a clearance value of 3 μm and a plunger pump volumetric efficiency of 0.8, the average flow rate of diesel fuel flowing through the clearance is calculated to be 11.5 m/s. According to the size of the *Re* value determined by the flow state judgment rule, the clearance diesel flow can be considered as a laminar flow process.

(2) System modeling assumption. Because of the complexity of the actual high-pressure common rail system, it is impossible to consider all the factors that affect the fuel supply and injection process during the establishment of the system model. Therefore, according to the characteristics of the high-pressure common rail system and the requirements of simulation calculations, and considering the characteristics of fluid flow in the clearance, the following basic assumptions are made in the simulation model:

a. In the whole calculation process, the change of the assembly itself and the fuel physical property by temperature are not considered.
b. Assuming that the pump cavity has a concentrated volume, the flow of fuel is not influenced by the pressure fluctuations and pulse disturbances.
c. In small time intervals, the flow of fuel through each orifice is constant and steady.
d. The elastic deformation of each part in the high-pressure system is not considered.
e. The friction resistance between the friction components is not considered.
f. Only the leakage caused by the fit clearance distribution change of the cylinder pair is considered; the leakage caused by the machining error of the sealing surface is not considered.

(3) Similarity theory for the analysis of the clearance flow field. When the model size is adjusted, all the dimensionless quantities must be consistent so that the model can be similar before and after. Namely, in the clearance flow field, the ratio of the axial length to the internal diameter π_1, the ratio of the inside diameter to the outside diameter π_2, and the Reynolds number π_3 are required to be the same.

a. Geometric similarity criterion. The gap is considered as a cylindrical geometry with a thickness of δ, an average radius of R, and a length of l; similar geometry by magnifying δ by k times has a thickness of $k\delta$, length of kl, and average radius of kR.

b. Kinematic similarity. When the flow rate is constant, the state of fluid flow through the clearance remains the same, and the time of flow through the clearance is

$$t = \frac{kl}{u} = kt_0 \tag{4.10}$$

In the formula, t is the time of fluid flow in the original clearance.

c. Dynamic similarity. Let Re be the Reynolds number of the fluid flow in the original clearance, and Re is the Reynolds number of the fluid flow at k times of the clearance amplification. In accordance with the similarity criterion: Re = Re,

$$Re = \frac{lu\rho}{\mu} \tag{4.11}$$

In the equation, ρ is the density of diesel fuel, μ is the kinematic viscosity of diesel fuel, and l is the characteristic length of fluid field. Thus

$$Re' = \frac{ulu'\rho'}{\mu'} \tag{4.12}$$

By kinematic similarity, it can be shown that $\mu = \mu'$; therefore

$$Re' = \frac{klu'\rho'}{\mu'} \tag{4.13}$$

If $\rho = \rho'$, then $\mu = k\mu'$ and $\mathrm{Re}/\mathrm{Re}' = 1$.

The conclusions can be drawn from the above derivation: under the condition of ensuring geometric similarity, kinematic similarity, and dynamic similarity, when the gap is magnified by k times, the term for the fluid flow state of a similar shape to remain the same is to increase its kinematic viscosity by k times or to reduce its density by k times.

The above conclusions are the basic starting point for the finite element analysis of fluid flow in a minimum clearance.

(4) Numerical analysis of the pressure field in the clearance flow. The continuous equation of fuel in the plunger cavity of a high-pressure pump is

$$Q_{VP} = Q_{PP} - Q_{P\to a2} - Q_{P0} - Q_P \tag{4.14}$$

In the equation:

Q_{VP} is the change rate of the compressed fuel caused by the change of the plunger cavity pressure,

Q_{PP} is the instantaneous plunger inflow, namely, the geometric fuel supply rate,

$Q_{P\to a2}$ is the flow rate of the plunger pump cavity to the one-way valve,

Q_{p0} is the flow rate to the low-pressure fuel passage,

Q_P is the leakage flow rate of the plunger cavity.

They can be calculated by the following equations:

$$\begin{cases} Q_{VP} = \frac{V_P}{E}\frac{\mathrm{d}P_P}{\mathrm{d}t} \\ Q_{PP} = F_P v_P \\ Q_{P\to a2} = \zeta(\mu F)_{\mathrm{Pr}}\sqrt{\frac{2}{\rho}|P_P - P_{a2}|} \\ Q_{P0} = \eta(\mu F)_{P0}\sqrt{\frac{2}{\rho}|P_P - P_0|} \\ Q_P = \frac{\pi d\delta^3 \Delta P}{12\mu_P LC} \end{cases} \tag{4.15}$$

V_P is the concentrated volume of the plunger cavity,

$$V_P = \frac{\pi d^2}{4} \times (H - h_p) \tag{4.16}$$

where

P_P is the pressure of the plunger cavity,
H is the height of the plunger at the lower dead point from the one-way valve,
E is the modulus of fuel elasticity, F_P is the sectional area of the plunger,
v_P is the speed of the plunger,
P_{rail} is the pressure of the common rail tube,
$(\mu F)_{Pr}$ is the flow area of the plunger cavity to the common rail tube,
ρ is the density of fuel in kg/m^3,
$(\mu F)_{P0}$ is the flow area from the plunger cavity to the low-pressure fuel passage,
P_0 is the pressure of the low-pressure fuel passage,
Q_P is the fuel leakage amount in the plunger cavity,
μ is the flow coefficient (for the thick wall orifice, the discharge coefficient is 1),
ξ, γ, η are step functions

$$\zeta = \begin{cases} 1 & \text{if } P_p \geq P_{a2} \\ -1 & \text{if } P_p < P_{a2} \end{cases}$$

$$\eta = \begin{cases} 1 & \text{if } P_p \geq P_0 \\ -1 & \text{if } P < P_0 \end{cases} \tag{4.17}$$

In Eqs. (4.14) to (4.16):
d is the plunger diameter,
μ_P is the dynamic viscosity of the diesel fuel,
L is the sealing length of the plunger,
δ is the radius clearance between the plunger and barrel,
C is the correction factor of the beginning part of laminar flow, where mostly $C\approx 1$,
ΔP is the pressure difference between the suction and discharge fuel, P_a,
v_p is the plunger speed.
Rearrange Eqs. (4.14) to (4.17) as

$$\frac{\mathrm{d}P_p}{\mathrm{d}t} = \frac{E}{H - h_p}\left[v_p - \frac{4}{\pi d^2}\left(\begin{matrix}\xi(\mu F)_{pr}\sqrt{\frac{2}{\rho}|P_p - P_{a2}|} + \frac{\pi d\delta_p^3}{12\mu_p L_p}P_p \\ +\eta(\mu F)_{p_0}\sqrt{\frac{2}{\rho}|P_p - P_0|}\end{matrix}\right)\right] \tag{4.18}$$

Assume the x-axis to be the axial direction of the plunger and the z-axis to be the circumferential direction. The Reynolds equation of fit clearance between the plunger and barrel is

$$\begin{aligned}&\frac{\partial}{\partial x}\left(\frac{h^3}{12\eta}\frac{\partial p}{\partial x}\right) + \frac{\partial}{\partial z}\left(\frac{h^3}{12\eta}\frac{\partial p}{\partial z}\right) - \frac{\partial}{\partial z}\left(\frac{u_1 + u_2}{2}h\right) \\ &-\frac{\partial}{\partial z}\left(\frac{\omega_1 + \omega}{2}h\right) + u_2\frac{\partial h}{\partial x} + \omega_2\frac{\partial h}{\partial z0} - v_2 + v_1 = 0\end{aligned} \tag{4.19}$$

$$\frac{\partial}{\partial x}\left(\frac{h^3}{\eta}\frac{\partial p}{\partial x}\right) + \frac{\partial}{\partial z}\left(\frac{h^3}{\eta}\frac{\partial p}{\partial z}\right) = 6u\frac{\partial h}{\partial x} \tag{4.20}$$

In the above equations: $\partial p/\partial x$ is the axial distribution of fuel pressure and $\partial p/\partial z$ is the circumferential distribution of the fuel pressure. Because $\partial p/\partial x >> \partial p/\partial z$ and the piston eccentricity is not considered, $\partial p/\partial z$ can be ignored. Thus,

$$\frac{\partial}{\partial x}\left(\frac{h^3}{\eta}\frac{\partial p}{\partial x}\right) - 6u\frac{\partial h}{\partial x} = 0 \tag{4.21}$$

Through the above model, the predicted pressure distribution in the fit clearance of the plunger and barrel can be obtained.

(5) Applying the boundary condition. When the boundary conditions are applied, the following three points should be considered:
   a. Try to set the boundary conditions of the problem in the place where it is known.
   b. Avoid setting the boundary near the areas of interest or sensitivity.
   c. Do not set the boundary in the place where the gradient of variable variations is large.

   The specific fluid field is chosen in the length of a stroke of the plunger and barrel assembly and in the enveloped geometric part of the plunger and barrel. (The difficulty of the method is that this geometric flow field must change all the time relative to the motion of the assembly.)

   At the upper end of the assembly, the gradient of the clearance fluid varies greatly, so that the flow field here needs to be properly adjusted and a length of the high-pressure cap is at the upper end of the clearance fluid. Considering the ultra-thinness of the gap, the radial grids are uniformly divided. From the consideration of accuracy, using grid mapping is more able to keep the boundary of the grid characteristic constant.

   The boundary conditions here include the velocity of the fluid, pressure, and other parameters of each end face of the wall. The upper end of the cross-section is the inlet boundary of the flow field and the leak at the lower part is the exit boundary. Then, according to the fuel dimensionless parameter Reynolds number, the working flow field in the laminar flow state can be ensured, without the need to activate the turbulence model. In the meshing of the flow field, since the near-wall part is an important region, the unstructured mapping grid is used. The flow field solving domain is bounded by four main parts: the inlet and outlet boundaries and the inner and outer wall boundaries. All boundary conditions need to be processed. Considering that the finite element mesh needs to be updated continuously, the boundary condition is loaded directly on to the solid model. The flow in this field can be regarded as a pressure-driven problem. The inlet boundary is the pressure boundary and the pressure is applied directly, so zero relative pressure is applied at the exit. Both the inner and outer wall boundaries are symmetrical boundaries, but the outer wall is considered as a stationary wall without sliding conditions. The interior wall is a moving wall and the velocity component is tangential to the wall surface, so the other velocity components are zero.

(6) Analysis of the finite element solution of the flow field. After FLOTRAN post-processing, the pressure and velocity distribution of the fluid field are extracted. Figure 4.3 is a nephogram of the pressure distribution in a similar initial clearance flow field. Figures 4.4 and 4.5 are nephogram profiles of the velocity distribution of the longitudinal flow field with similar initial clearances. The following conclusions can be obtained

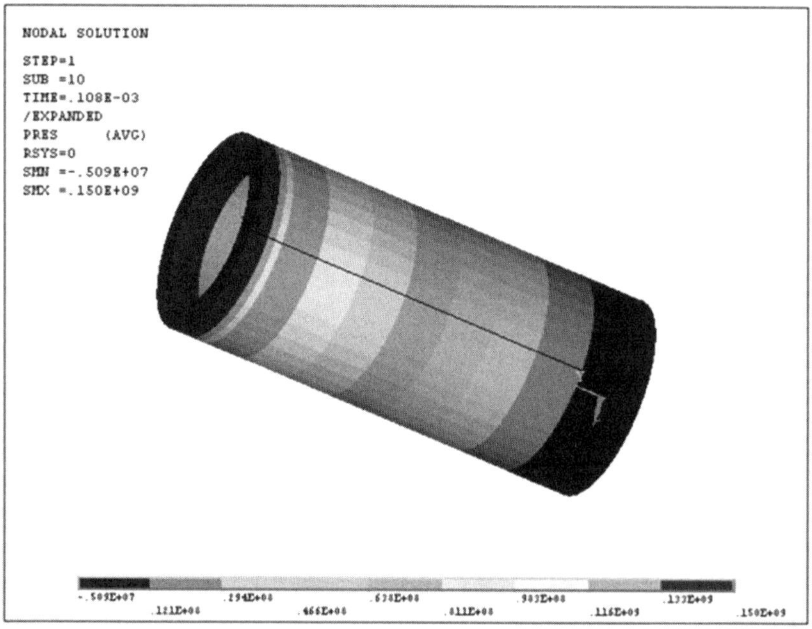

Figure 4.3 Nephogram of the pressure distribution in a similar initial clearance flow field.

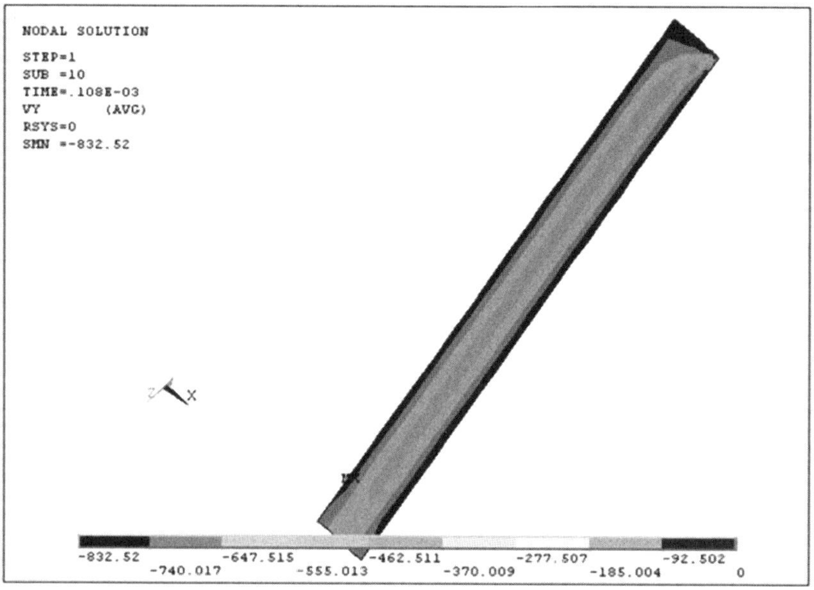

Figure 4.4 Enlarged nephogram profile of the velocity distribution of the longitudinal flow field with similar initial clearances.

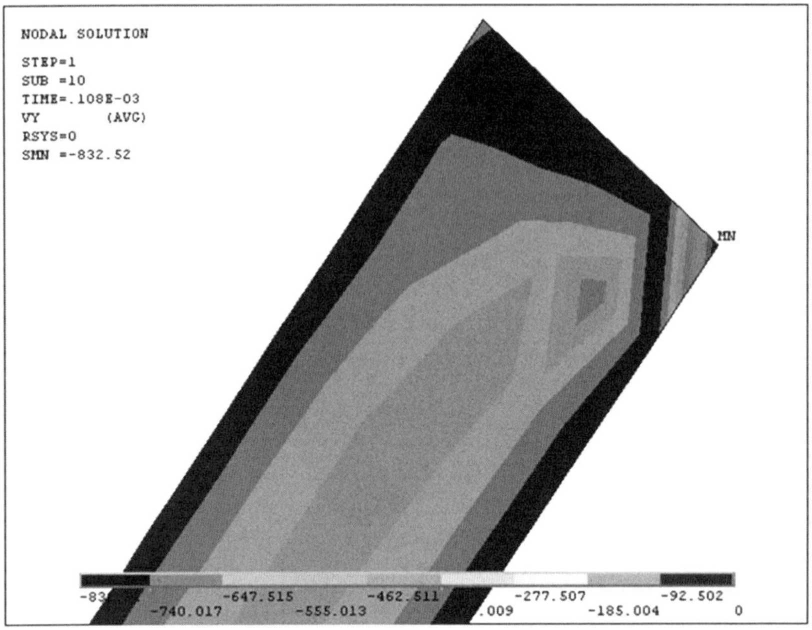

Figure 4.5 Nephogram profiles of the velocity distribution of the longitudinal low field with similar initial clearances.

The pressure gradient in the axial direction of the fluid field is relatively large, while the radial direction is basically the same.
In the velocity distribution map from the axis direction, at the entrance, because of the throttle effect, speed changes greatly, the middle and lower part develops uniformly, and the speed basically remains unchanged; from the radial direction, because of the two-wall viscous effect, the speed appears to be of a dragnet shape, small on both sides and large in the middle.

4.1.2 Finite Element Analysis of the Plunger and Barrel Assembly Structure

4.1.2.1 Three-dimensional Solid Finite Element Model

In this section, a bottom-up and solid modeling method is proposed. The material of the plunger and barrel assembly is GCr15, the elastic modulus is E = 206 GPa, the Poisson ratio is $\mu = 0.3$, and the density is =7.9 g/cm^3. The original size of plunger was diameter 7 mm, length 60 mm, initial clearance width 3 μm, barrel diameter 25 mm, and height 45 mm. Taking into account its considerable influence on the actual oil film, this method takes it as the key point of modeling and as the difficulty of grid division. The stroke guide at the lower part of the plunger has little influence on the clearance analysis, so this part is omitted. Because the structure and load of the plunger and barrel are symmetrical about the x–y plane, the finite element model of 1/2 is adopted to save computing time and occupying computer resources. The 8 node 3D body element solid 45 unit is used to divide the plunger into 14 057 units and 2603 nodes for the first time. The barrel is divided into 29 083 units and 5004 nodes. The model is shown in Figure 4.6.

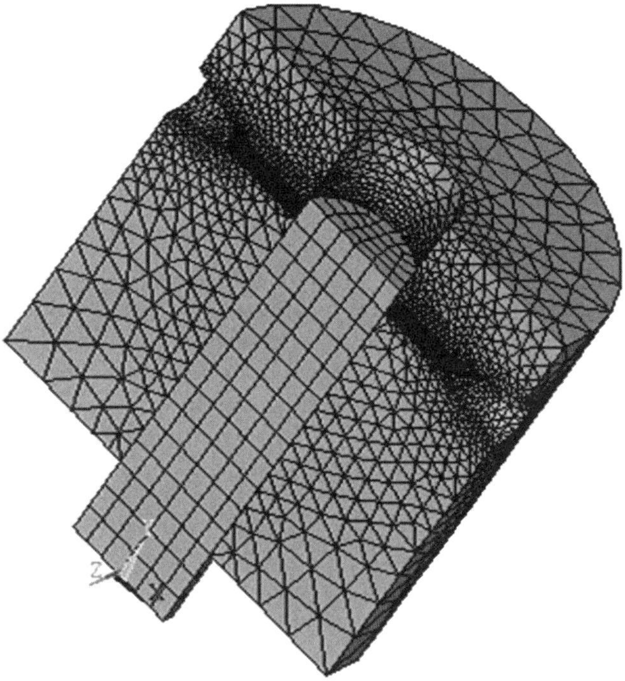

Figure 4.6 Structural mesh diagram of the plunger and barrel assembly.

4.1.2.2 Constraint Condition of Structure Field

(1) The plunger barrel and the plunger, respectively, restrain the three-direction displacement of its symmetrical surface. The lower end of the plunger is held by the cam and the lower end of the plunger restrains UY = O; the plunger barrel is supported by the pump body and the lower end restrains UY = O.
(2) Camshaft driving force. The plunger rises under the action of the cam and because of the relatively great impact caused by the roller and cam touching, the plunger axis is slightly skewed to one side, which means that the force of the cam on the plunger is not on the central axis of the plunger, but on one side. Meanwhile, the plunger compresses the fuel in the upper space of the plunger with P force, and the high-pressure fuel pressure in the plunger cavity has a counterforce to the plunger. This counterforce, due to the deviation of the plunger axis and the outlet and inlet of the oil orifice on the plunger axis, does not affect the other side of the axis and skews at another side, which means the plunger is bent and deformed under the action of these two forces.
(3) Fuel pressure at the end of the plunger. In the upward movement of the plunger, the fuel pressure gradually increases with the decrease in the volume. The barrel inner cavity is affected by the fuel pressure and undergoes a compression and bending combined deformation, so that the inner hole of the barrel protrudes and contracts. The specific ram cavity pressure loading is determined by test.
During the working process, several forces received by the plunger and barrel assembly, the pressure from the oil outlet valve, the support reaction force of the support surface, and the driving force of the cam driven part, can be obtained according to

the force balance condition, and the fuel pressure is subjected to numerical loading based on the initial estimated distribution.

(4) The supporting reaction of the opposite support side of the barrel. The supporting reaction between the barrel and the support surface of the pump body is balanced by the pressing load at the end of the barrel, and the forces can be thought to be evenly distributed on the supporting surface and are symmetrically distributed:

$$q_4 = \frac{q_3({r_3}^2 - r_5^2)}{r_3^2 - r_4^2} \tag{4.22}$$

In the formula, r_4 is the inner radius of head support surface of the barrel.

(5) Pre-tightening force of a bolt in the outlet oil valve. To ensure the tightness of the barrel mating surface, the end of the barrel must bear the pressing force of the compression bolt, which can be thought to be evenly distributed as pressure on the end face of the barrel is symmetrically distributed:

$$q_3 = \frac{T}{k\pi d({r_3}^2 - {r_5}^2)} \tag{4.23}$$

In the formula,

q_3 is the barrel bolt compression force (MPa),
k is the constant coefficient ($k = 0.1 \sim 0.3$),
d is the pitch diameter of thread, 17 mm,
r_3 is the outer radius of barrel head, 25 mm,
r_5 is the inner radius of barrel, 7 mm.

4.1.2.3 Structural Field Solution

The direct sparse matrix solution is adopted to solve the structure field, and in the automatic iteration option setting, the iteration accuracy is determined by the convergence. The structure field can be solved step by step.

By post-processing, the pressure deformation nephogram of the plunger and barrel assembly is obtained as shown in Figure 4.7. The following conclusions can be obtained:

(1) Under the simulation of a 150 MPa static working condition, the barrel expands in a radial direction and the piston shrinks in the radial direction. The contraction range of the plunger is obviously larger than that of the piston sleeve.
(2) The deformation distribution of the plunger and barrel assembly in the axis direction is uneven, decreasing from the top dead point to the lower dead point.

4.1.3 Structural Optimization of the Plunger and Barrel Assembly

4.1.3.1 Analysis of the Preliminary Simulation Result

By extracting the final restrained result of the coupling solution under different working conditions of the plunger and barrel assembly, the final deformation distribution of the corresponding assembly can be obtained. Figures 4.8–4.11 show the deformation nephograms of the original assembly at 150 and 180 MPa.

On the basis of Figures 4.8–4.11, the clearance radial deformation distribution of the plunger and barrel assembly, under the different working pressures of the plunger volume cavity, is calculated when the plunger is in the top stop stroke of the plunger. The results are summarized as shown in Figure 4.12, where the horizontal clearance of the

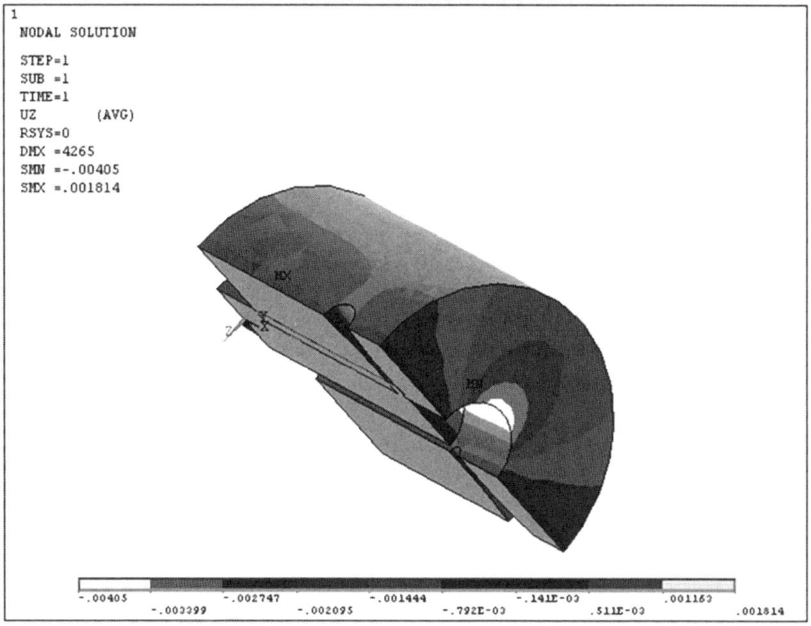

Figure 4.7 Initial load deformation nephogram of the structure field.

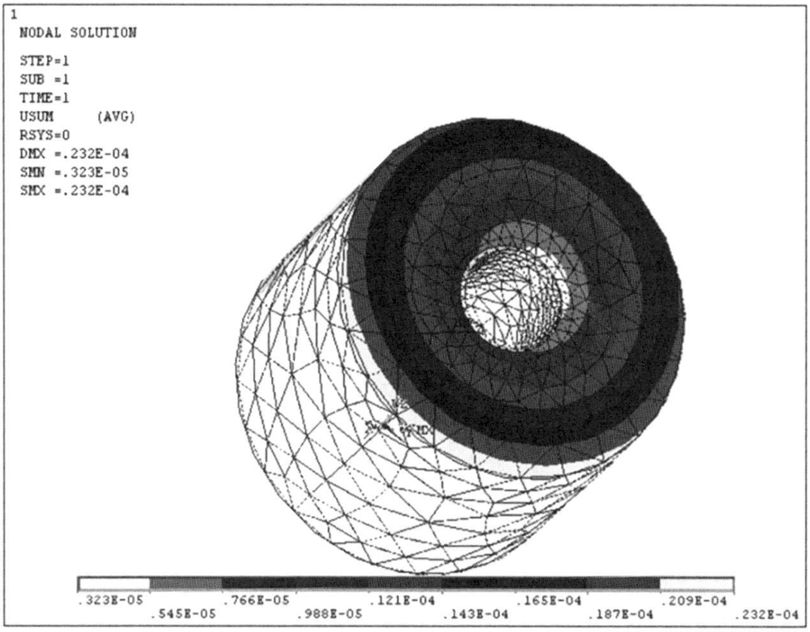

Figure 4.8 Simulation of the deformation of the original barrel at 150 MPa.

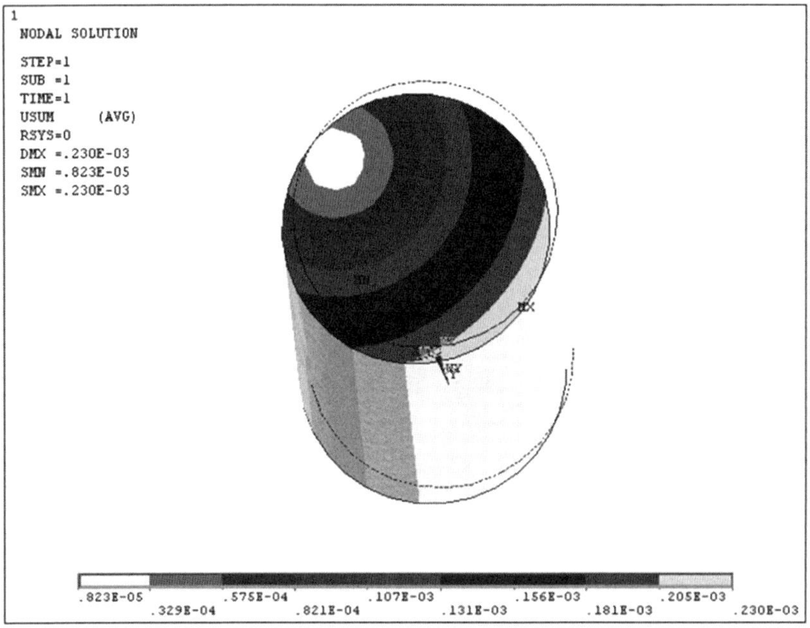

Figure 4.9 Simulation of the deformation of the original plunger at 150 MPa.

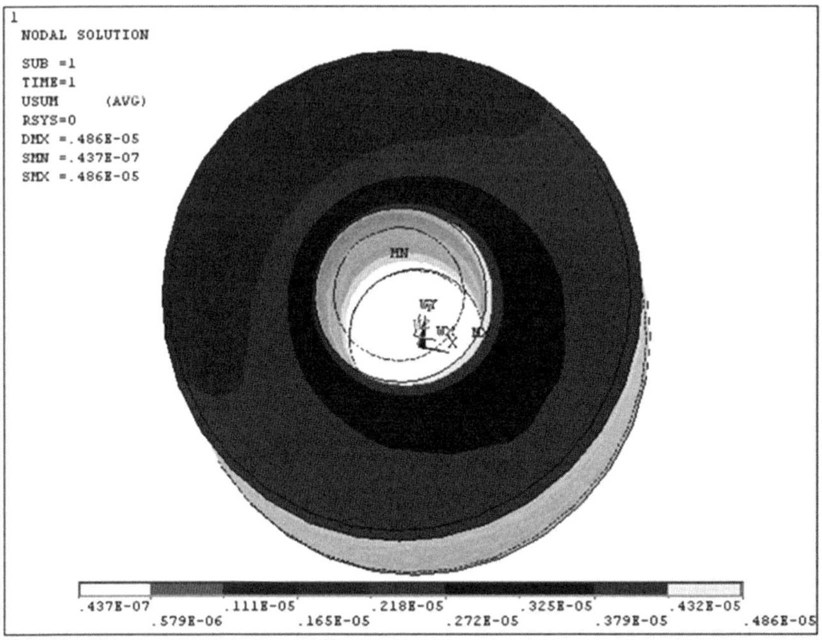

Figure 4.10 Simulation of the deformation of the original barrel at 180 MPa.

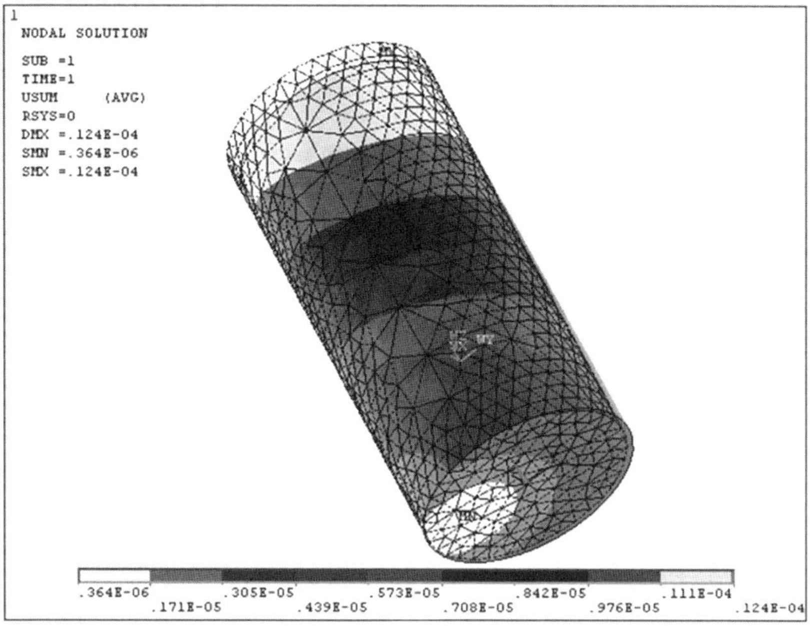

Figure 4.11 Simulation of the deformation of the original barrel at 180 MPa.

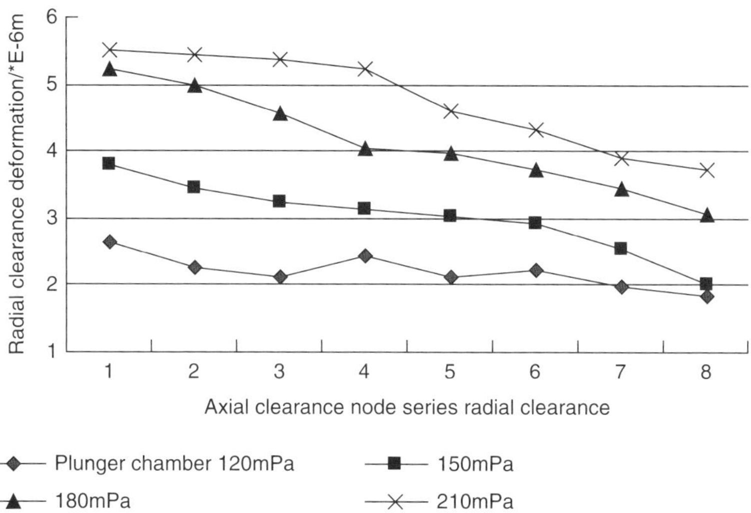

Figure 4.12 Simulation of the radial clearance of the plunger and barrel assembly under different pressures.

axial joint in the figure is equal to the axial part between the oil inlet orifice of the barrel and the upper dead center.

Figure 4.12 shows that the reason why the axial gradient is small under the action of high-pressure plunger barrel uniform expansion is that the barrel wall is thick, the ability to withstand the radial expansion pressure is strong, while the ability to bear the axial support force and the support reaction force is relatively small. In contrast, under the full working pressure of the plunger cavity, the axial shrinkage of the plunger head is obvious, while a larger pressure distribution of the plunger axial gradient in the clearance field is the reason why the plunger has a slight taper deformation. The above two points determine the low working sealing performance of the original plunger and barrel assembly under an ultra-high-pressure condition. In order to enhance the working performance under an ultra-high pressure, static consideration of reducing the fit clearance size between parts of the assembly is of no avail; to solve the sealing performance of the plunger and barrel assembly under a no-sticking condition, it is necessary to consider the deformation distribution and its reasons for the assembly performance under ultra-high pressure.

4.1.3.2 Deformation Compensation Optimization Strategy

The main reason for the plunger and barrel assembly high-pressure leakage is the deformation of the sealing clearance. In the optimization stage of the structural design of the plunger and barrel assembly, while considering the increase of the structure processing strength, the structure of the assembly itself can be used as a deformation compensation design, which means that the flexible variable cross-section deformation compensation technique is adopted. This technique starts with axial non-uniformity of the initial deformation of the sealing clearance, where the axial non-uniformity is mainly caused by the axial pressure gradient of the interstitial flow field.

The specific technology is that, based on the pressure distribution, the plunger is structurally compensated in reverse, and an urn-type compensation groove with a conical degree is trenched at its head. Therefore, in this working process of the assembly, the high-pressure fuel that enters the compensation groove of the plunger head produces a pressure wedge to expand itself, and causes the plunger to produce a certain axial gradient radial expansion; in this way, the axial deformation of the plunger caused by the extrusion of the fuel in the assembly clearance is partly neutralized. Eventually, the fitting clearance of the plunger and barrel assembly is kept in the appropriate numerical range. In this method, the greatest feature is that all fit clearance deformations caused by the compensation groove and the pressure gradient of the interstitial fluid are positively correlated with the pressure in the pressure cavity. If the structure is well handled, the full pressure range of deformation compensation can be achieved.

The sealing performance of the plunger and barrel assembly under a no-sticking condition can be featured by the ratio of the fuel leakage to the geometric fuel supply in several working cycles, or by the deformation and deformation gradient of the plunger and barrel assembly clearance.

4.1.3.3 ANSYS Optimization Analysis

In addition to the general structural stress analysis and dynamic system simulation, the ANSYS finite element analysis software can also be used for an optimization design. The main reason is that the most optimized design function is added to ANSYS, which is fully functional and very simple to use. There is no difference between the steps of the ANSYS optimization design and the work of the general finite element analysis software. Only the design variables, the objective function control, and the number of executions are demanded by the most optimized design in the post-processing phase. The ANSYS optimization design function can not only be used to solve the general function for optimum solution but can also be used to produce the structure of the geometric optimization design and shape optimization of the physical structure.

(1) ANSYS optimization steps. The mathematical model of optimization can be expressed as follows:

$$\text{Objective function}: f = f(x_1, x_2, \dots, x_N) \tag{4.24}$$

$$\text{Design variables}: x_{Li} \leq x_i \leq x_{Ui}\ (i = 1, \dots, N) \tag{4.25}$$

$$\text{Constraints}: g_{Lj} \leq g_j(x_1, x_2, \dots, x_N) \leq g_{Uj}\ (j = 1, \dots, N) \tag{4.26}$$

In the formulas, $f(x)$ represents the objective function, x_{Li}, x_{Uj} represent the upper and lower bounds of the design variables, N in Eq. (4,25) represents the number of design variables, g_{Lj}, g_{Uj} represent the upper and lower bounds of the constraints, and N in Eq. (4.26) represents the number of constraints.

The objective function is a scalar function of the design variable. It can be divided into single objective optimization and multiobjective optimization according to the number of functions. Usually, the objective function of the optimum design in mechanical engineering is divided into two categories: the first kind of objective function is to solve the optimum distribution shape of the stress value, and only the size of the shape design variables is considered as a constraint, so it is a linear constraint condition; the second kind of objective function is for the area or volume of the object, and the constraint condition is the maximum stress or deformation, which is the non-linear constraint condition.

The design variables are mainly used to describe the design direction and can set the range to achieve the desired target value. Design variables are independent variables that are often used as geometric variables, such as length, thickness, radius, or node. According to the number of design variables, they can be divided into a single design variable and a multiple design variable.

The constraint is also called a state variable. While seeking solutions to optimum design problems, designers are often limited by many design conditions, such as time, space, material, cost, etc. However, if the optimum design is limited by these constraints, the optimum solution cannot be found at all. For example, the reduction of material and the decrease of stress are often not able to be satisfied at the same time. These constraints are called state variables, which are one of the design variables. If such constraints are added to the optimum design problem, the problem becomes a constrained optimization design problem. The general constraints are divided into equality constraints and inequality constraints.

(2) Optimization parameter and its range settings. The optimization target of the plunger and barrel assembly is determined when the theoretical volume efficiency is the largest and the volumetric efficiency is the smallest with a fluctuating pump output pressure. Considering the complexity of a multiobjective parameter

optimization, the volumetric efficiency is considered as a constraint for solving optimization problems.

When setting, X is the set of geometric shapes for the compensation orifice of the plunger upper cavity and P is the output backpressure of the pump. $\Delta\eta_P$ is the variation function of the volumetric efficiency of the pump with the backpressure. Therefore:

$$\Delta\eta_P = f(X, P) \tag{4.27}$$

Optimization objective: $\Delta\eta_P = \min$

Optimization constraints :

$$\begin{cases} n = 300 \sim 500\ \text{rpm} \\ \eta_{\min} = 0.83 \\ P = 150s(s = 0.3 \sim 1.2) \\ d = 7\ \text{mm} \\ h_e = 7.5\ \text{mm} \\ \delta \geq 2\mu\text{m} \end{cases} \tag{4.28}$$

In the above equation:

n is the range of the plunger pump camshaft speed,
$\eta_{\min}$ is the minimum volume efficiency of the pump design,
P is the pump output backpressure,
d, h_e are the plunger effective diameter and effective stroke,
δ is the thickness of the plunger and barrel assembly clearance.

Under less stringent convergence criteria, the design variable range is initially locked. Furthermore, in this variable range, a set of schemes that fits the present level of machining is selected. Then a set of optimized sized schemes is selected by comparing and calculating the finite element. It should be noted that during the opening of the taper groove of the plunger head, the axial symmetry must be strictly observed, and a tiny deviation will cause a large pressure distribution disorder, resulting in the work stopping due to wear. Specific schemes are as follows:

Scheme A: inner groove of the plunger head is a cylindrical inner orifice with diameter 3 mm, depth 8 mm,
Scheme A: inner groove of the plunger head is a cylindrical inner orifice with diameter 3 mm, depth 10 mm,
Scheme A: inner groove of the plunger head is a cylindrical inner orifice with diameter 4 mm, depth 8 mm,
Scheme A: inner groove of the plunger head is a cylindrical inner orifice with diameter 4 mm, depth 10 mm,
Scheme E: plunger head groove is an inverted cone inner orifice with diameter 4 mm, bottom diameter 3 mm, depth 8 mm,
Scheme E: plunger head groove is an inverted cone inner orifice with diameter 4 mm, bottom diameter 2 mm, depth 8 mm,
Scheme E: plunger head groove is an inverted cone inner orifice with diameter 4 mm, bottom diameter 2 mm, depth 10 mm.

4.1.3.4 Evaluation of the Optimization Result

(1) Evaluation of clearance deformation. The nephogram of clearance deformation can directly reflect the effect of each optimization scheme. The final calculated convergent nephogram of each group is extracted and compared horizontally. Figures 4.13 and – 4.14 show the deformation nephogram of the plunger and barrel assembly in Scheme F at 180 MPa working pressure.
   Analyze the above seven schemes with coupling analysis, and calculate the clearance deformation of each working stroke. Then divide the plunger stroke position by n, mark these state points, and select a set of m nodes in the axial direction of the flow field and it is easy to find that the radial deformation of the clearance flow field corresponding to the axial position of the jth node is the algebraic sum of σ_{1j} and σ_{2j}, which are the radial deformations of the plunger and barrel assembly near the two walls of the flow field. Take the nth state point, record the amount of deformation in the radial diffusion of the outer wall of the flow field as positive and. conversely, the radial deformation of the inner wall of the flow field as positive.
   According to the distribution of clearance deformation, Schemes E and F are better.
(2) Evaluation of the stroke leakage rate. The stroke leakage rate at the lower end of the flow field is calculated as follows. According to the plunger rate curve, take the characteristic velocity value v_t of the plunger at each state point and calculate the flow rate q_i at the clearance outlet of each state point:

$$q_i = \frac{\pi v_m}{4}[(D + \sigma_{2m})^2 - (d + \sigma_{1m})^2] \tag{4.29}$$

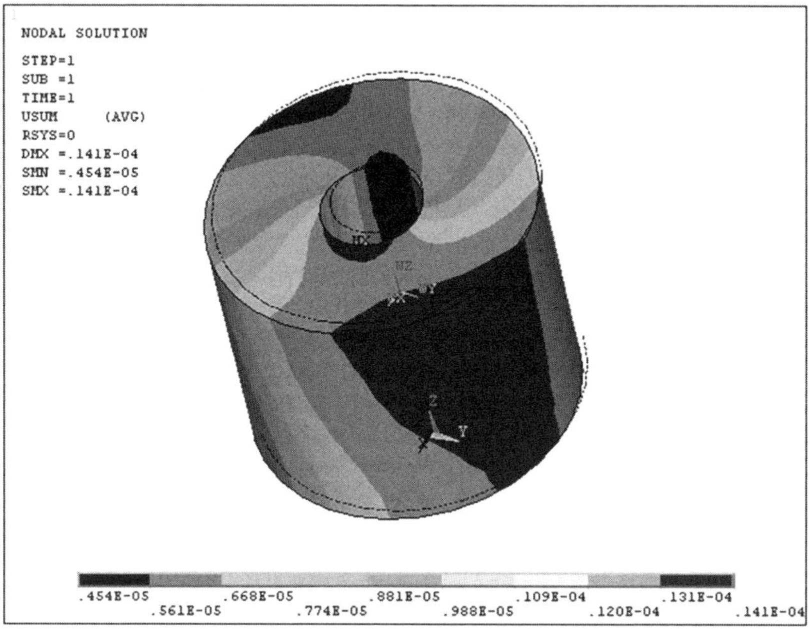

Figure 4.13 Simulated deformation nephogram of the plunger barrel in Scheme F.

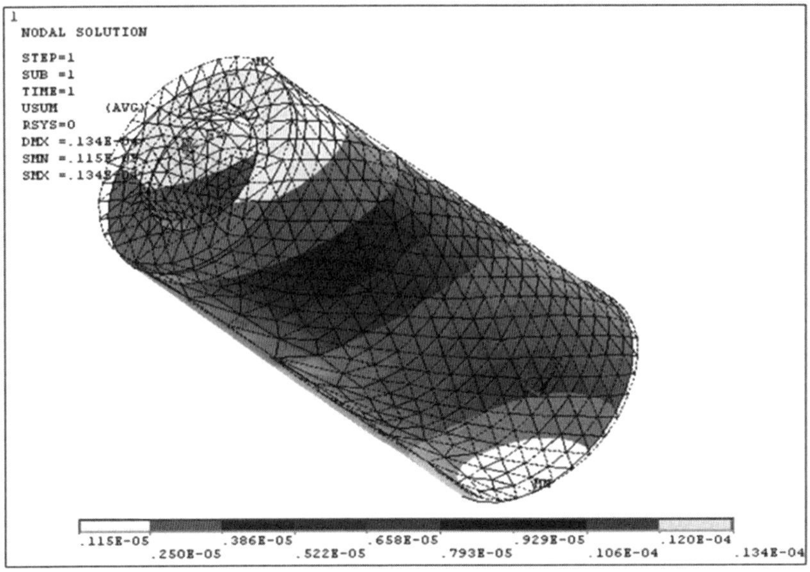

Figure 4.14 Simulated deformation nephogram of the plunger in Scheme F.

In the formula,

D is the original clearance outer diameter of the assembly,

d is the original clearance inside diameter of the assembly,

σ_{1m} and σ_{2m} are the radial deformations of assembly near the two walls of the flow field,

v_m is the outlet rate of the fluid,

subscript i corresponds to the i th state point,

subscript m corresponds to the m th node, which is the selected nearest node at the exit.

According to the stroke leakage rate evaluation, Schemes D and E are relatively close and relatively minimum. By summing up the two indexes, it can be concluded that Scheme E is the best simulation optimization structure.

4.1.4 Experimental Study on the Deformation Compensation Performance of the Plunger and Barrel Assembly

According to the results of the study, the deformation compensated plunger and barrel assembly was tested. The assembly was installed in a special tool for the sealing test of the assembly and the sealing performance and deformation of the assembly were tested on the special test bench.

4.1.4.1 Test for the Sealing Performance of the Plunger and Barrel Assembly

The used experimental device is shown in Figure 4.15. The seven improved schemes are used together with the original plunger and barrel assembly for the static pressure leak test. Every scheme has four sets of assembly and is tested one by one. By comparison, a group with the best leakage performance is selected in each group to participate in a transverse comparison.

Figure 4.15 Special test devices for the sealing test of the plunger and barrel assembly.

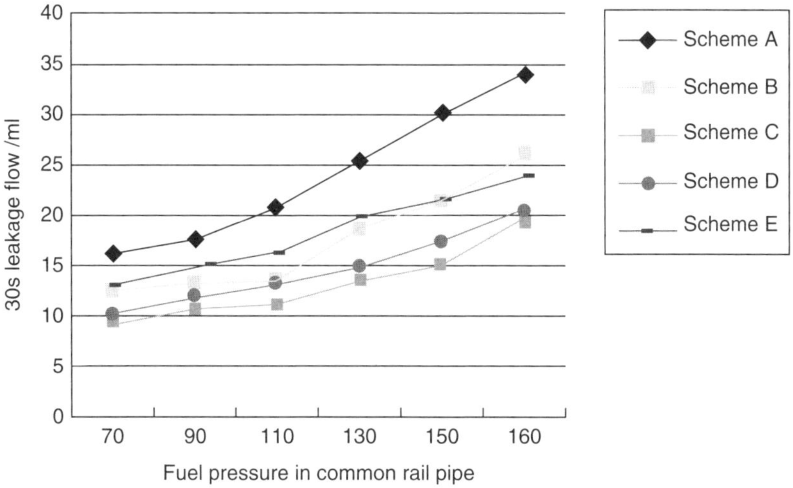

Figure 4.16 Leakage rate of different assembly schemes.

Figure 4.16 shows the timing leakage at special stroke positions at different assembly schemes under different common rail cavity pressures. It can be seen that the leakage of the improved assembly is obviously reduced before the improvement, and among them, the leakage amount of Scheme E is the optimum among several optimization schemes. Figure 4.17 gives the leakage compared situation of the plunger of different relative positions in the effective stroke in a selected group of assemblies in Scheme E. It can be seen that the change of the leakage flow of the stroke is relatively uniform, and meets the requirement of sealing under high pressure.

Static pressure leakage can only be used as a relative reference to evaluate the effect of structural optimization. For a further experiment, eight groups of representative scheme

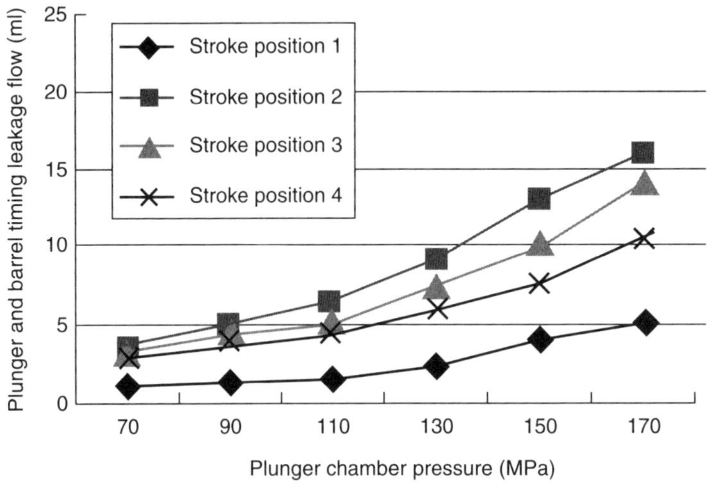

Figure 4.17 Leakage at different stroke positions of the plunger.

Table 4.1 The fuel supply of the assembly in Scheme D under different track pressures.

Rotational speed (r/min)	30s Theoretical displacement (ml)	*P* (MPa)	Flow rate *Q* (ml)	Volume efficiency (%)
160	23.58	0	26.1	110.69
		30	24.5	103.90
		50	24.2	102.63
		70	23.8	100.93
		90	23.4	99.24
		110	22.8	96.70
		130	21.9	92.86
		150	20.8	88.21

Note: Calculation of theoretical displacement per minute of plunger: $Q = (\pi d^2/4)h_e \times n$, where the cam speed is 160 r/min.

assemblies, including the original plunger and barrel assembly, were selected in the static pressure leakage test.

The seven schemes are used together with the original 135 pump assembly to test the volumetric efficiency. Table 4.1 shows the fuel supply of the Scheme D assembly at the cam rotation speed of 160 rpm. Figure 4.18 shows how the volumetric efficiency of the improved Scheme D and the original 135 pump varies with the fuel supply pressure. It is not difficult to draw the following conclusions:

(1) The high-pressure pump that uses the assembly in Scheme D can satisfy the design requirements of high volumetric efficiency under backpressure between 0 and 150 MPa. When the fuel supply pressure rises to 150 MPa, the volumetric efficiency can reach more than 88%, and is able to improve. Compared with the original 135 pump, it greatly improves the working performance of a high-pressure pump.

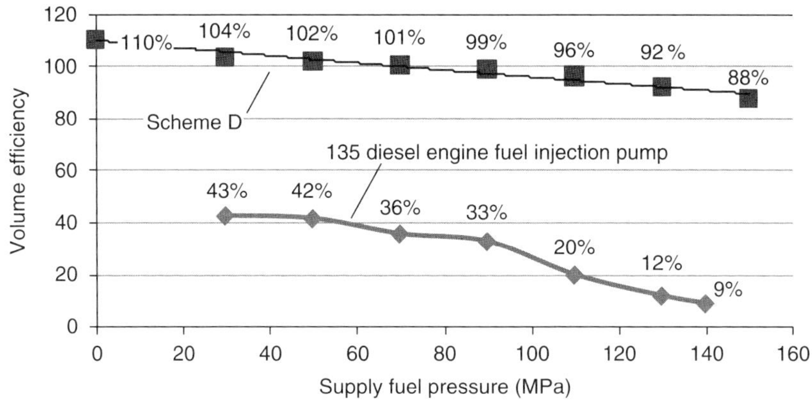

Figure 4.18 The volume efficiency of the original 135 pump and Scheme D under different supply pressures.

(2) The volumetric efficiency decreases gradually with the pump output pressure, and the scope that volume efficiency of the high-pressure pump of the improved Scheme D decreases greatly with the decrease of the rotational speed and is obviously smaller than that of the original 135 pump. When the fuel supply pressure reaches 140 MPa, the volumetric efficiency of the original 135 pump is only 9%, which cannot meet the pressure requirements of the ultra-high pressure jet.
(3) Before the outlet pressure of the high-pressure pump of the improved Scheme D exceeds 70 MPa, its volumetric efficiency exceeds 100%, which may be caused by the clearance effect of the low-pressure inlet orifice of the plunger and barrel assembly.
(4) After running the Scheme D test assembly with no sticking, the plunger and barrel assembly has no damage on the surface. Therefore, they achieve better motion characteristics and higher working reliability.

4.1.4.2 Plunger and Barrel Assembly Deformation Test

The resistance strain gauge sensor is used to obtain the compressed deformation of assembly outer wall and convert it to the resistance change of strain gauge itself. Then, it is transformed into a weak voltage signal by bridge circuit, and the required parameters are acquired through the signal acquisition and processing system. Finally, according to the principle of structural mechanics, the deformation of the coupling, the pressure distribution of the gap fluid and the load characteristics of the plunger cavity during working process are obtained. The system structure is shown in Figure 4.19.

In order to facilitate the arrangement of strain gauges and main lead wires, the key structures of the high-pressure pump are simplified. A special test piece of assembly is made, and the structure is assembled as shown in Figure 4.20. The plunger and barrel assembly is fixed by an upper and a lower pressure block bolt. The upper end is hermetically connected with the high-pressure connector and the lower end is supported by the plunger lift adjustment bolt. The strain gauge is attached to the outer wall of the plunger barrel and a specific wiring terminal of the measuring bridge circuit is exported by the contact lead wire.

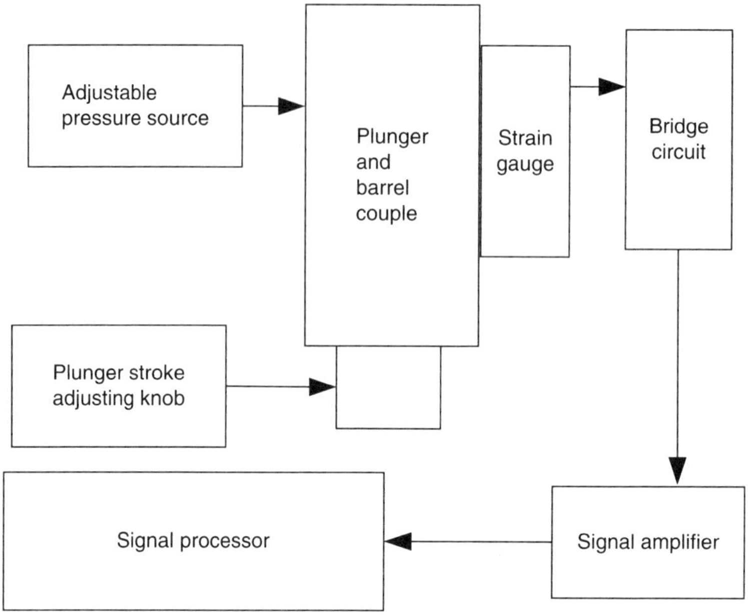

Figure 4.19 Structure design of the static pressure leakage measurement system.

In the eight schemes of processing plunger and barrel assembly, the shape of the plunger barrel remains the same, which is convenient for determining the number of patches and the location (circumferential and axial position) of the strain gauges. In the experiment, the fifth strain gauge is used to measure the radial deformation of the outer wall of the barrel. The plunger barrel is chosen far from the oil outlet orifice and four measuring points are evenly selected along the axial direction, which are distributed on the upper and down stroke of the plunger. The fuel pressure in the test is between 50 and 170 MPa.

In accordance with the above test plan, the assembly is gradually pressurized in the pressure test and the corresponding readings of five sets of strain gauges of the barrel outer wall are recorded. Through the related conversion, the deformation variable values of the eight sets of pairs at different positions in the axial direction of the plunger barrel under different pressures can be obtained. The results show that there is a good linear relation between the amount of deformation and the working pressure. Figure 4.21 shows the relative deformation of eight pairs of packages at 150 MPa working pressure.

As shown in Figure 4.21, the clearance deformation of the original Scheme A is large, and the axial deformation is uneven. The clearance deformation of the plunger in reciprocating motion is large. In comparison, the clearance pressure deformation of Scheme E is smaller, the axial variation is smaller, and the dynamic variation is uniform, which coincides with the static pressure seal test results. The change of clearance between the plunger and barrel assembly consists of the plunger deformation and the plunger barrel deformation. The deformation of the plunger barrel is the result of the combined action of the flow field pressure in the clearance and the pre-tightening force of the upper and

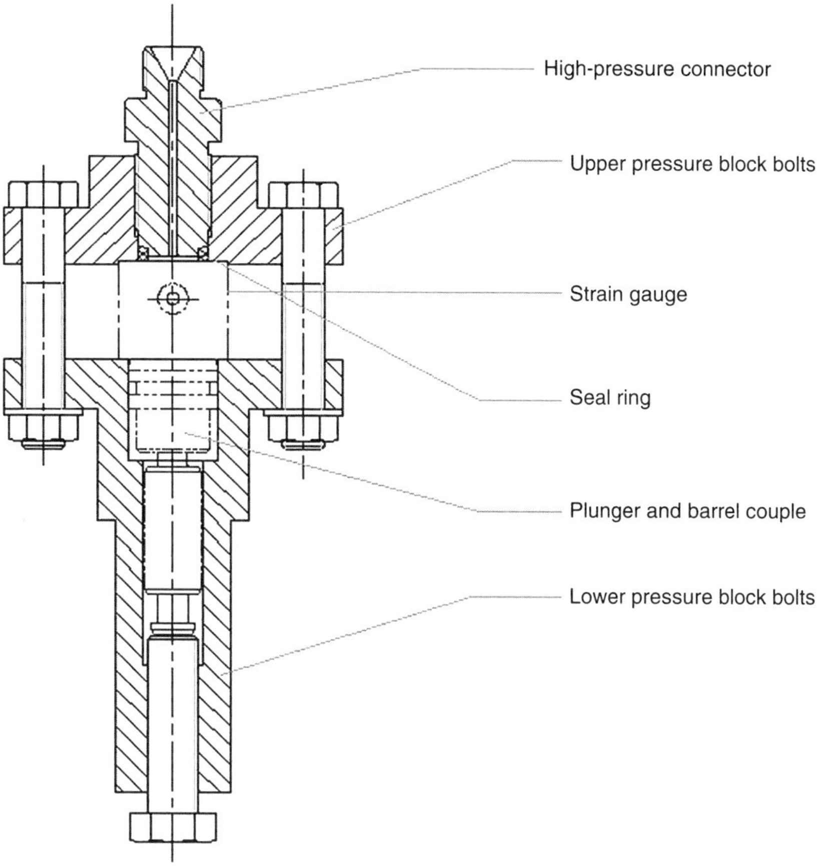

Figure 4.20 Special test piece for the static pressure deformation test.

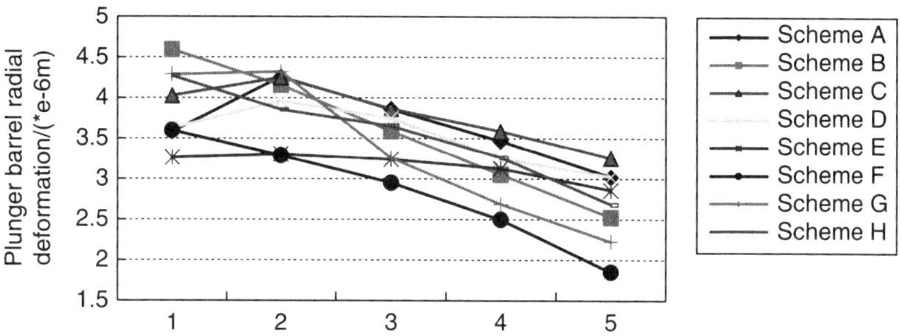

Figure 4.21 Radial deformation of the plunger barrel with different schemes under 150 MPa.

lower pressure blocks. The deformation of the plunger is caused by both the fuel pressure of the upper cavity of the plunger and the flow field pressure in the clearance.

Figure 4.22 is the simulation and comparison result of the radial deformation of the plunger barrel of Scheme E. The test results show that there is good consistency between the simulation and results.

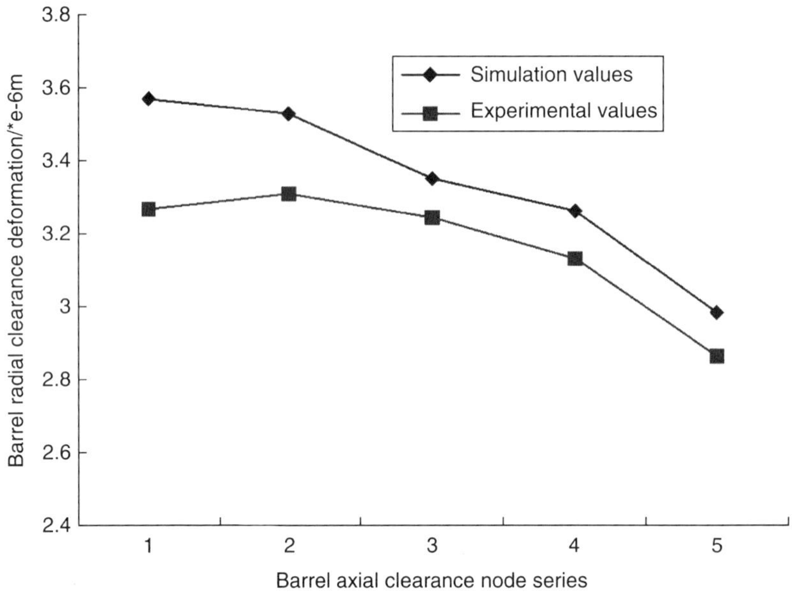

Figure 4.22 Comparison between the simulation and experiment.

4.2 Strength Analysis of the Cam Transmission System for a High-pressure Fuel Pump

The mechanical load on the high-pressure pump cam–roller mechanism is mainly the pressure of the upper plunger cavity. When the high-pressure fuel pump is working, the periodic alternating load suffered by the camshaft has the obvious impact characteristics. Therefore, the dynamic characteristics must be considered in the study of the strength of the cam–roller pair.

ADAMS (Automatic Dynamic Analysis of Mechanical Systems) is a virtual prototype analysis software developed by MDI (Mechanical Dynamics Inc.). At present, ADAMS has been adopted by hundreds of major manufacturers all over the world. ADAMS uses an interactive graphical environment, unit library, constraint library, and force library, to create a fully parametric geometric model of the mechanical system, and its solver uses the Lagrange algorithm in the dynamics theory of the rigid multibody system, to establish system kinetic equations, to analyze the virtual machine system with statics, kinematics, and dynamics, and to give the outputs of displacement, velocity, and acceleration and the reaction curve. ADAMS simulation can be used to predict the mechanical system performance, motion range, collision detection, peak load, and calculate the input load of the finite element. On the one hand, ADAMS is the application software of a virtual prototype analysis: users can use this software to conveniently carry out statistics, kinematics, and dynamics analysis of the virtual mechanical system. On the other hand, it is a virtual prototype analysis and development tool; its open program structure and various interfaces can be used as a secondary development tool platform for special types of virtual prototype analysis for special industry users. ADAMS consists of five modules: basic module, expansion module, interface module, professional domain module, and toolbox. Users can use general modules to simulate general mechanical systems, and use special modules to quickly and effectively model and simulate the problems in specific industrial application areas.

4.2.1 Dynamic Simulation of the Cam Mechanism of a High-Pressure Pump

4.2.1.1 Solid Modeling

By using the three-dimensional modeling software ProeWildfire2.0 to build solid modeling of the cam transmission mechanism and using ADAMS to build a rigid body, a variety of motion pairs is added in ADAMS; the results of the model, which is basically completed, are shown in Figure 4.23.

It can be calculated after the cam and the force part are loaded by the pressure of the high-pressure fuel cavity at the upper part of the plunger and the driving moment of the fuel pump. After the cam mechanism is loaded, as shown in Figure 4.24, the calculated cam contact pressure change is shown in Figure 4.25 and the friction change of

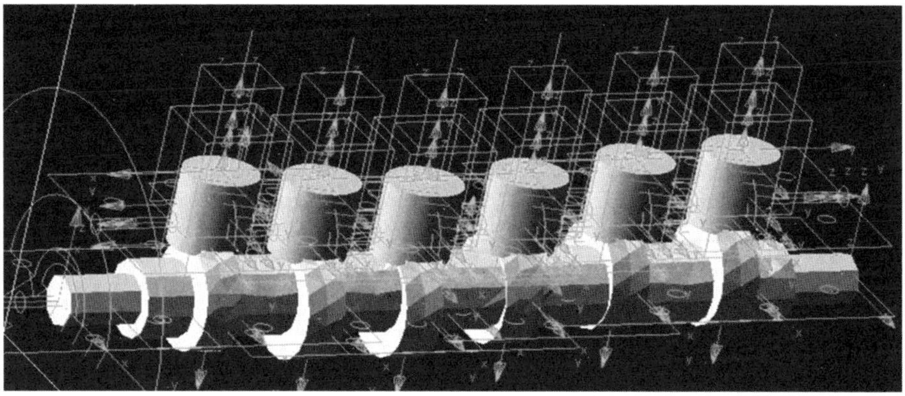

Figure 4.23 ADAMS model of the cam.

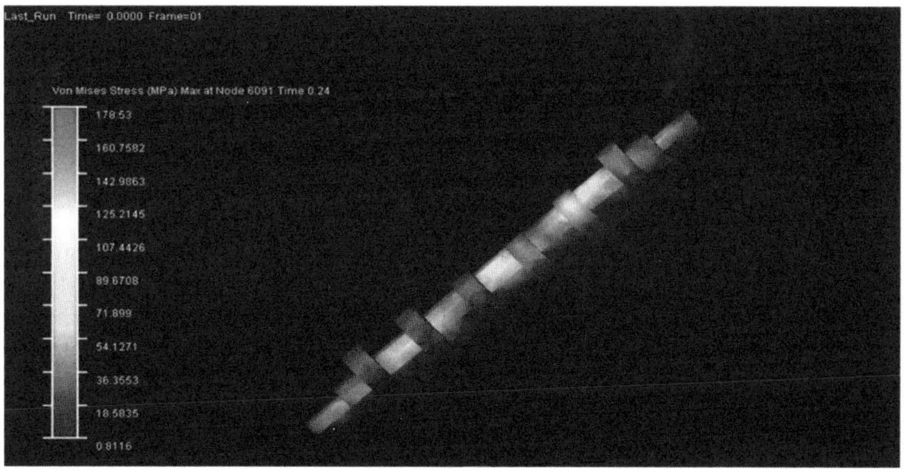

Figure 4.24 Loading diagram of the cam mechanism.

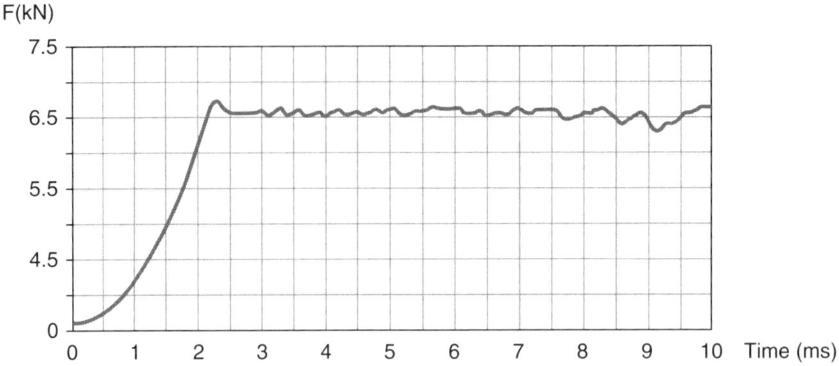

Figure 4.25 Cam contact pressure diagram.

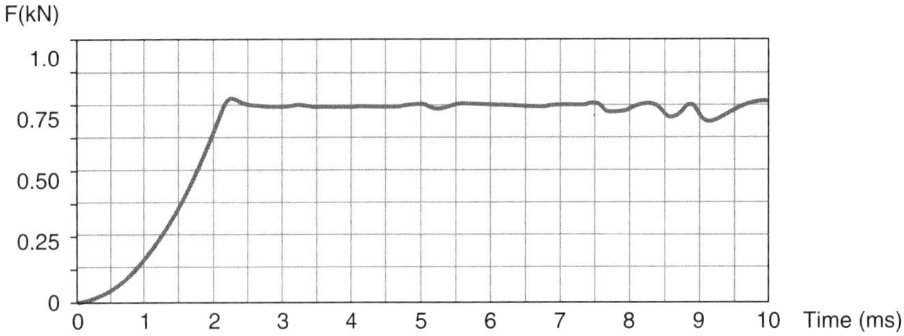

Figure 4.26 Friction diagram of the cam contact pair.

the cam roller contact surface is shown in Figure 4.26. It can be seen from the diagram that the friction force of the cam roller contact surface increases with the compression of the plunger cavity diesel fuel, and the friction power has negative effects on the work of the cam–roller pair.

4.2.1.2 Rigid–Flexible Hybrid Modeling and Simulation of the Camshaft Mechanism

MSC.Patran is an integrated parallel frame finite element analysis simulated system for front and rear processors. Its unique direct geometry access (DGA) technology provides a perfect integration environment for the geometric model communication between various CAD/CAM software systems and seamless connections of various analysis models.

The camshaft model built in Proe is imported into the finite element analysis software MSC.Patran to perform the finite element analysis and the units are set to mm, kg, N, and s. The mesh divide is processed, the connection point is defined, and the locations of the fixed points at both ends are selected. RBE2 is added near the two points; the surrounding points are regarded as dependent points and the fixed end is regarded as the independent point. Then the material, input material density, elastic modulus, Poisson's ratio, and other corresponding parameters are defined. The MNF file output of ADAMS, after confirmation, enters the Nastran module to calculate and obtain the corresponding flexible body file.

When the flexible body is built, the input position of the flexible body is not necessarily in the desired position, and it is not convenient to define the kinematic pair and the load on the flexible body. In order to facilitate the operation, ADAMS developed a tool to replace the rigid body directly with the flexible body or replace the flexible body with another flexible body. After the replacement, the moving pair and load on the rigid body or flexible body will automatically transfer to the flexible body, and the Maker point on the rigid body or flexible body will be transferred to the node closest to the Maker point on the flexible body. In addtion, the new flexible body also inherits some characteristics of the rigid or flexible body, such as color, icon, size, initial speed, modal displacement, etc., so that it is convenient for operation.

In the main menu, select "Build – Flexible Bodies – Rigid to Flex" and after the dialog box with the replacement of rigid flexible body pops up, pick the rigid body to be replaced in the "Current Part" input box. In the "Mnf File" input box, enter the Link.mnf path of the Mnf file to replace the rigid body component and click Apply to complete the process of replacing the rigid body with a flexible body. The rigid–flexible mixed model is established, as shown in Figure 4.27. The membrane is calculated and simulated, and the results are shown in Figure 4.28.

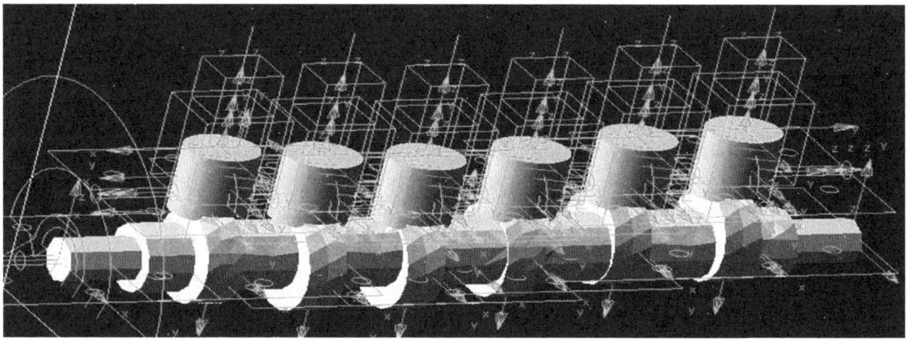

Figure 4.27 Mnf model of the cam mechanism.

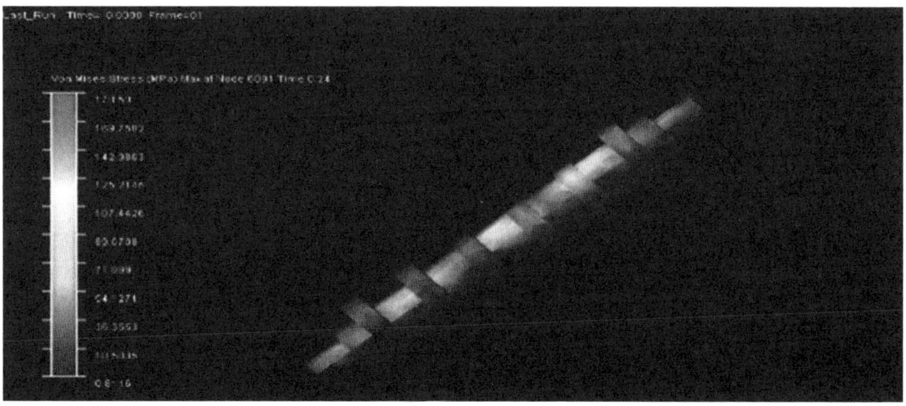

Figure 4.28 Stress nephogram.

From the figure it can be seen that the maximum stress point is in node 6091 and the maximum stress value is 179 MPa. The minimum stress value of the node is 0.8 MPa, the stress amplitude is 178.02 MPa, and the formula is calculated using the Gotaverken–Soderberg strength calculation:

$$n = \frac{1}{\left(\frac{\sigma_m}{\sigma_b}\right) + \left(\frac{\sigma_a}{\sigma_x}\right)}$$

In the formula, σ_m is the maximum stress value, σ_a is the stress amplitude, σ_b is the fracture limit, and σ_x is the endurance limit. Taking the parameters $\sigma_b = 400$ MPa and $\sigma_x = 600$ MPa of A3 steel, the fatigue strength safety factor of the camshaft is $n = 1.34$.

4.2.2 Stress Analysis of the Cam and Roller Contact Surface

The study of the stress on the contact surface is one of the most complex and difficult fields in the theory of elasticity. In essence, it is the problem of the transition from the classical linear elastic theory to the non-linear elastic (elastic plastic) theory. The earliest calculation of the contact stress based on the linear elastic theory was proposed by Hertz. Because of the complexity of the problem, in more than 100 years there was little outstanding progress achieved in elasticity analysis, although from the beginning of the 1960s many research workers have made outstanding contributions in this field, such as Merwin and Johnson, who put forward the contact stress of the elastic–plastic solution is an approximation method. Many conditions have been assumed in order to simplify the problem. For example, in the calculation process, it is assumed that the plastic deformation is equal to the elastic deformation, which is only reasonable in the case of a low friction coefficient ($f < 0.5$). Because the plastic zone occurs at a certain distance below the surface, surrounded by the elastic zone, the total stress will not be significantly different from the elastic strain. However, in the case of a high friction coefficient, the plastic zone rises to the surface and the larger plastic deformation will occur near the surface; in such a case the calculation results will have larger errors. Therefore, the elastic–plastic analysis of contact stress is far less convenient than the Hertz theory to form a complete set of calculation methods.

Later the theoretical and experimental results showed that in most parts of the machine working condition, according to Hertz theory, the contact stress calculation error compared with the actual situation is not too large, so this simple calculation method is still widely used now.

Because of the complexity of the contact problem, Hertz made the following assumptions:

(1) The two contact materials are absolutely homogeneous and isotropic.
(2) All the deformations occur within the elastic limit without any residual deformation.
(3) The two contact surfaces are absolutely smooth.
(4) There is no tangential load, which means there is no friction.
(5) The contact area is much smaller than that of the whole contact object.
(6) There is no lubricant film between the surfaces.

The calculation of contact stress can be carried out after the above assumptions are made.

4.2.2.1 Contact Stress Calculation Method

According to the shape characteristics of two contact bodies, contacts are mainly divided into point contact (such as roller bearing) and linear contact (such as gear). Because under the contact load all materials will have elastic deformation, a certain size of contact area must be formed in the contact area. In the case of point contact, the contact surface is round or oval; in the case of linear contact, the contact surface is rectangular.

Firstly, take point contact: if the two surfaces contact at the O point (see Figure 4.27), then M_1, M_2 are the two points corresponding to each other on the surface, which means they have the same coordinate (x, y). The elastic deformation of the contact zone will occur under the action of the load. If ω_1, ω_2, respectively, represent the displacement of points M_1, M_2 at the undeformed part along direction z_1, z_2, and $\omega_1(o), \omega_2(o)$ represent the displacement of point o along direction z_1 and z_2, the distance between M_1 and M_2 is reduced by $[\omega_1(o) - \omega_1] + [\omega_2(o) - \omega_2]$.

If M_1, M_2 eventually arrive at the same contact surface due to local compression,

$$z_1 + z_2 = [\omega_1(o) - \omega_1] + [\omega_2(o) - \omega_2] = [\omega_1(o) + \omega_2(o)] - [\omega_1 + \omega_2] = \delta - (\omega_1 + \omega_2)$$

In the formula, δ represents the distance between two contacts that are approaching each other:

$$\omega_1 + \omega_2 = \delta - (z_1 + z_2) = \delta - c \tag{4.30}$$

If the two contact bodies are spheres whose radii are R_1 and R_2,

$$c = z_1 + z_2 = \frac{r^2}{2R_1} + \frac{r^2}{2R_2} = \frac{r^2(R_1 + R_2)}{2R_1R_2} \tag{4.31}$$

In the formula, r represents the distance between point o and M_1, M_2.

If the two contact bodies are curved surfaces with constant curvature,

$$c = z_1 + z_2 = Ax^2 + By^2 \tag{4.32}$$

In the formula, $x, y \rightarrow M_1, M_2$ point in x, y on the plane coordinate, so
$A, B \rightarrow$ coefficients are positive.

Because the contact bodies are much larger than the contact areas, Hertz considered it possible to replace the original contacts with two semi-infinite elastic bodies and leave the contact area of the ellipse to bear the load. In this situation, $\omega_1 = k_1 \int_F \frac{pdF}{r} = k_1 \int_0 d\varphi_1 \int_{1l_1} pdr$

$$\omega_2 = k_2 \int_F \frac{pdF}{r}$$

In the formula,

$p \rightarrow$ represents normal contact stress,
$2l_1 \rightarrow$ represents the chord length through M_1 $(BC = 2l)$,
$r \rightarrow$ represents the distance from the union area dF to point M_1:

$$k_1 = \frac{1 - \mu_1}{2\pi G_1}, k_2 = \frac{1 - \mu_2}{2\pi G_2} \tag{4.33}$$

In the formula, μ_1, μ_2 represent Poisson ratios and G_1, G_2 represent shear modulii of elasticity. Therefore,

$$\omega_1 + \omega_2 = (k_1 + k_2)\int_F \frac{pdF}{r} = \left(\frac{1-\mu_1}{2\pi G_1} + \frac{1-\mu_2}{2\pi G_2}\right)\int_F \frac{pdF}{r} = k_0 \int_F \frac{pdF}{r} \tag{4.34}$$

and therefore

$$G = \frac{E}{2(1+\mu)}$$

where E represents the elastic modulus. Therefore

$$k_0 = \frac{1}{\pi}\left[\frac{1-\mu_1^2}{E_1} + \frac{1-\mu_2^2}{E_2}\right] = \frac{\eta}{\pi} \tag{4.35}$$

Now the expression of p must be found to satisfy

$$k_0 \int \frac{pdF}{r} = \delta - c \tag{4.36}$$

It is obvious that the center o point on the elliptical contact surface has the maximum displacement, and therefore it will bear the maximum compressive stress. Thus the distribution function p of the stress on the contact surface must be a semi-ellipsoid with vertex p_0 and at the bottom of the elliptical contact surface boundary (Figure 4.29):

$$\left(\frac{x_1}{a}\right)^2 + \left(\frac{y_1}{b}\right)^2 + \left(\frac{\zeta}{c}\right)^2 = 1 \tag{4.37}$$

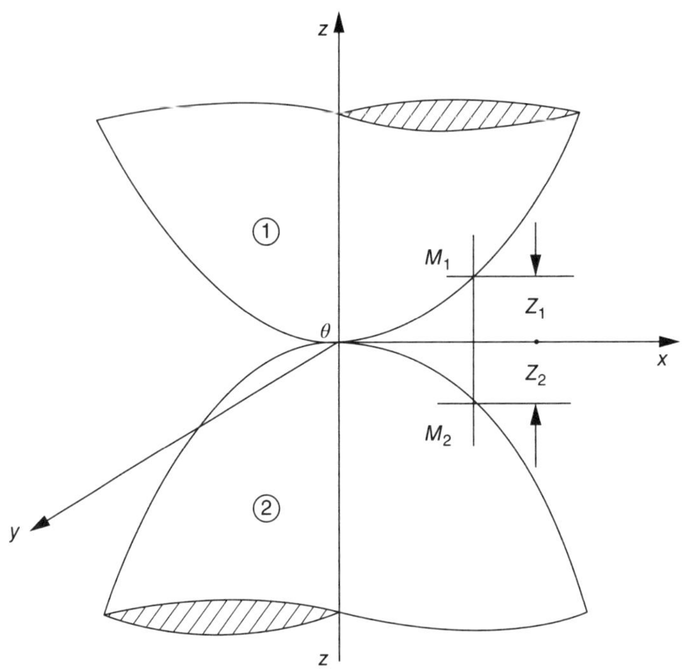

Figure 4.29 Sketch of the contact diagram of two curved surfaces.

In the formula, ζ represents the vertical coordinates of a semi-ellipsoid at x_1, y_1:

$$p = p_0 \times \frac{\zeta}{c} = p_0\sqrt{1 - \left(\frac{x_1}{a}\right)^2 - \left(\frac{y_1}{b}\right)^2} \tag{4.38}$$

$$p = \int_F pdF = \frac{p_o}{c}\int_F \zeta dF \tag{4.39}$$

Therefore,

$$\int_F \zeta dF = \frac{2}{3}\pi abc$$

and

$$p_o = \frac{3}{2} \times \frac{P}{\pi ab}$$

Now the values of a, b are required to obtain p. When the contact surface between the camshaft and roller is equivalent to that of two parallel cylindrical contacts, which is a rectangle as shown in Figure 4.30, it can be thought that $a \rightarrow \infty$ and the load distribution of unit length is a semi-ellipse area with the bottom of the $2b$:

$$\left(\frac{y}{b}\right)^2 + \left(\frac{z}{c}\right)^2 = 1 \tag{4.40}$$

$$P = \int_{-b}^{b} pdy = \frac{p_o}{c} \times \frac{\pi bc}{2}, p_0 = \frac{2P}{\pi b} \tag{4.41}$$

where

$$p_o = 0.5642\sqrt{\frac{P}{\eta}\left(\frac{R_1 + R_2}{R_1 R_2}\right)}$$

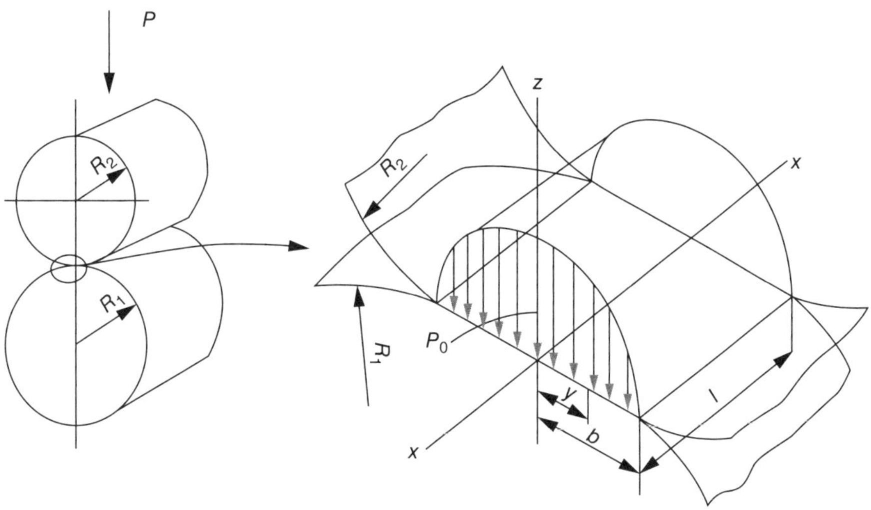

Figure 4.30 Hertz stress distribution during linear contact.

If $\mu_1 = \mu_2 = 0.30$,

Then $p_0 = 0.418\sqrt{\frac{PE(R_1+R_2)}{R_1R_2}}$, $b = 1.522\sqrt{\frac{P}{E} \times \frac{R_1R_2}{(R_1+R_2)}}$

The distribution of the contact stress along the depth can be calculated by calculating the values of p_o and a, b. For the case of linear contact between two cylinders, the stress at one point on the z-axis can be calculated using the following formulas:

$$\sigma_x = -p_o \times 2\mu\left[\sqrt{1+\left(\frac{z}{b}\right)^2} - \frac{z}{b}\right], \sigma_y = -p_0\left[\frac{1+2\left(\frac{z}{b}\right)^2}{\sqrt{1+\left(\frac{z}{b}\right)^2}} - 2\frac{z}{b}\right],$$

$$\sigma_x = -p_o\frac{1}{\sqrt{1+\left(\frac{z}{b}\right)^2}},$$

$$\tau_{\max} = \frac{1}{2}(\sigma_x - \sigma_y)$$

4.2.2.2 Calculation of Contact Stress under the Combined Action of Normal and Tangential Loads

Since Hertz first established the contact theory of elastic body, the effect of tangential force has not been considered in the calculation of contact stress for more than 50 years. By 1953, Smith and Liu published an article about the stress produced by tangential and normal loads on an elastic body and its applications to some contact stress problems. This is the most complete set of contact stress calculation methods for an elastic body following the Hertz theory.

Firstly, the stress produced by the normal load on the contact surface is calculated by the ellipse distribution:

$$\sigma_{xy} = -\frac{p_{n\max}z}{\pi}\left[\frac{b^2+2y^2+2z^2}{b}\overline{\psi} - \frac{2\pi}{b} - 3y\psi\right] \tag{4.42}$$

$$\sigma_{zn} = -\frac{p_{n\max}z}{\pi}[b\overline{\psi} - y\psi] \tag{4.43}$$

$$\tau_{yzn} = -\frac{p_{n\max}}{\pi}z^2\psi \tag{4.44}$$

In Eqs. (4.42) and (4.43), b represents contact width. Also

$$\psi = \frac{\pi}{k_1}\frac{1-\sqrt{\frac{k_1}{k_2}}}{\sqrt{\frac{k_2}{k_1}}\sqrt{2\sqrt{\frac{k_1}{k_2}} + \left(\frac{k_1+k_2-4b^2}{k_1}\right)}} \tag{4.45}$$

where

$$\overline{\psi} = \frac{\pi}{k_1}\frac{1+\sqrt{\frac{k_2}{k_1}}}{\sqrt{\frac{k_2}{k_1}}\sqrt{2\sqrt{\frac{k_2}{k_1}} + \left(\frac{k_1+k_2-4b^2}{k_1}\right)}} \tag{4.46}$$

and $k_1 = (b + y)^2 + z^2$

$$k_2 = (b - y)^2 + z^2$$

Then the stress distribution of the tangential load, which is also elliptical, is obtained:

$$\sigma_{yt} = -\frac{p_{t\max}}{\pi}\left[(2y^2 - 2b^2 - 3z^2)\psi + 2\pi\frac{y}{b} + 2(b^2 - y^2 - z^2)\frac{y}{b}\overline{\psi}\right] \tag{4.47}$$

$$\sigma_{zt} = -\frac{p_{t\max}}{\pi}z^2\psi$$

$$\tau_{yzt} = -\frac{p_{t\max}}{\pi}\left[(b^2 + 2y^2 + 2z^2)\frac{z}{b}\overline{\psi} - 2\pi\frac{z}{b} - 3yz\psi\right] \tag{4.48}$$

where $p_{t\max} = f \times p_{n\max}$ and f represents friction ecoefficiency. If $f = 1/3$, then $p_{t\max} = 1/3p_{n\max}$.

The total contact stress can be obtained by adding the stresses generated by the normal and tangential loads:

$$\sigma_y = -\frac{p_{n\max}}{\pi}\left\{(b^2 + 2y^2 + yz^2)\frac{z}{b}\overline{\psi} - 2\pi\frac{z}{b} - 3yz\psi + \right.$$

$$\left.\frac{1}{3}\left[(2y^2 - 2b^2 - 3z^2)\psi + 2\pi\frac{y}{b} + 2(b^2 - y^2 - z^2)\frac{y}{b}\overline{\psi}\right]\right\} \tag{4.49}$$

$$\sigma_y = -\frac{p_{n\max}}{\pi}\left[b\psi - y\psi + \frac{z}{3}\psi\right] \tag{4.50}$$

$$\tau_{yz} = -\frac{p_{n\max}}{\pi}\left\{z^2\psi + \frac{1}{3}[b^2 + 2y^2 + 2z^2]\frac{z}{b}\overline{\psi} - 2\pi\frac{z}{b} - 3yz\psi\right\} \tag{4.51}$$

When $z = 0$, the stresses on the contact are

$$\sigma_{y(z=0)}\begin{cases} = -\frac{2}{3}p_{on}\left(\frac{y}{b} - \sqrt{\frac{y^2}{b^2} - 1}\right) & \text{when } y \geq b \\ = -p_{on}\left(\sqrt{1 - \frac{y^2}{b^2}} + \frac{2}{3}\frac{y}{b}\right) & \text{when } |y| \leq b \\ = -\frac{2}{3}p_{on}\left(\frac{y}{b} + \sqrt{\frac{y^2}{b^2} - 1}\right) & \text{when } y \leq -b \end{cases} \tag{4.52}$$

$$\sigma_{z(z=0)}\begin{cases} = -p_{on}\sqrt{1 - \frac{y^2}{b^2}} & \text{when} |y| \leq b \\ = 0 & \text{when } y \geq a \text{ and } y \leq -b \end{cases} \tag{4.53}$$

$$\tau_{yz(z=0)}\begin{cases} = -\frac{1}{3}p_{on}\sqrt{1 - \frac{y^2}{b^2}} \text{ when } |y| \leq a \\ = 0 \text{ when } y \geq a \text{ and } y \leq -a \end{cases} \tag{4.54}$$

The formula of plane stress is given above. In the case of plane strain, the third normal stresses $\sigma_x \neq 0$ are

$$\sigma_x = \mu(\sigma_y + \sigma_z) = -\frac{2\mu}{\pi}p_{on}\left\{\left[(b^2 + y^2 + z^2)\frac{z}{b}\overline{\psi} - \frac{\pi z}{b} - 2yz\overline{\psi}\right]\right.$$

$$\left. + \frac{1}{3}\left[(x^2 - b^2 - z^2)\psi + \frac{\pi y}{b} + (b^2 - y^2 - z^2)\frac{y}{b}\overline{\psi}\right]\right\} \tag{4.55}$$

When $z = 0$,

$$\sigma_x/z = 0 \begin{cases} = -\mu p_{on}\dfrac{2}{3}\left(\dfrac{y}{b} - \sqrt{\dfrac{y^2}{b^2} - 1}\right) & \text{when } y \geq a \\ = -\mu p_{on}\left(2\sqrt{1 - \dfrac{y^2}{b^2}} + \dfrac{2}{3}\dfrac{y}{b}\right) & \text{when } |y| \leq a \\ = -\mu p_{on}\dfrac{2}{3}\left(\dfrac{y}{b} + \sqrt{\dfrac{y^2}{b^2} - 1}\right) & \text{when } y \leq -a \end{cases} \tag{4.56}$$

Three principal stresses can be obtained by using Mohr's circle:

$$\left.\begin{aligned} \sigma_1 &= \frac{\sigma_y + \sigma_z}{2} + \sqrt{\left(\frac{\sigma_y - \sigma_z}{2}\right)^2 + \tau_{yz}^2} \\ \sigma_2 &= \frac{\sigma_y + \sigma_z}{2} - \sqrt{\left(\frac{\sigma_y - \sigma_z}{2}\right)^2 + \tau_{yz}^2} \\ \sigma_3 &= \mu(\sigma_1 + \sigma_2) \end{aligned}\right\} \tag{4.57}$$

Finally, the maximum shear stress and the maximum eight body shear stress, τ_{max} and τ_{Gmax}, can be obtained by

$$\tau_{max} = \frac{1}{2}(\sigma_1 - \sigma_3)$$

The maximum eight body shear stress is given by

$$\tau_{Gmax} = \frac{1}{3}\sqrt{(\sigma_1 - \sigma_2)^2 + (\sigma_2 - \sigma_3)^2 + (\sigma_3 - \sigma_1)^2}$$

According to the above calculation steps, the distribution of the whole stress field can be obtained. Figures 4.31 and 4.32 show the distribution of the perpendicular shear stress and the maximum shear stress of the cam contact surface along the depth calculated using this method. It can be found that due to the tangential force (friction), the maximum principal stress is increased by 39%, the maximum shear stress is increased by 43%, and the maximum eight shear stress is increased by 37%. Besides, the maximum shear stress moves from the depth of $0.78b$ to the surface. If the f is greater than a certain value, the location of the maximum shear stress always stays on the contact surface.

Thus, if the friction between the cam and roller can be reduced, the surface stress of the cam mechanism can be effectively reduced, thus improving the service life of the cam mechanism.

4.2.2.3 Analysis of the Cam Working State

According to the rolling theory, there is an empirical relationship between the fatigue life L (expressed by the number of rolling cycles) of the rolling pair and the load N acting on the rolling pair:

$$NL^{1/p'} = \text{constant} \tag{4.58}$$

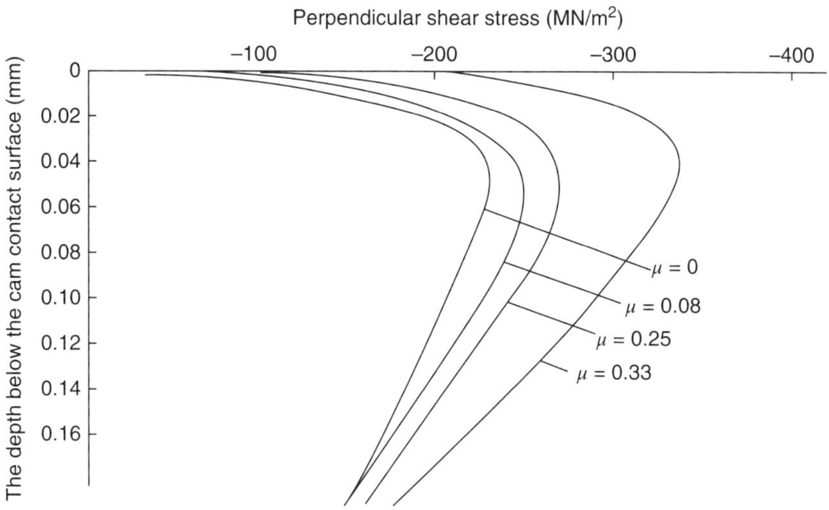

Figure 4.31 The distribution of the perpendicular shear stress along the depth below the cam contact surface under different friction coefficients.

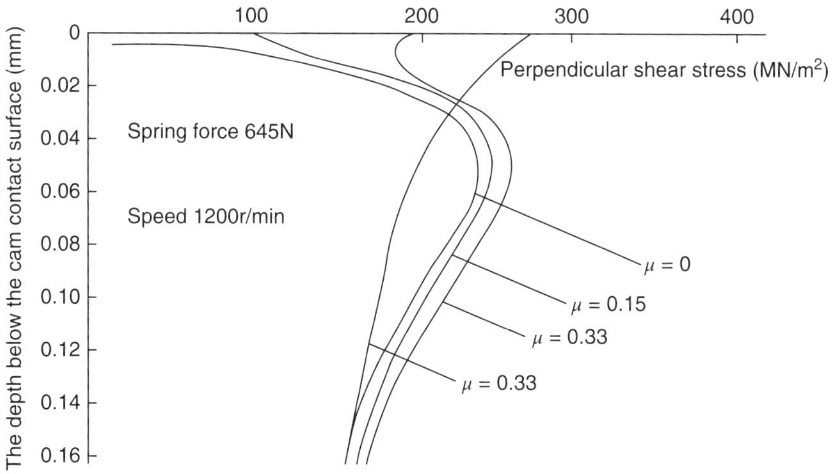

Figure 4.32 The maximum shear stress distribution on the cam surface under different friction coefficients.

It has been pointed out in some literatures that the fatigue life between the two rolling pairs is proportional to the 1/9th power of the Hertz stress. For the cam and roller pair, the fatigue process of the cam is similar to that of the cam pair. Certainly, the experimental conditions, structural parameters, and material properties have an impact on the exponential $p\prime$ and the constant term in the fatigue life equation. According to the constant running test on the cam and the roller by the Austria Friedmann & Maie Company, the relationship between the Hertz stress σ and the life of the cam is shown in Figure 4.33, and the relation can be expressed by the following cam life equation:

$$\sigma L^{0.118} = 1896 \tag{4.59}$$

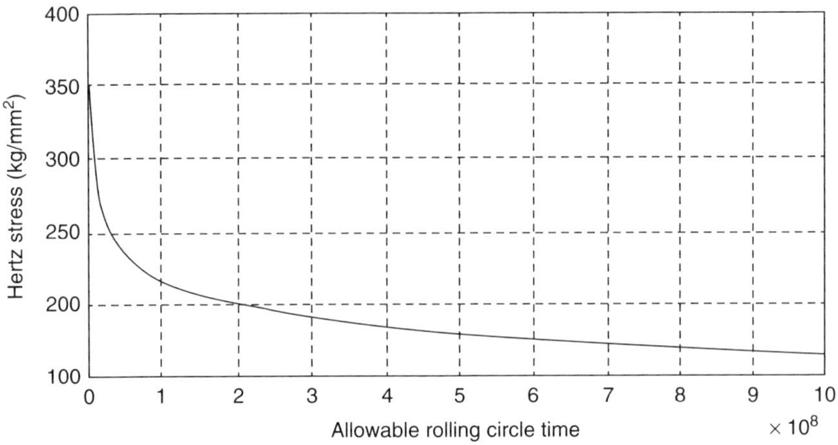

Figure 4.33 The relationship between the Hertz stress and service life.

Thus, it is very important to reduce the Hertz stress of the cam to improve the fatigue life of the cam. In order to find a way to reduce the Hertz stress of the cam, it is necessary to analyze some factors affecting this cam stress.

(1) The influence of the cam load. The total load on the cam is determined by the fuel pressure, the inertia force of the moving parts, the spring force, and the pressure angle.
The fuel pressure is the main component of the cam load, and the maximum fuel pressure on the cam surface can approach more than 100 MPa, which has an important influence on the Hertz stress. The fuel force depends on the diameter of the plunger, the cam speed, the size of the outlet valve, and other structural parameters. By reducing the quality of the moving parts and reducing the pump cam acceleration, the inertia force can be reduced. The change of acceleration is related to the profile of the selected cam. Although in most cases the inertial force is limited and has little effect on the Hertz stress, in some special circumstances, such as cam acceleration mutation, the acceleration value tends to infinity, and in this case a large Hertz stress will be produced, which leads to the total mechanism of shock and vibration. Therefore, it is very important to choose the cam profile carefully, which is why the circular cam is selected.
The magnitude of the spring force is related to the stiffness and compression of the spring. Because the two factors are finite values, they have little effect on the Hertz stress.
Because the final effect on the Hertz stress of the cam is the force F_t along the normal contact direction between the roller and the cam, the size of the F_t is related to the pressure angle α. The smaller the pressure angle α, the smaller the F_t, thus reducing the Hertz stress value of the cam. In order to reduce the maximum pressure angle, the cam base circle radius should be properly chosen, but its range of adjustment is also limited.
(2) The influence of the cam width. Widening the cam width is beneficial to reducing the Hertz stress of the cam. However, in order to avoid the stress concentration in the cam and roller width edge, generally the cam and roller are required to have

equal width, so the increase in the width of the cam is actually limited by the same size of pump body. Therefore, it is often limited in its ability to increase the width of the cam. Thus, in many practical situations, to avoid the unfavorable condition of a Hertz stress surge caused by the partial contact and even point contact between the roller and cam, more attention is often needed to ensure the straightness of the generatrix of the rollers and cams and the parallelism between the central lines of the rollers and cams in the process of assembly.

(3) The influence of elastic modulus of the cam and roller material. The elastic modulus depends mainly on the atomic nature, lattice type, and lattice constant. Some important methods to change the mechanical properties of metals, such as alloying, heat treatment, and hot and cold processing, have little effect on the lattice constants. For general carbon steel or alloy steel, the positive modulus of elasticity at room temperature is almost equal to 206 GP, so there is no room for change in the Hertz stress of the cam. Because of the difference in heat treatment performance, the combination of different cam and roller materials has an effect on the service life.

(4) The influence of oil film thickness. The basic forms of cam damage are strain, pitting, and wear, which are related to the conditions of oil film forming:

a. Strain. When the oil film lubrication is insufficient, or the oil film is damaged, the metal surface often appears as a strain, which occurs when the lubrication is in the boundary lubrication or mixed lubrication. Especially, the friction heat caused by the roller sliding along the cam makes the situation more serious, resulting in local surface welding, material failure, shear stress, and an increase in thermal stress. At this time, fatigue will be rapidly formed and can cause complete damage to the surface of the cam. In this respect, the thermal properties of the contact zone are important because the thermal properties change the physical and chemical properties of the lubricant and the surface, thus affecting the stability of the lubricating oil film. Dyson and Naylor put forward the theory of a flash point temperature. It is considered that once the oil film temperature at the contact point exceeds a critical value, the strain will occur. Later, Bell and Dyson concluded that the critical value of strain mainly depends on three factors, which are oil film temperature, oil film thickness, and friction work at the contact point.
b. Pitting. Pitting is related to the operation time and Hertz stress condition. It is the result of fatigue of the lower surface of the metal surface. This occurs because when the surface is rough, after the rough part of the rolling element contacts repeatedly, the situation appears like a system with a large number of μHz, whose severity depends on the surface load properties, surface roughness, and the convex shape of the starting point. The plastic deformation occurs near the rough peak when the load is not too severe and a rough surface with a circular peak is formed after fatigue at these rough peaks. At this time, the load is carried at a lower stress level, and a better load bearing surface can be formed under this working condition. However, if it is in a serious load condition, the load on each rough point of the cam surface will cause plastic shear deformation at the root of the rough part. This plastic shear deformation causes a crack to form at the rough surface, which then causes the surface of the cam to become more roughened. This process occurs repeatedly until the surface is completely destroyed. Of course, the pitting can be solved by designing the cam reasonably, reducing the

load, improving the surface quality, and improving the material. Another effective way to improve surface fatigue and prevent pitting is to separate the surfaces with oil film. This is also why it is important to form the theory of compressible fluid oil film.

The calculation of oil film thickness between the cam and roller is generally carried out by using the simplified formula proposed by Dowson:

$$h_{\min} = 2.65\frac{R^{0.43}(\eta v)^{0.7}\alpha^{0.54}}{W^{0.13}E^{0.03}} \tag{4.60}$$

In the equation:
$h_{\min}$ is the minimum thickness of the oil film,

$R = (R_1^{-1} + R_2^{-1})^{-1}$ is the equivalent cylinder radius,
E is the equivalent elastic modulus of the cam and roller materials,
η is the lubricating oil viscosity entering the contact zone,
α is the viscosity index of lubricating oil pressure,
v is the average velocity, which is $(v_1 + v_2)/2$ (where v_1 and v_2 are, respectively, the surface velocities of the two rolls),
W is the load on a cam with a unit width.

According to the existing pump condition, the rotational speed is 750 r/min, $R_1 = 11.5$ mm, $R_2 = 21$ mm, only pure rolling occurs between the roller and the cam, $\eta = 1.57 \times 10^{-2}$ N s/m^2, $E = 2.1 \times 10^{11}$ Pa, and $\alpha = 1.8 \times 10^{-8}$ m^2/N. It can be calculated that $W = 6623/15 \times 10^{-3} = 441\,000$ N/m and the load is 6623 N, the contact width is 15 mm, and the minimum thickness of the oil film is calculated to be 0.137 μm. Because the cam damage is closely related to the oil film thickness and surface quality, it can be considered that the ratio of oil film thickness to surface roughness is a function of surface fatigue; this ratio is usually referred to as the ratio of film thickness to roughness film thickness λ:

$$\lambda = \frac{h_{\min}}{\sum \text{roughness of surface}} \tag{4.61}$$

In the formula, $\sum$ roughness of surface $= \sqrt{\Delta_1^2 + \Delta_2^2}$, where Δ_1 and Δ_2, respectively, are the roughness of two contact surfaces, expressed by the arithmetic mean or root mean square value.

The related literature pointed out that when the ratio of film thickness to roughness oil film thickness of the cam is calculated at a range of 0. 6–0.8, only a small part of the load transfers by the oil film, while most of the load is received by the rough part of the contact area between the cam and the roller. Under such conditions, the plastic deformation of the rough part of the cam contact area may lead to the destruction of the cam surface. Therefore, the ratio of the film thickness to roughness thickness of the oil film should be increased to more than one if possible. In general, the arithmetic average of the surface roughness of the cam is 0.4 μm. Thus, when the cam is easily damaged during the critical ratio of the film thickness to roughness oil film thickness λ is 0. 6–0.8, the corresponding critical allowable oil film thickness is

$$h_{\min} = 0.4 \times (0.6 \sim 0.8) = 0.24 \sim 0.32\ \mu\text{m} \tag{4.62}$$

Therefore, obviously it is difficult for the cam with the minimum oil film thickness of 0.137 μm to endure a high load operation for a long time. Considering the theory of elastohydrodynamic lubrication, to increase the reliability of the injection pump cam, the thicker ratio of film thickness to roughness is needed, which can generally be achieved by increasing the thickness of the oil film and reducing the surface roughness of the cam and the roller surface.

The thickness of the oil film is a function of the equivalent contact radius of the cam and the roller, the acting load on the cam, the oil film speed, and the viscosity of the lubricating oil. If the cam load is constant, the thickness of the oil film mainly depends on the radius of the compound equivalent curvature radius. In the shaped line design of the cam, it is important to increase the thickness of the oil film by increasing the radius of curvature in order to improve the fatigue life of the cam surface. On the other hand, if the machining quality of the cam surface is improved and the roughness of the cam surface is reduced, the increase of the thickness of the ratio of film thickness to roughness of the oil film is obvious. If the roughness of the cam surface is raised one level, the thickness of the ratio of film thickness to roughness of the oil film can be doubled. Therefore, in the actual production, to ensure the machining quality of the cam surface is a very important condition to prevent fatigue damage of the cam.

The minimum oil film thickness is a function of the equivalent contact radius of the cam and the roller follower, the acting load on the cam, the oil film velocity, and the viscosity of the lubricating oil. The smaller the composite equivalent curvature radius, the larger is the cam load and the lower the viscosity of the lubricating oil, while the slower the oil film speed, the thinner is the oil film. If the cam load is constant, the thickness of the oil film depends mainly on the radius of the composite equivalent curvature radius. Therefore, when working in the same Hertzian stress, it is important to increase the radius of curvature to increase the thickness of the oil film and to improve the fatigue of the cam surface.

4.2.3 Experimental Study on Stress and Strain of the High-Pressure Fuel Pump

4.2.3.1 Test and Analysis of the Pressure of the Plunger Cavity

In order to study the pressure change regulation of the plunger cavity, a method that measured the plunger strain to obtain the pressure change of the plunger cavity was used. Through this experiment, the change regulation of the pressure of the plunger cavity is obtained. The resistance strain gauge sensor is used to sense the compression deformation of the outer wall of the plunger and the deformation transforms into the transformation of the strain gauge's self-resistance. Then through the bridge path, it is converted to a weak voltage signal and the required parameters are obtained through signal acquisition and the processing system. Finally, according to the principle of structural mechanics, deformation of the plunger is obtained. The system structure is shown in Figure 4.34 and the actual system is shown in Figure 4.35.

(1) Resistance strain gauge. The resistance strain gauge is made of resistance wires with a diameter of 0.025 mm and high resistivity. The strain gauge consists of a sensitive gate, a base, a lead, a cover, a binder, an electrode, and other parts. The strain gauge is widely used because it has many advantages: the strain measurement has high sensitivity and accuracy, is stable, and has a reliable performance. It can measure 1–2 μm

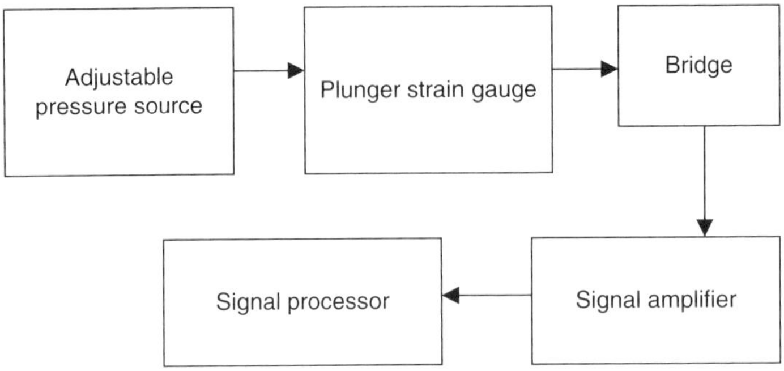

Figure 4.34 Structural diagram of the plunger edge measuring system.

Figure 4.35 Actual view of the test system.

and the error is less than 1%. The strain gauge is small in size, light in weight, simple in structure, easy to use, and fast in measurement. There is almost no influence on the working state and the stress distribution of the measured parts. The range of measurement is large: it can be used both in static measurement and in dynamic measurement. The adaptability is strong: it can measure both the elastic deformation and the plastic deformation. The deformation range can be from 1 to 2%. It can be used in harsh environments such as high temperature, ultra-low temperature, high pressure, underwater, in a strong magnetic field, and in an area of nuclear radiation. It is convenient for multipoint measurement, long distance measurement, and telemetry. It is cheap, comes in a variety of forms, is mature, convenient for selection and use, and can measure various physical quantities.

When the strain gauge with an initial resistance value R is pasted on the surface of the specimen, the surface strain induced by the specimen will be transferred to the sensitive gate of the strain gauge, resulting in the relative resistance change. The experiments have shown that in a certain range of strain, there are the following relationships:

$$\Delta R/R = K\varepsilon_X \tag{4.63}$$

In the formula, ε_X is the axial strain of the strain gauge and K is the sensitivity coefficient of the strain gauge and represents the ratio of the relative resistance change ($\Delta R/R$) of the strain gauge mounted on the tested part when it suffers uniaxial stress on its axis, to the axial strain ε_X of the specimen caused by the uniaxial stress.

Under constant temperature and constant load conditions in the strain gauge attached to the test piece, the characteristic of the variation of the stress variable with time is called creep. When the specimen initially has an empty load, the phenomenon that the strain gauge will still change with time is called zero drift. The creep reflects the stability of the strain gauge in a long period of time and usually requires $\theta < 3 \sim 15\,\mu\text{s}$. The main reason that causes creep is the slippage between the wire grid and the base, especially the adhesive layer, which is caused by internal stress generated during the production of the strain gauge and the shear stress in operation. Selecting the binder and base material with larger modules of elasticity, properly thinning the film adhesive layer and base, and solidifying them are beneficial to improve the creep property.

Foil strain gauges were selected in this experiment. Those selected in this chapter were made of modified phenolic substrates and copper foil, with a closed structure. The sensitivity coefficient is 1.95, the strain limit is $\pm 2\%$, and the typical resistance value is 300 Ω. The temperature self-compensation and creep self-compensation can both be realized simultaneously.

The strain gauge is bonded to the measured piece by a binder. The adhesive formed by the binder must quickly transfer the strain of the measured piece to the sensitive gate. The performance of the binder and the quality of the bonding process directly affect the working characteristics of the strain gages, such as zero drift, creep, lag, and sensitivity. Therefore, it is obvious that the selection of binder and the correct bonding process are very important to the measurement accuracy of the strain gauge.

(2) Strain amplifier. The TS3821 strain amplifier is a high gain and broadband signal adjuster. It can be matched with a resistive strain sensor or strain gauge to measure force, pressure, torque, displacement, speed, acceleration, and other physical aspects with a swift change to the static. When this instrument is used as a DC voltage amplifier, it can measure the microvolt voltage signal and can be matched with various voltage output sensors to process multiparameter measurements; a built-in bridge box is convenient to use, where the bridge is automatically balanced, the bridge pressure is adjustable, and it can be used with multiple channels.

(3) Strain measurement circuit. The concrete bridge connection method is shown in Figure 4.36. The characteristic of a static strain measurement is that, between two loads, the strain of the specimen is stable for a certain time. In this way, a set of reading bridges and amplifiers can be used to connect with the various measuring bridges, and then the strain at each point can be measured one by one. Therefore,

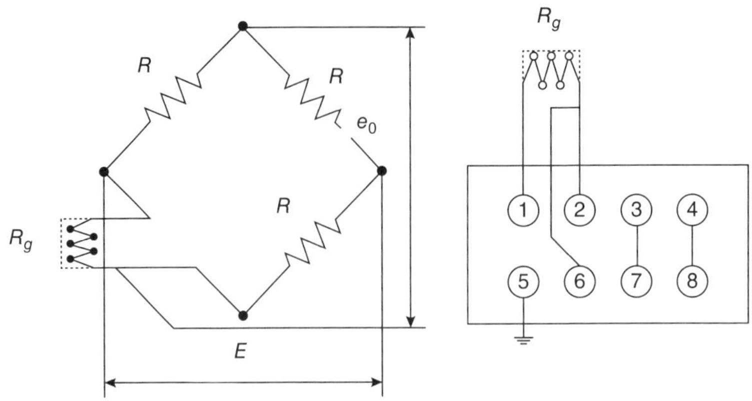

Figure 4.36 Bridge connection method of the strain gauge.

the pre-conditioning balance box is used to solve the problem of multipoint strain measurement.

When multiple points are tested, because the resistance of each strain gauge is not the same as that of the connection wire, the pre-set balancing box has a resistance balance potentiometer, which can balance all the points before loading the test piece. After loading, by pre-adjusting the switches of the balance resistance box, each point on the measurement bridge can be connected to the strain gauge.

(4) The relation between strain and output voltage. The relation between strain and output voltage is

$$e_0 = \frac{E}{4} Ks\, \varepsilon_0 \tag{4.64}$$

In the formula, E is the supply bridge voltage, Ks is the strain gauge sensitivity coefficient, e_0 is the bridge circuit output voltage, and the ε_0 is the strain value. If the bridge pressure is 1 V and the strain coefficient is 2.00, then $e_0 = 0.5\varepsilon_0$. For example, the strain value is 500 μV/V = 0.5 mV/V.

(5) Analysis of test results. The patch is selected to paste on the parts exposed to the outside of the plunger at the lower part of the plunge, as shown in Figure 4.37. The test range of pressure is selected from 50 to 150 MPa and the speed is 450 r/min. Firstly, a static calibration is made for the variable plate. The cam is at rest. The upper spring check valve of the plunger cavity is removed and the plunger cavity is connected to the common rail cavity, pressurized to 150 MPa. Then the unloading pressure is obtained and the calibration curve is as shown in Figure 4.38.

CH1 is the strain gauge deformation map and CH2 is the voltage signal output from the common rail cavity pressure sensor. The sensor is calibrated in advance and the relation between the voltage output and pressure, $P = 40$ U, where P is the pressure corresponding to the output voltage. The empirical formula of change of strain gauge pressure and pressure of the plunger cavity is obtained through data fitting, where $P = 106$ U. In this way, the curve of the pressure change of the plunger cavity can be obtained through the deformation curve of the lower strain gauge of the plunger in the dynamic case, as shown in Figures 4.39 and 4.40.

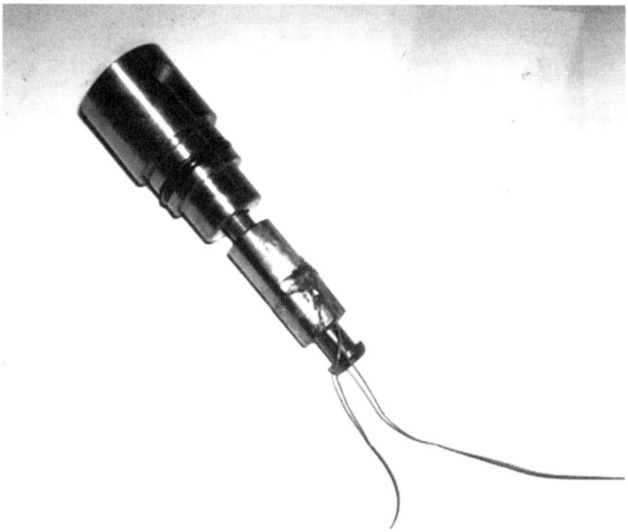

Figure 4.37 Position of the plunger strain gauge.

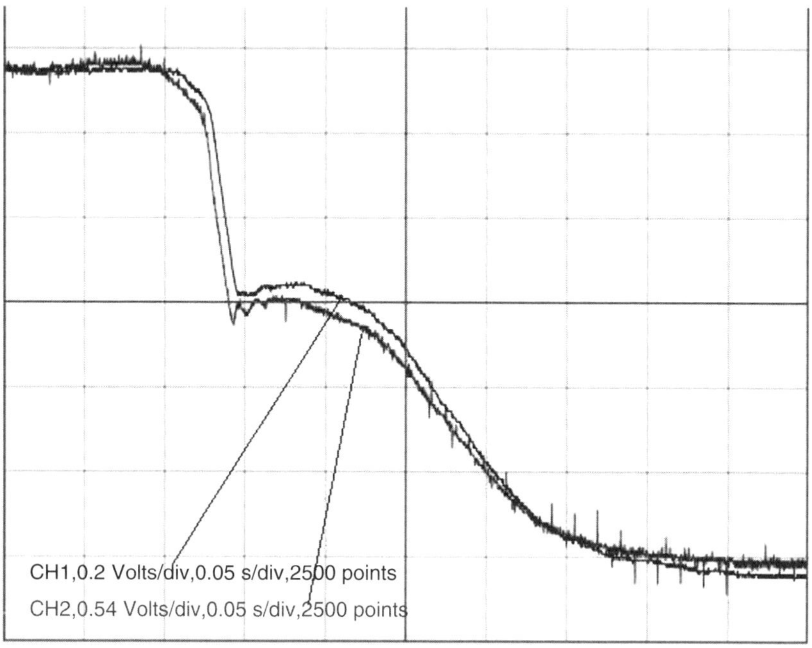

Figure 4.38 Oscilloscope calibration diagram.

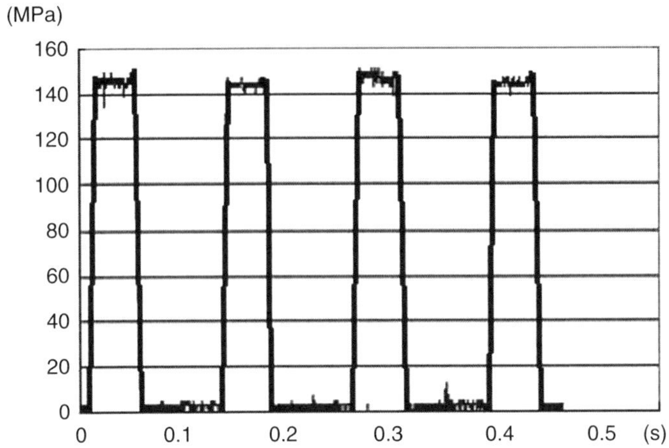

Figure 4.39 Pressure of plunger cavity at 150 MPa.

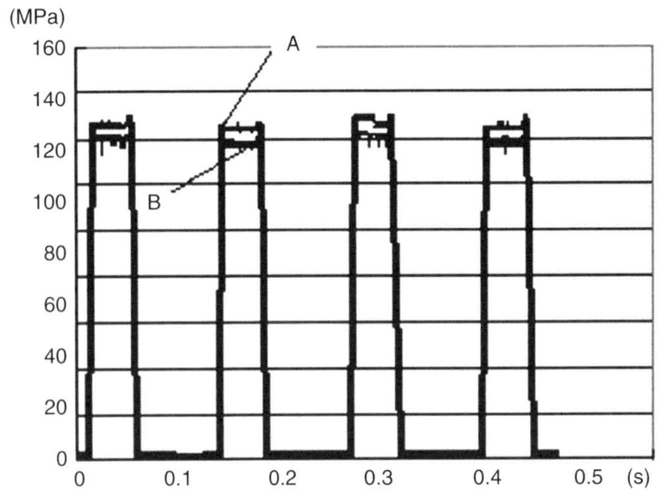

Figure 4.40 Pressure of the plunger cavity with different valve hole diameters.

From the comparison between the test results and the corresponding simulation analysis, it can be seen that the correctness of the simulation analysis is verified by the test results: the peak pressure of the plunger cavity decreases with the increase of the diameter of the fuel outlet valve.

4.2.3.2 Stress Test and Analysis of the Camshaft

In order to test the static contact area between the cam and roller and the stress near the contact point of a cam, and to test the bending stress in the middle part of the shaft, the experimental study is carried out by using the resistance strain gauge sensor to sense the camshaft's compression deformation.

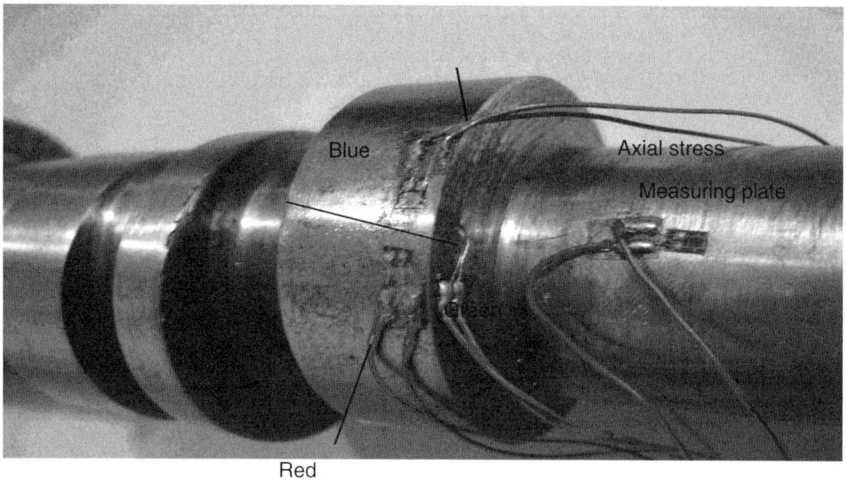

Figure 4.41 Stress patch.

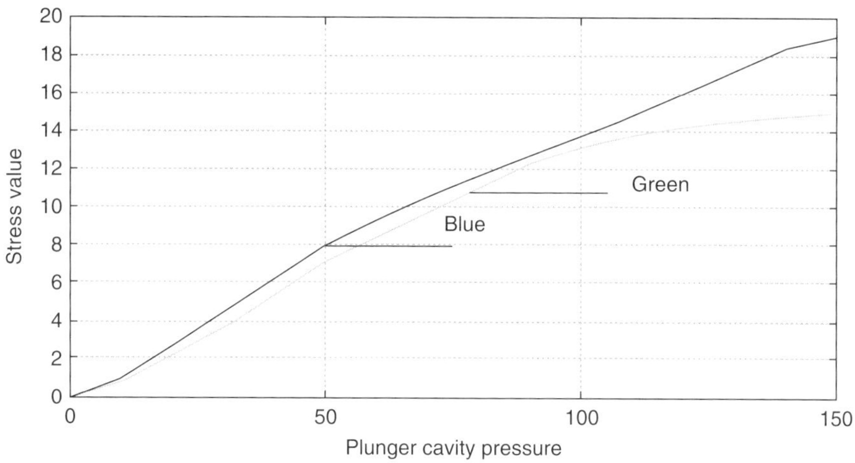

Figure 4.42 Surface stress of the cam.

The deformation principle of the strain sheet has been described and will not be repeated here. If deformation is measured by the strain resistance, the stress will be calculated by formula $\sigma = E\varepsilon$, of which the amount of elastic film is $E = 206$ GPa.

For the test, the strain gauges are installed in the camshaft, as shown in Figure 4.41. By using the previous stress and strain test method, the maximal static stress of the camshaft is measured. The cam and the roller contact parts are red, blue, and green strain gauges. The blue line, and green line stress changes are as shown in Figure 4.42. Figure 4.43 shows the camshaft stress, while Figure 4.44 shows the calculation result of injection cavity stress.

The contact is a line contact. During the static pressure measurement, the cam is first smeared with red pigments and the contact width is measured as 0.45 mm. The contact width $2b = 0.36$ is calculated according to the formula (4.45). There is a certain deviation

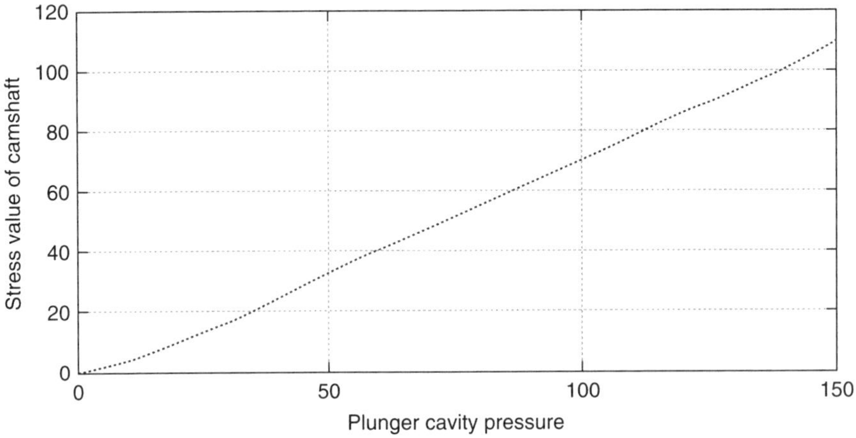

Figure 4.43 Axial stress.

Figure 4.44 Corresponding stress simulation diagram of the single plunger cavity.

from the measured value error. By analysis, the reason is that the contact stress is generated on a very small contact area. It belongs to the local stress and occurs in a smaller range. The required accuracy of the measurement method is higher.

4.3 Research on Common Rail Pressure Control Technology Based on Pump Flow Control

In the high-pressure common rail system, in order to control the rail pressure stability under the target pressure and to ensure a certain precision (usually the pressure fluctuation ≤3–5%), the fuel supply of the high-pressure fuel pump and fuel consumption must be controlled with certain adaptability and with dynamic accuracy. Usually there are two ways to achieve the above goals. One is to use rail cavity overflow to control the stability, which means that the high-pressure pump cycle of fuel production remains unchanged and the sum of fuel consumption of the injector and the fuel spillage in the common rail cavity is equal to the amount of fuel supply. The above process ensures the stability of the pressure cavity, like the early BOSCH CR common rail system. Another process is to change the fuel supply of the high-pressure pump to control, which means in a short time interval (such as one cycle) the fuel supply is in a dynamic balance with the fuel consumption of the injector, and then the stability of the rail pressure is ensured, like the ECD-U2 common rail system. Obviously, the former is not energy saving and the latter has a certain advantage. Therefore, the latter is the most important form of realizing pressure control of the rail cavity.

4.3.1 Design Study of a High-Pressure Pump Flow Control Device

4.3.1.1 Overview of a High-Pressure Pump Flow Control Device

The flow control device of common rail pressure regulation mainly controls the common rail pressure changes and fluctuations in the common rail pressure. The former provides a wider control space for optimized Map control, and helps to achieve the optimization of diesel engine performance. The latter influences the stability of fuel injection and the stability of the diesel engine running speed.

(1) Fuel suction stroke adjustment type. In this section, the electronic control high-pressure pump introduced is a straight column type plunger pump, with six plunger pairs arranged axially along the pump body. Each set does not use the plunger slanting groove type to adjust the fuel quantity, but the flow control device is installed on each set. The basic structure of the flow control device of the electronically controlled high-pressure pump is shown in Figure 4.45. The control valve is composed of two one-way valves and one solenoid valve, and the control of the inlet volume of the high-pressure pump is completed. The operating process of the electronically controlled high-pressure pump is described as follows:

   a. The process of sucking fuel. When the plunger is down, the solenoid valve opens and the plunger working cavity enlarges and the fuel pressure decreases. When the fuel pressure is less than the output pressure of the fuel pump, the one-way valve A opens and the fuel enters the plunger cavity through the solenoid valve and the one-way valve A.
   b. The process of pressurizing fuel. With the high-pressure pump cam rotation (around base circle), the plunger can go upwards, the plunger cavity volume becomes smaller, the fuel is compressed, the pressure increases, the one-way valve A is closed, and low-pressure fuel closes the plunger cavity to form a closed space. When the fuel pressure exceeds the sum of the one-way valve B and back pressure and spring force, the one-way valve B opens and the fuel enters the common rail tube from the plunger cavity. When the plunger lift reaches the top stop, the fuel supply cycle is completed.
   c. Fuel supply of high-pressure pump and regulation of common rail pressure. The fuel output of the fuel pump goes through a solenoid valve into the plunger cavity, and the input flow amount is decided by the conduction time of the solenoid. The solenoid valve is controlled by the ECU pulse signal control, so the fuel flow in the plunger cavity can be controlled by controlling the duty cycle of the pulse

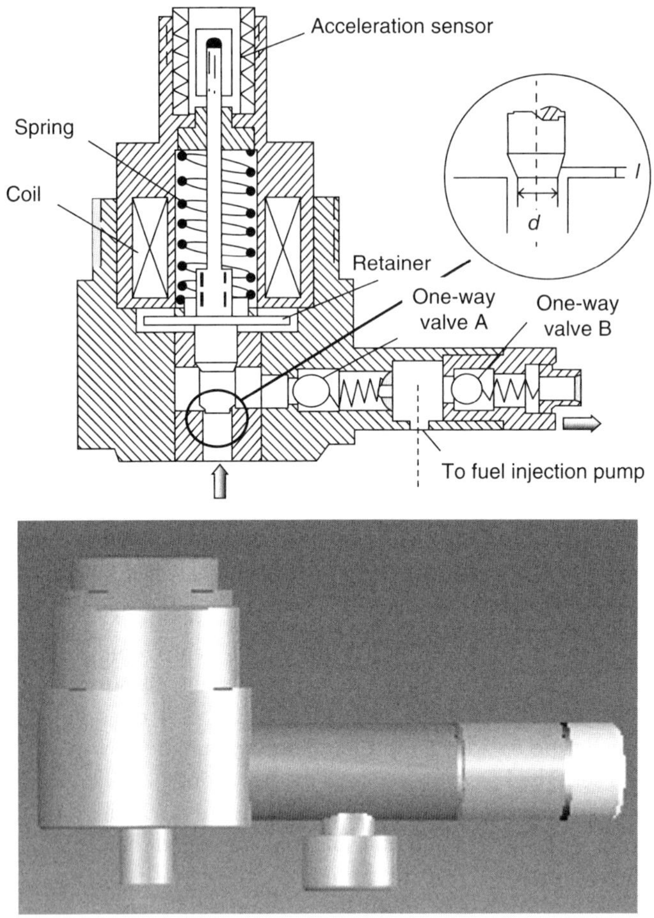

Figure 4.45 Sketch of the high-pressure pump flow control device.

signal and by controlling the fuel supply of the high-pressure pump using the flow control device. Finally, the control of the common rail pressure can be realized. As shown in Figure 4.46, the actual sequence of the working process is controlled by the effective fuel suction stroke of the high-pressure pump.

d. Work characteristics. The single inlet fuel suction stroke volume can be adjusted directly by constantly opening and closing the control solenoid valve. The solenoid valve realizes the separation of the high-pressure fuel due to the function of the one-way valve A. Therefore, the opening and closing of the electromagnetic force required is greatly reduced, the design difficulty of the solenoid valve is reduced, and the solenoid valve seat shock force is reduced, which improves the working environment, thus reducing the solenoid valve sealing requirements. By controlling the low-pressure fuel amount, the common rail pressure is adjusted and the energy is saved without loss of the high-pressure fuel.

(2) Fuel pressurization stroke adjustment type. The fuel pressurization stroke flow regulation device is composed of an electromagnet and two one-way valves. The electromagnet and a one-way valve are arranged in the axial direction, coupled to one valve, and the other one-way valve is arranged longitudinally. The basic structure is shown in Figure 4.47. The working process of the electrically controlled high-pressure pump of the pressurization stroke adjustment type is described as follows:

a. The process of the sucking fuel. When the plunger is down, the plunger cavity increases and the fuel pressure decreases. When the fuel pressure is less than that of the output of the fuel pump, the one-way valve B opens and the fuel is sucked into the plunger cavity through the inlet A.
b. The process of pressurizing fuel. With the rotation of the fuel pump cam (rotating the base circle), the electromagnet is energized. The push rod of the guide electromagnet pushes the one-way valve B. Though the plunger goes up and the volume of the plunger cavity is smaller, the fuel is not compressed and flows back to the low-pressure oil cavity through the A hole. When the target amount of fuel is left in the plunger, the electromagnet is cut off, the one-way valve B is closed, the low-pressure fuel circuit is closed, and the plunger cavity forms a closed space. When the fuel pressure exceeds the sum of the C spring force and the backpressure of the check valve, the one-way valve C opens and the fuel enters the common rail pipe from the plunger cavity.
   Regulation of the fuel supply of the high-pressure pump and common rail pressure. The fuel flow rate of the plunger cavity into the common rail cavity is determined by the electrified time of the electromagnet. The opening and closing of the electromagnet is controlled by the ECU pulse signal. Therefore, the fuel flow into the rail pipe can be controlled by controlling the duty cycle of the pulse signal, and control of the rail pressure can be achieved. In the sketch of the flow control device, as shown in Figure 4.48, the actual sequence of the working process is controlled by the effective fuel pressurizing stroke of the high-pressure pump.
c. Work characteristics. (i) The electromagnet uses a spiral tube electromagnet to directly control the opening and closing of the one-way valve B and to indirectly control the amount of fuel supply. (ii) When the electromagnet is not in operation, the fuel supply of the full load of the high-pressure pump can be realized and the common rail pressure can be quickly established, which saves the energy consumption and reduces the heat dissipation. (iii) The formation control of fuel pressurization is at the beginning of the pressurization stroke, so the fuel pressure of the plunger cavity is low and the force required for the valve B opening is small, which reduces the difficulty of the electromagnet design. If the fuel pressurization stroke control is changed to the end of the stroke, it will be difficult to design the electromagnet. (iv) Controlling the amount of low-pressure fuel to regulate the common rail pressure can save power consumption of the high-pressure pump.

The electromagnet is one of the most basic and key components used in the flow control device. For the electronically controlled high-pressure pump of the diesel engine, to realize the flexible control of the common rail pressure, the fuel supply

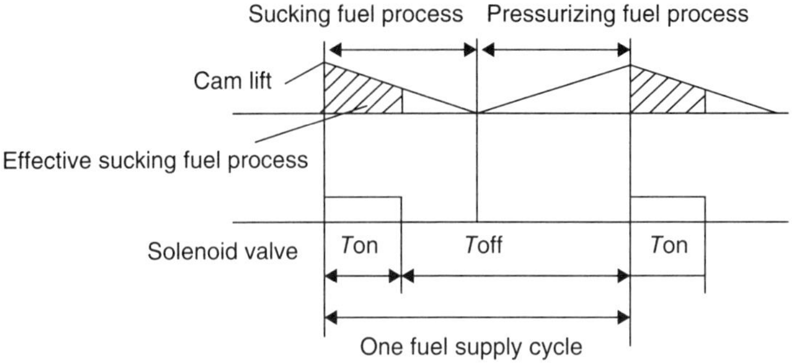

Figure 4.46 Sketch of the fuel suction stroke adjustment type working.

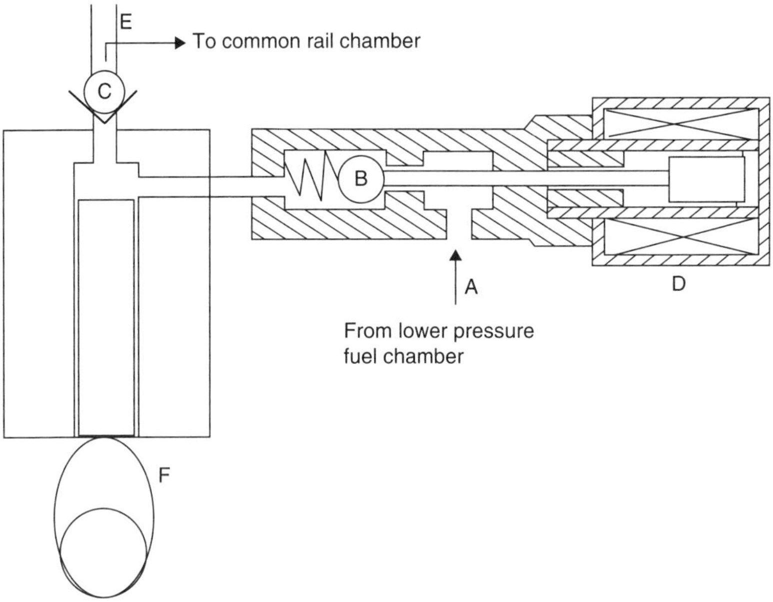

Figure 4.47 Working diagram of the fuel pressurization stroke adjustment control device.

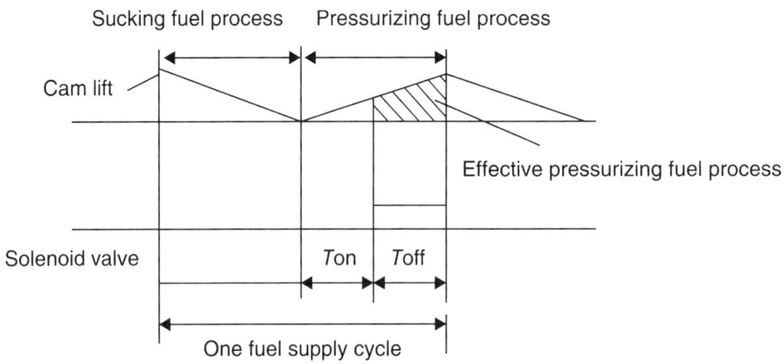

Figure 4.48 Working diagram of the fuel pressurization stroke adjustment type.

must be adjusted flexibly. The traditional mechanical device cannot satisfy the flexible and accurate control of the fuel supply and the adjustment of rail pressure, especially of a small amount of fuel supply, due to its inertia, driving torque, control precision, and other restrictions. However, the required high-speed response performance is impossible. The high-speed electromagnet can totally make up for the defects of the mechanical actuator because of its fast switching response, low power consumption, and good consistency characteristics. When the high-speed fuel supply response characteristic of the high-pressure pump is matched with the high-speed injection characteristic of the injector, the pressure fluctuation of the rail chamber can be controlled in the minimum range. The opening and closing of the electromagnet has an important influence on the precision of the common rail pressure regulation.

The high-speed solenoid, as the interface between the fuel transfer pump and the high-pressure pump, directly controls the fuel supply process and plays a decisive role in controlling the pressure fluctuation of the rail cavity.

4.3.1.2 Structure and Working Principle of the High-Speed Solenoid Valve

The application of solenoid valves in many industrial fields has a long history and a wide range of research. The solenoid valve for a high-pressure pump flow control device of the diesel engine common rail system, due to the requirements for fuel supply characteristics of high-pressure pumps by diesel engines, has the following special requirements: fast response (the response time is required to be matched with the highest rotational speed of the cam, and must guarantee that during every work cycle of the plunger the solenoid valve can work at once – the faster the response, the higher must be the accuracy of the rail pressure to adjust), small size, good heat dissipation, good fuel sealing performance, and good environmental adaptability and reliability.

Based on the above characteristics, three main solenoid structural types, as shown in Figure 4.49, have been presented globally in the past 10 years or so: the spiral tube type,

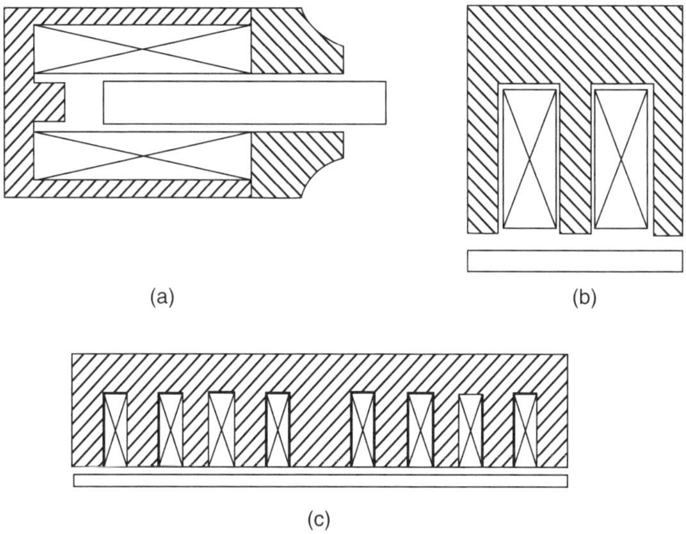

Figure 4.49 Three types of solenoid valve.

the "E" type, and the multipole type. The multipole solenoid valve can be considered as a combination of multiple "E"-type electromagnets. The electromagnetic force generated by it is the sum of the electromagnetic force of all "E"-type electromagnet units, and the working time is the same as the single unit "E"-type electromagnet.

The solenoid valve in the high-pressure pump flow control device of a high-pressure common rail system has different structural modes due to the different working mode, the restrained installation position, and the different structural factors of the solenoid valve.

The structure factor of the solenoid valve was first proposed by an American scholar, Roters, and later was used in many electrical science articles as a coefficient based on the selection of structural types for the design of the solenoid valve. The definition of the structural factor is as follows:

$$K_\phi = \frac{\sqrt{F_c/10}}{\delta_c} \tag{4.65}$$

In the formula, F_c is the electromagnetic suction (N) under the initial stroke and δ_c is the initial stroke (cm).

Under the same load conditions, different structural types of solenoid valves have different work efficiencies. Namely, under a certain initial stoke δ_c and electromagnetic attraction F_c, there will be a most suitable technical and economic index in all kinds of solenoid valves. Any kind of structural type of solenoid valve must also have the most appropriate technical and economic indicators of the initial stroke and electromagnetic attraction force range. However, at the beginning of a design, what can be known is only that the initial stroke and electromagnetic attraction, and the relation between the structural size and them, is unknown. However, these data are often needed to be known when the technical and economic indicators are determined.

Under the initial gas clearance δ_c, the electromagnetic attraction of the solenoid valve should be calculated using the energy balance formula:

$$F_x = \frac{1}{2}(IN)^2 \left|\frac{\mathrm{d}G_\delta}{\mathrm{d}\delta}\right| \approx \frac{1}{2}(IN)^2 \frac{\mu_0 S}{\delta^2} \tag{4.66}$$

For the DC exciting solenoid valve with constant magnetic potential, the IN is constant, so its electromagnetic attraction F_c is proportional to the cross-section area of the core column, Sz. Because the core of the DC solenoid valve is mostly cylindrical, it can be considered that $\sqrt{F_c}$ is proportional to the diameter of the core column dz, which is just one of the main structural parameters of the solenoid valve.

The length of the core column, lz, is also a main structural parameter of the solenoid valve. Under certain conditions, lz mainly depends on the length of the coil, and the latter rates to excitation ampere IN. When the attraction force F_c and the cross-sectional area Sz of the core are unchanged, the initial air gap δc is roughly proportional to IN. Therefore, it can be considered that the length of the core column lz is proportional to the initial stroke δ_c.

In this way, the initial condition and the main structure size are connected by the structure factor $K\phi$. This is the physical meaning of the structural factor.

The computing and experimental results show that every kind of structural type of solenoid valves has a certain range of the value of the structure factor $K\phi$. In this range, the solenoid valve will have a higher economic weight index, and Table 4.2 is

Table 4.2 The reasonable range of K_{ϕ} of different types of constant magnetic potential solenoid valves.

Structure types of solenoid valves	Range of K_{ϕ}
1. Disc-type solenoid valve with a flat-type armature	>93
2. Spiral tube-type solenoid valve with a plane shape armature with end seat	$93 \sim 16.5$
3. Clapping-type solenoid valve	$26 \sim 2.6$
4. Spiral tube-type solenoid valve with a cone-shaped armature with end seat	$16.5 \sim 5$
$\alpha = 45°$, $2\alpha = 60°$	$5 \sim 1.2$
5. Spiral tube-type solenoid valve with a staircase-shaped armature	$4 \sim 1.85$
6. Spiral tube-type solenoid valve with no end seat	<0.2

the reasonable range of $K\phi$ values under different types of constant magnetic potential solenoid valves.

The disc-type solenoid valve has no solenoid force, so has the steepest attraction characteristics and the maximum $K\phi$ value. This valve is suitable for small and large suction strokes. On the contrary, the electromagnetic attraction of the spiral tube-type solenoid valve without a stop seat is a spiral tube force, and it hardly decreases with the change of air clearance value. Therefore, the suction characteristic is the flattest and the $K\phi$ value is also the smallest. This type of solenoid valve should be used for a large stroke and small suction.

The stroke of the solenoid valve in the fuel suction stroke adjustment is short and requires a quick response of the opening and closing, so the single pole "E"-type structure is adopted. The "E"-type structure has the more compact structure, a smaller quantity of moving parts and a relatively larger suction compared with other structural-type solenoid valves. Through the optimization design of the system, better dynamic response characteristics and a satisfactory control effect can be obtained. The main function of the solenoid valve in the fuel pressurization stroke adjustment mode is to promote the opening and closing of the one-way valve. Therefore, a guide rod is needed and the solenoid-type solenoid valve is selected here to meet the design requirements.

4.3.1.3 Simulation of the Static Characteristic of the Solenoid Valve

The boundary value problem of the electromagnetic field is actually the problem of solving Maxwell equations under given boundary conditions and other partial differential equations evolved by the equation set. From the technical means of solving the problem, it can be divided into two categories: analytic solution and numerical solution. For a simple model, the analytical solution of the equation can sometimes be obtained. If the complexity of the model is increased, it is often difficult to obtain the analytical solution of the model. The finite element method is a commonly used numerical method and corresponding electromagnetic software is available. Ansoft's Maxwell 2D/3D is a very excellent electromagnetic analysis software.

Maxwell 2D/3D is an important part of the electromechanical system design solution. Maxwell 2D is a powerful and accurate two-dimensional electromagnetic field finite element analysis software. It is also easy to use. It includes an electric field, static magnetic field, eddy current field, transient field, and temperature field analysis module. It can be used to analyze the characteristics of static, steady state, transient, normal working conditions and fault conditions of electromagnetic devices such as motors, sensors, transformers, permanent magnet devices, exciters, and so on. It includes the execution user interface from top to bottom, advanced adaptive mesh generation technology, user defined material library, and other characteristics, making Maxwell 2D much easier to use than any other software. Maxwell 2D has the capability of a high-performance matrix solver and multi-CPU processing, which provides the fastest solution speed.

(1) The design of the armature stroke of the solenoid valve. The following is an analysis of the structure design and static electromagnetic force of the fuel suction stroke adjustment flow control device.

The high-pressure pump used in this study is a straight column pump without an inclined slot. The maximum amount of working fuel required per cycle can be calculated according to the following formula:

$$V_p = \frac{\pi \times D^2}{4} \times H$$

In the formula, D is the diameter of the plunger (mm) and H is the geometric stroke of the plunger (mm).

The maximum fuel supply of the high-pressure pump is 307 mm^3/cycle. In addition, considering the solenoid valve, the one-way valve throttling, the plunger coupling leakage, and so on, then

$$V_S = \frac{V_{p\max}}{\eta_k}$$

where $\eta_k = 0.80$.

When the speed of the high-pressure pump is 1000 r/min, the amount of working fuel required for a single cylinder per cycle solenoid valve is $V_S = 383.75\,\text{mm}^3$/cycle = 0.056 L/s.

The flow rate calculation formula at the valve port of the solenoid valve cone valve is as follows:

$$Q = C_d A \sqrt{\frac{2\Delta P}{\rho}}$$

In the formula:

Q is the flow through the port,
C_d is the flow coefficient at the port (selected to be 0.7, the empirical value),
A is the geometric circulation area of the solenoid valve,
ΔP is the pressure difference between the front and back of the valve port.

The expression of the circulation area at the valve port of the solenoid valve is as follows:

$$A = \pi l \sin\theta \left(\frac{1}{2}d + l \sin\theta \cos\theta\right)$$

In the formula, d is the middle diameter of the valve core, l is the valve core stroke, and θ is the conical cone angle, selected to be 45°. The matching values of different d and l

Table 4.3 Matching values of different *d* and *l*.

d(mm)	21.5	10.6	6.9	5.0	3.8
l (mm)	0.1	0.2	0.3	0.4	0.5

are shown in Table 4.3. In order to take account of both the small motion quality and the valve core stroke at the same time, the diameter of the valve core is 4 mm and the valve core stroke is 0.5 mm.

(1) Optimum design of the iron core size for the solenoid valve. Because the structure of the solenoid valve is axisymmetric, its two-dimensional form can be used to reflect the three-dimensional situation. Therefore, in order to reduce the computation and improve the computation accuracy, the analysis of the solenoid valve magnetic field is carried out by using the 2D axisymmetric model. Maxwell is a large electromagnetic finite element analysis software launched by Ansoft Company. Under the Maxwell 2D solver, the 2D model is established, the specific copper material property is assigned to the coil, the specific steel-1010 material property is assigned to the armature, the balloon boundary condition is specified for the background, and the 3600 A source current is applied to the winding window. Figure 4.50 is a sketch of the process of solving the electromagnetic force using Maxwell 2D. Figure 4.51 is a two-dimensional finite element mesh model of the solenoid valve in which the most outsourced layer is air.

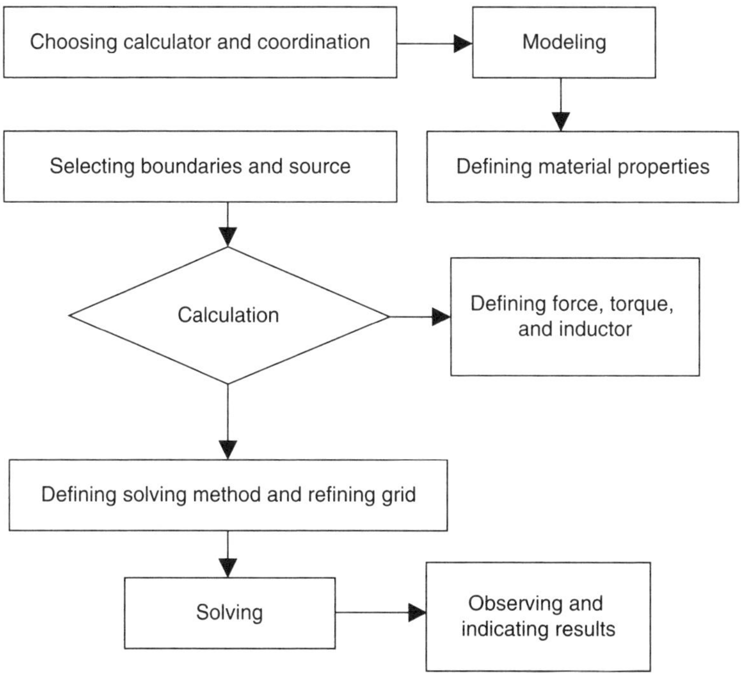

Figure 4.50 Flow chart of the finite element analysis program.

Figure 4.51 Gridding diagram of the solenoid valve.

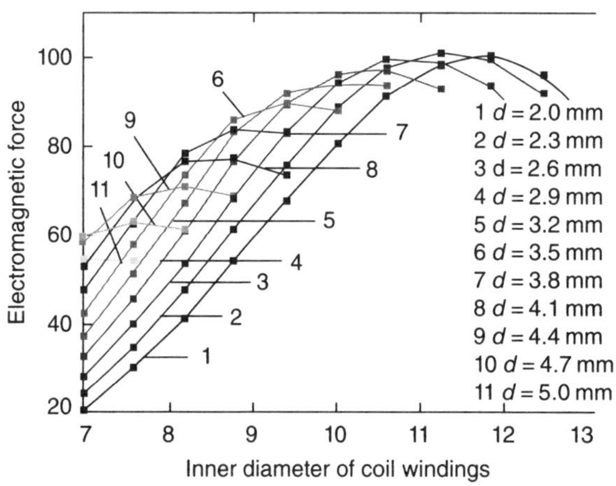

Figure 4.52 Electromagnetic force simulation results of the cross-section and inner diameter of different coil windings.

The electromagnetic force simulation results of the cross-section width and inner diameter of different coil windings are shown in Figure 4.52.

Figure 4.52 shows that if the inner diameter of the winding is too small, the reluctance of the magnetic circuit will increase, and the magnetic field strength will be weakened. If the inner diameter of the winding increases, the magnetic field strength will increase and the area of magnetic force will decrease. When the section area of the winding is constant, there is an optimal shape of the coil section, which can make the same size magnetic field the maximum. Figure 4.53 is the magnetic field line distribution of the magnetic field with width of the same winding section and the inner diameter of the different windings. Figure 4.54 is the distribution of magnetic field lines with the inner diameter of the same winding and different widths of the cross-section of the winding. After a comprehensive comparison, the inner diameter of the solenoid valve is selected to be 10.6 mm, and the width of the coil winding is 2.6 mm. At this time, the electromagnetic force is 97.2 N, which can meet the electromagnetic force required for the solenoid valve to open.

(2) Optimal design of armature thickness. In theory, the smaller the armature thickness, the greater is the magnetoresistance of the armature and the greater the magnetic potential, the less is the magnetic potential of the air gap part and the smaller the electromagnetic force. If the thickness of the armature is greater, the opposite situation will happen. Figure 4.55 shows the calculation results of the electromagnetic force for different armature thicknesses. After the thickness of armature increases, the motion quality of the spool will increase, resulting in a slower response of the spool, and if the armature thickness increases to a certain value, the increase of the electromagnetic force will not be obvious. Therefore, the thickness of the armature should be selected in the middle, and the large electromagnetic force and the small motion quality of the valve core should be taken into account to ensure the fast response characteristic of the solenoid valve. The thickness of the armature is 1.5 mm in the control device of the fuel suction control adjustment type.

(3) The influence of electromagnetic material on the electromagnetic force.

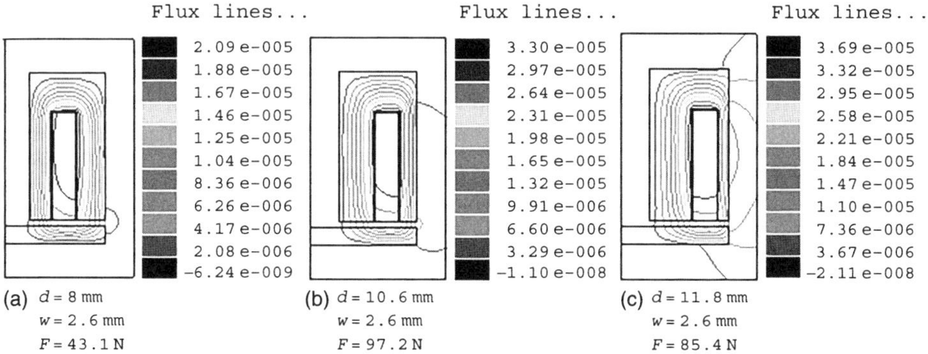

Figure 4.53 Magnetic field line distribution of the magnetic field with the same width of winding section and a different inner diameter of the winding.

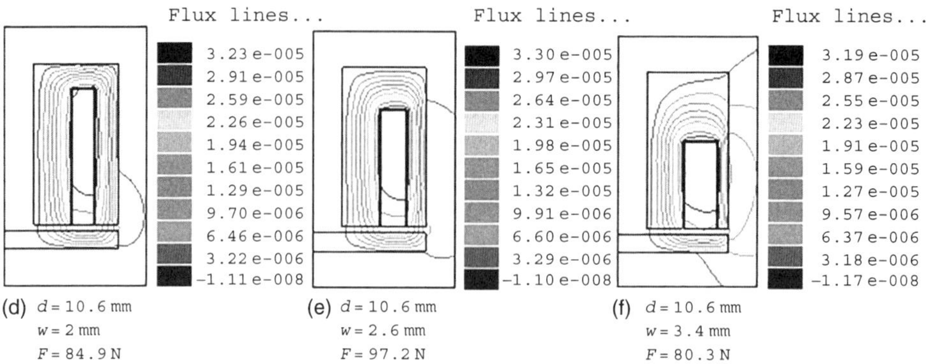

Figure 4.54 Magnetic field line distribution of the magnetic field with different widths of the winding section and the same inner diameter of the winding.

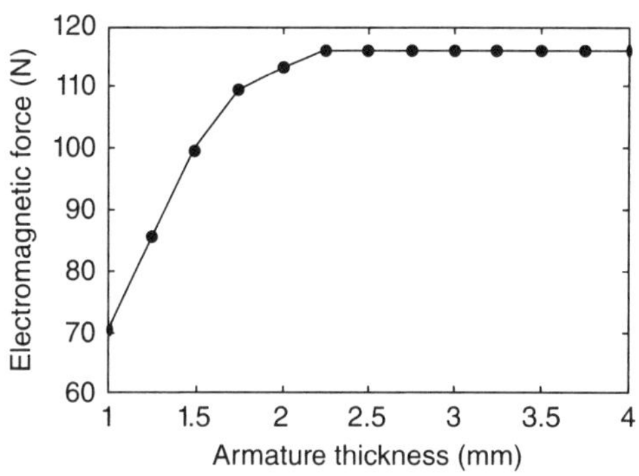

Figure 4.55 Electromagnetic force under different armature thicknesses.

Table 4.4 The structural parameters of the electromagnet.

Material	Coil diameter (mm)	Coil turns (*n*)	Resistance (Ω)	Inductance (H)	Stroke (mm)
Industrial pure iron	0.06	1500	18.0	0.022	1.0

Figure 4.56 is the effect of two kinds of solenoid valves made of industrial pure iron or iron nickel alloy on the opening response time under different excitation voltages. It can be seen from the diagram that the industrial pure iron has a faster response than the iron nickel alloy under the same voltage excitation. Therefore, industrial pure iron is more suitable as the iron core material. The structural parameters of the electromagnet are shown in Table 4.4.

4.3.1.4 Simulation of Dynamic Characteristics of the Solenoid Valve

(1) The effect of spring stiffness on the dynamic response time of the solenoid valve. When undertaking a study of the effect of a spring pre-load on the response time of the electromagnetic valve, it has been found that in the opening stage, because the coil current can reach a peak in a very short time, the electromagnetic force can also reach a larger value. The influence of the pre-load force on the opening time is not

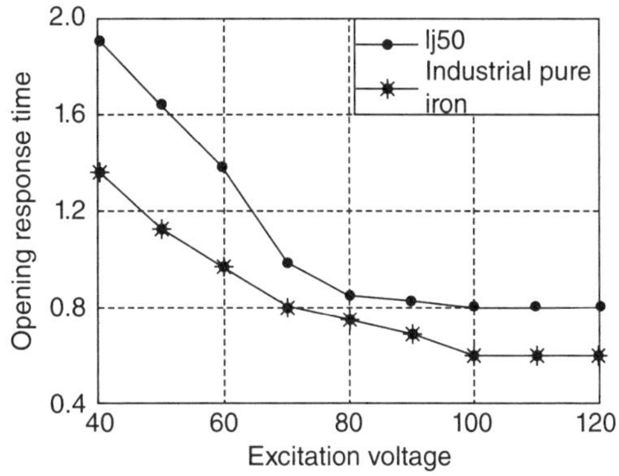

Figure 4.56 Response time curve of different electromagnetic materials.

related to the closing time. Therefore, in the discussion of the total response time, its effect on the closing time of the solenoid valve should receive more attention and a large pre-tension value should be taken.

Figure 4.57 are electromagnetic valve stroke curves obtained from four different spring stiffnesses ($k = 2$, 7, 12, and 17 N/mm). It can be seen from the graph that the influence of spring stiffness on the opening time is tiny, so it can be neglected. Under the condition that the solenoid valve can be opened, the larger spring stiffness should be taken.

(2) The influence of the driving voltage on the dynamic response time of the solenoid valve.

Figure 4.58 is the effect of the driving voltage of the solenoid valve on the dynamic response time when the spring force pre-tightening force of the solenoid valve is 32 N and the driving voltage is 24 ~ 110 V, respectively.

As shown in Figure 4.58, when the low voltage is driven, the response time of the solenoid valve increases, the dynamic time is increased, and the 24 V drive solenoid valve is opened for 2.0 ms. With the increase in driving voltage, the opening time of the solenoid valve decreases and the dynamic performance improves. However, when the driving voltage increases to a certain value, the driving voltage and the opening time gradually slowly decreaase. The reason is that due to the influence of magnetic saturation and other factors, increasing the driving voltage of the solenoid valve makes the influence of a smaller opening response time decrease. Due to power outages, the valve is closed under the effect of pre-load of the spring, so increasing the driving voltage has little influence on the closed solenoid valve. The increase of driving voltage will have an adverse effect on the power consumption of the drive circuit and will heat the coil. On the premise of satisfying the dynamic response characteristics, the driving voltage should be reduced.

(3) The influence of the mass of the moving parts on the dynamic response time of the solenoid valve.

From Newton's second law of motion $F = ma$, it can be seen that a smaller mass of m can obtain a larger acceleration under a certain force F. Therefore, the reduction of the mass of the moving parts can improve the response speed of the closing and opening of the solenoid valve. Figure 4.59 is the dynamic simulation results of the response time when the mass of moving parts is 10, 20, 30, and 40 g. It is clear that the smaller the mass, the better the response characteristics that can be obtained.

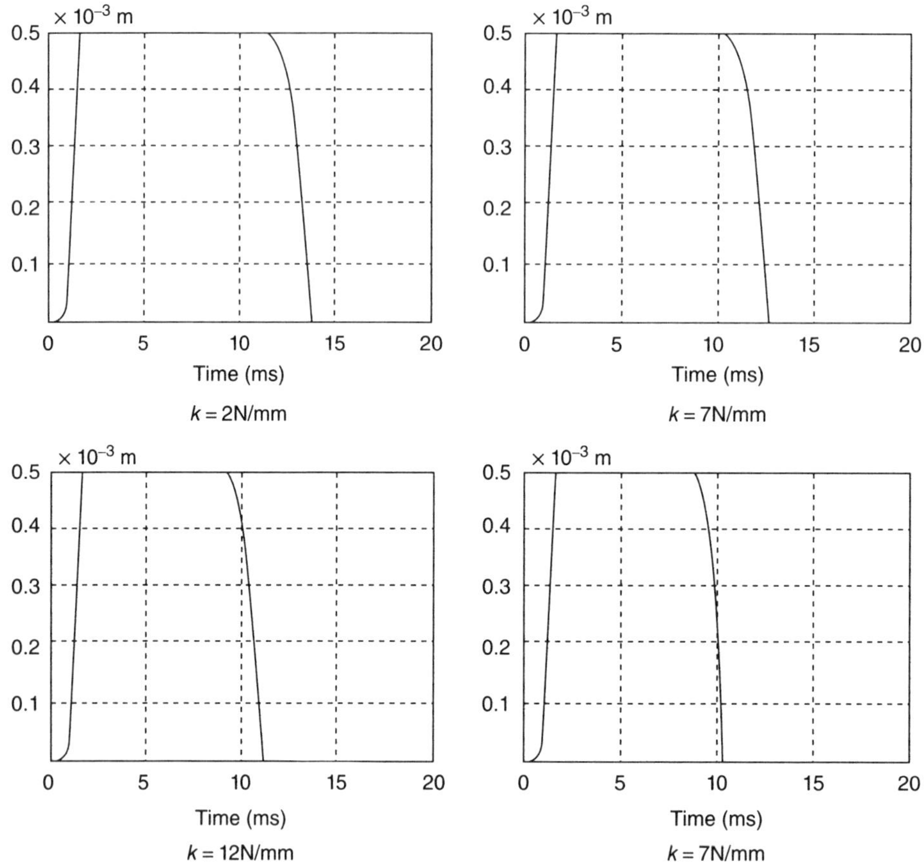

Figure 4.57 The effect of spring stiffness on the response of the solenoid valve.

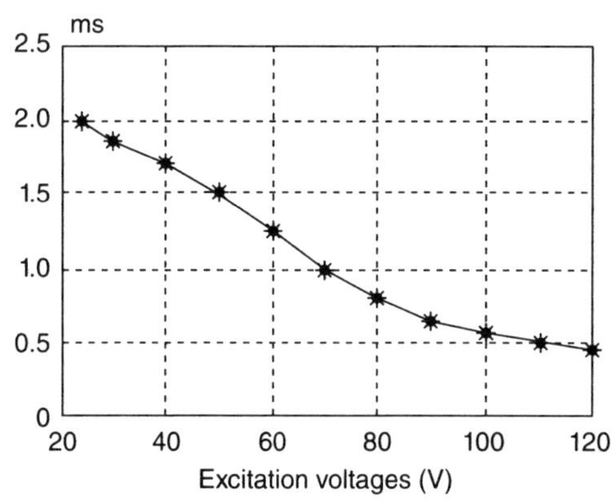

Figure 4.58 The effect of driving voltage on the response time of the solenoid valve.

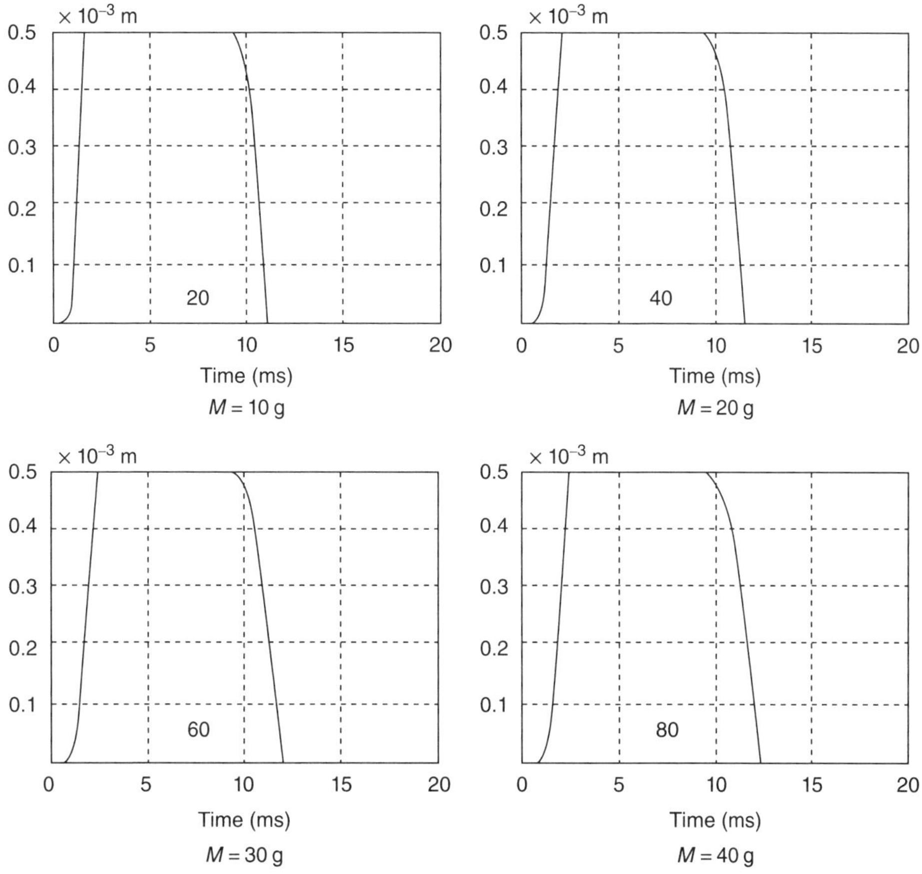

Figure 4.59 The effect of the mass of the moving parts on the response time of the solenoid valve.

4.3.1.5 Design and Optimization of the One-Way Valve

A one-way valve is a directional control valve that ensures the fluid flow through the valve is passed only in one direction and does not flow in reverse. The requirements of a one-way valve are mainly that the pressure loss of the liquid flow is small, the sealing ability is good when the reverse cut-off is closed, the action is swift, and there is no impact and noise when it is working.

The direct one-way valve can be divided into two kinds: the steel ball type and the cone valve type. The structure of the steel ball check valve is simple, but the sealing property is not as good as that of the cone valve type. In addition, because the steel ball has no guiding part, it is easy to generate vibration when working, and is usually used when there is less flow. The cone valve type is more popular. Although the structure is more complex than the steel ball type, its guidance is better and the sealing performance is more reliable.

The spring of the one-way valve can ensure that under the premise that it can reset after it overcomes the valve core friction and gravity, the stiffness of the spring is as small as possible, by which the pressure loss of the one-way valve is reduced. Generally, the opening pressure of the one-way valve is 0.035 ~ 0.05 MPa and the pressure loss through the rated flow should not exceed 0.1 ~ 0.3 MPa.

The designed flow control device is installed on the 3CB-0.8/160 pressure test pump. The plunger diameter is $d = 7$ mm, the plunger lift is $h = 8$ mm, the single stroke full load fuel consumption is 0.307 ml, and the flow rate is small. The steel ball one-way valve can meet the requirement.

The one-way valve A is subjected to high pressure when the plunger rises and the one-way valve B is also subjected to high pressure after the common rail tube is set up. Therefore, the two one-way valves are high-pressure one-way valves. For the convenience of designing and processing, the two one-way valves can be designed as one-way valves with the same structural parameters. This paper focuses on the one-way valve A, which is connected to the low-pressure end.

The initial design of the one-way ball valve has the following parameters: base diameter $D = 4$ mm, steel ball diameter $d = 6$ mm, spring stiffness $k = 0.5$ N/mm, spring pre-load $f = 2.5$ N, steel ball opening stroke $x_{max} = 1$ mm. Under the fuel suction stroke control mode, the solenoid valve is fully open, or under the fuel pressurization stroke control mode, the solenoid valve is always closed, and the plunger supplies fuel with a full stroke. Namely, when the fuel supply is at a maximum, the effect of the solenoid valve can be neglected, and it can be simplified where there are only two one-way valves at work. When the cam speed is 600 r/min, the pressure of the fuel pump is 5 MPa and the common rail pressure is 80 MPa. The one-way valve simulation model is established as shown in Figure 4.60 and the stroke curve of the one-way valve is obtained as shown in Figures 4.60 and 4.61.

The complete opening time of the one-way valve is 2.3 ms and the complete closing time is 0.7 ms. The flow difference, the pressure characteristic curve, is shown in Figure 4.62. It can be seen that the simulation flow rate difference – the pressure characteristic curve and the target curve of the one-way valve – has a gap. It is necessary to optimize the five parameters of the one-way valve. The conjoint simulation of AMESim and Matlab is carried out to optimize the design of one-way valve parameters.

Figure 4.63 is an optimized structure model of a one-way valve. Figure 4.64 shows the calculated flow rate difference – the pressure characteristic curve. The actual flow rate difference of the one-way valve after the optimization, the pressure characteristic curve, is very close to the expected curve and the optimization result is very good (Figure 4.65).

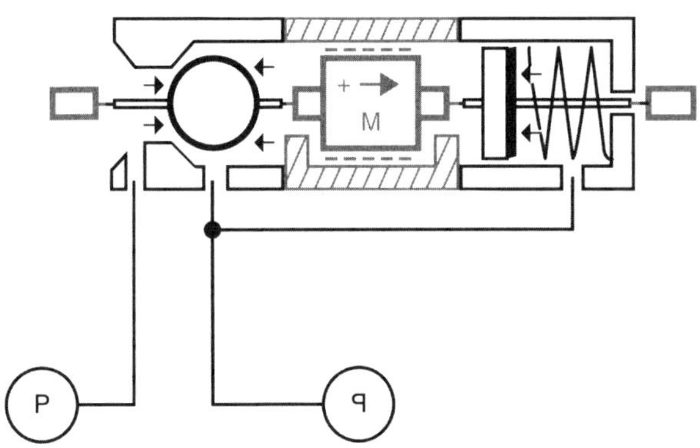

Figure 4.60 The structure diagram of the one-way valve before optimization.

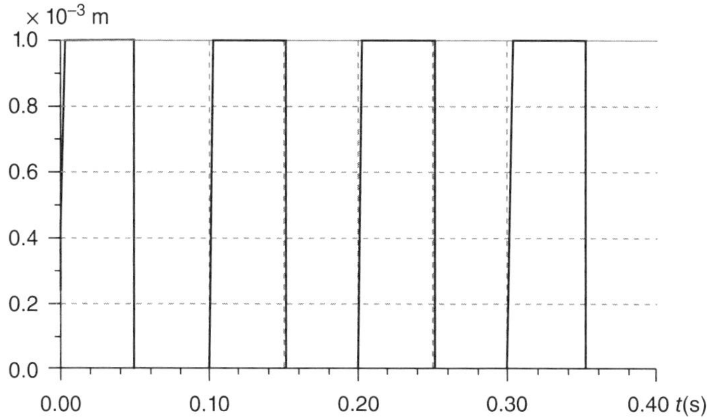

Figure 4.61 The opening characteristics of the one-way valve before optimization.

Figure 4.62 Simulation flow rate–pressure difference characteristic curve before optimization.

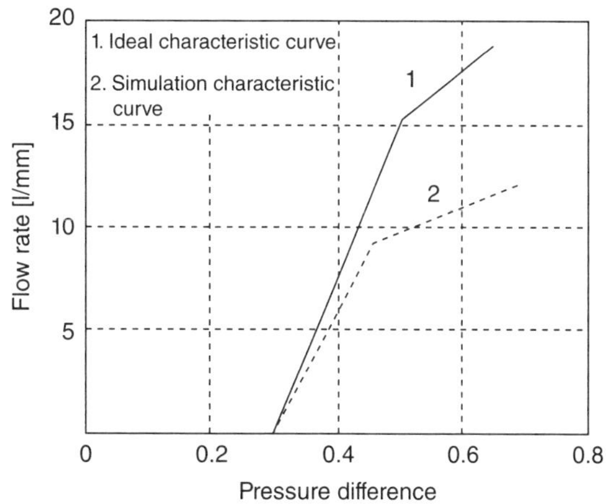

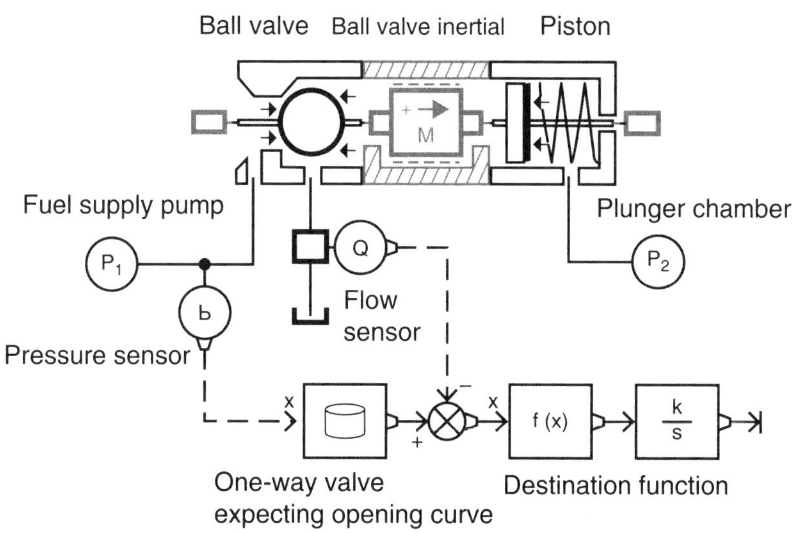

Figure 4.63 Optimal structure model of the one-way valve.

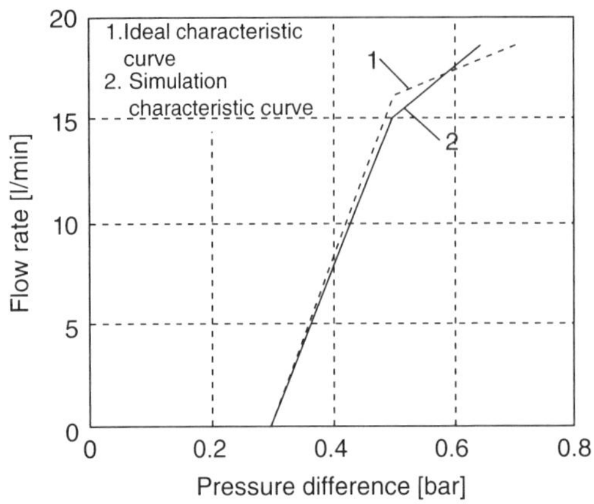

Figure 4.64 Optimized flow rate–pressure difference characteristic curve.

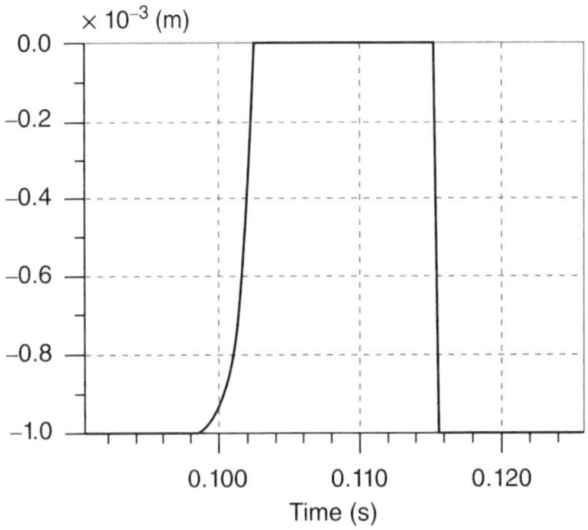

Figure 4.65 Displacement of the one-way valve.

4.3.2 Conjoint Simulation Analysis of a Flow Control Device and the Common Rail System

The flow control device of the high-pressure pump is the actuator of the common rail chamber pressure control system. The control target is the rail chamber pressure, and the control object is the high-pressure fuel pump. Namely, ECU controls the circulating fuel supply of the high-pressure pump through the flow control device to realize control of the pressure of the common rail chamber.

4.3.2.1 Simulation of the Flow Control Device

(1) Simulation model of the flow control device. The existing hydraulic simulation software is mostly based on the system level simulation, but the simulation function of the component level is weak, especially the structure simulation of the component. The AMESim developed by IMAGINE company can carry out system level simulation, and can design and optimize the structure of hydraulic components by using the hydraulic components design library provided by this software.

In this section, the AMESim software will be used to establish the physical simulation model of the fuel suction stroke adjustment flow control device and simulate the working process.

According to the physical structure and working principle of the flow control device, the flow control device model is shown in Figure 4.66:

Component 1 is a control signal, which is the simulation of the control pulse emitted by a microcontroller. Changing the step time of 1 can provide a different duty ratio of the control signal.

Components 2, 3, and 5 are the simulations of the solenoid valve in the flow control device.

Component 4 is the simulation of the low-pressure tank.

Components 6 and 7 are simulations of the low-pressure one-way valve A.

Components 8, 9, 10, 11, and 14 are simulations of the high-pressure pump.

Components 12 and 13 are simulations of the high-pressure one-way valve B.

Component 15 plays the role of limiting the flow.

Component 16 is a speed reference.

The setting speed of the cam is $n = 600$ r/min and the pressure of the high-pressure end is 100 MPa. The duty signal of component 1 is 30% and the simulation results are shown in Figures 4.67 and 4.68.

(2) Simulation of the common rail pressure control system

a. Establishment of a computational model. The conjoint simulation of the Simulink control algorithm and HYDSIM can be realized by using the HYDSIM and Matlab interface. In a high-pressure common rail system, according to the operating conditions of the diesel engine, ECU sets the optimum rail pressure target to be P_0, and programmed the control program with Simulink to compare the difference ΔP between the target rail pressure P_0 and the actual rail pressure P. The

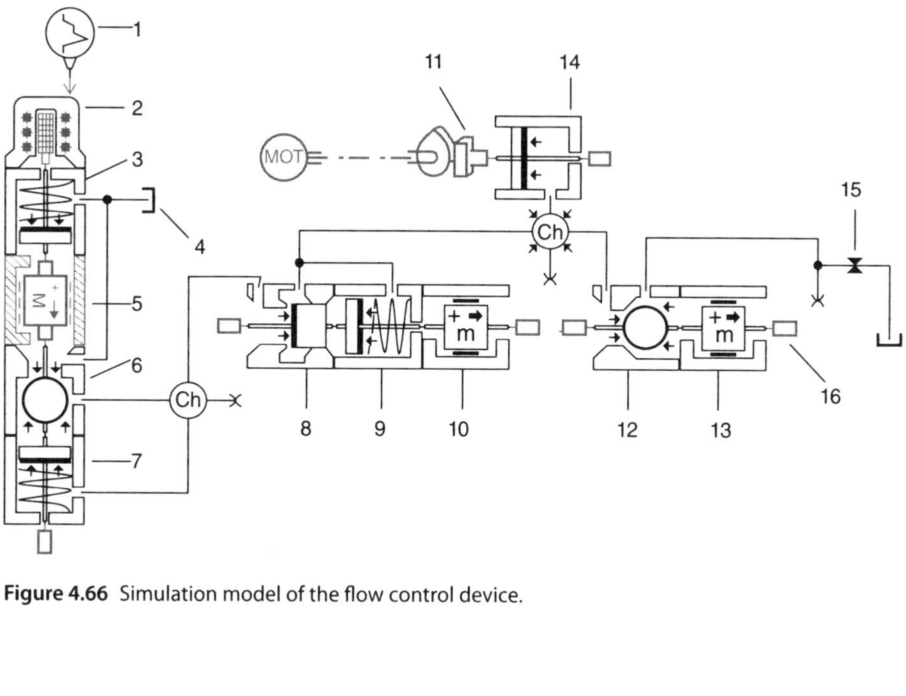

Figure 4.66 Simulation model of the flow control device.

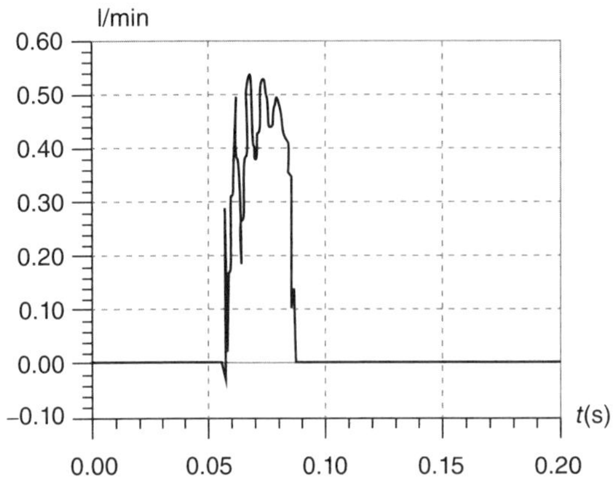

Figure 4.67 Flow rate of the low-pressure one-way valve.

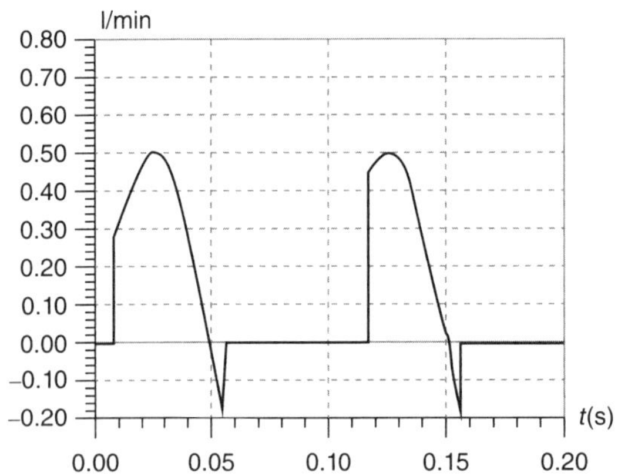

Figure 4.68 Flow rate of the high-pressure one-way valve.

process of regulating the rail pressure by Simulink is shown in Figure 4.69 and the conjoint simulation model of the PID control and the common rail system are shown in Figure 4.70.

4.3.3 Analysis of Simulation Results

Set the cam speed to be 600 r/min, the initial pressure of the common rail to be 130 MPa, the rail chamber volume to be 15 cm^3, the high-pressure pump plunger diameter to be 7 mm, the plunger stroke to be 8 mm, and the width of the solenoid valve pulse of the electronically controlled injector to be 0.8 ms. After setting the PID parameters, the simulation model is operated to get the pressure fluctuation of the high-pressure common rail cavity, as shown in Figure 4.71a.

In the comparison graph of the common rail pressure fluctuation shown in Figure 4.71a, the simulated pressure and the measured pressure fluctuate between 126 and 134 MPa, and the fluctuation range is within 5%. It can ensure the stability of the rail pressure and meet the requirements of fuel injection. However, the fluctuation amplitude of the measured pressure fluctuation is larger than that of the simulation result and the simulation result is slightly lagging in the phase. The fluctuation amplitude error is not more than 2% and the phase lag is 0.4 ms. This error is caused

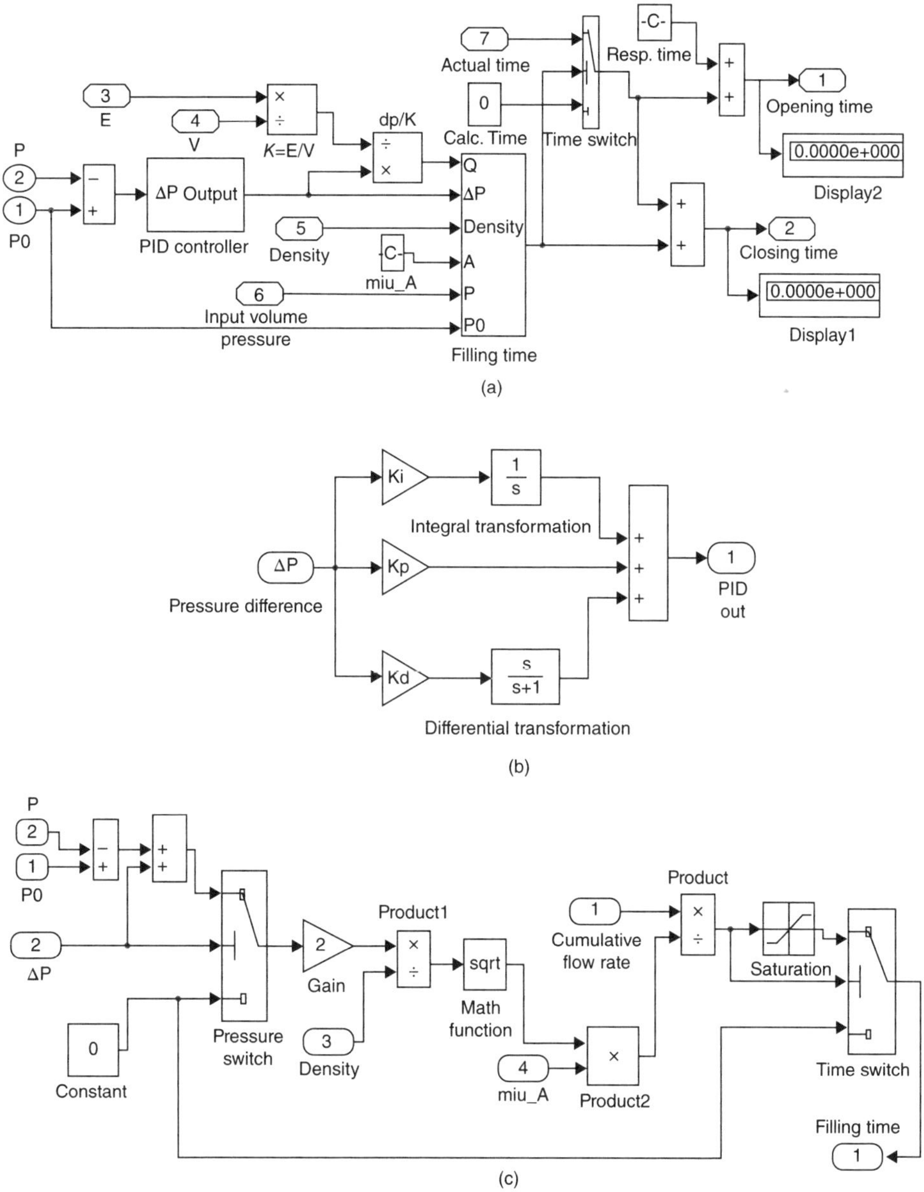

Figure 4.69 PID control program of the common rail system.

mainly because in the simulation model, the impact state, the variation of fuel property parameters with temperature of the one-way valve, and the installation location of the pressure sensor are not considered. However, the deviation between the simulation result and the measured value is small and is acceptable. Figure 4.71b is the pressure fluctuation of the common rail chamber under the PID control algorithm using this flow control device, when the speed of the high-pressure pump is 600 r/min and the target rail pressures are, respectively, 70, 100, and 135 MPa.

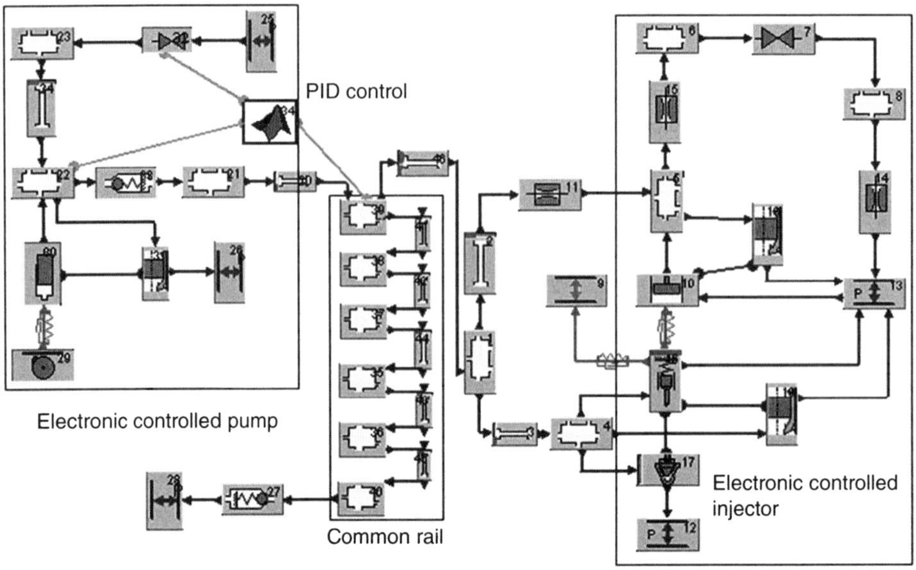

Figure 4.70 Simulation module of the common rail system.

Figure 4.72 is the time needed for the fuel pressure of the common rail system to reach the set value after the common rail system is started. If the cam speed is 392 r/min and the target pressure value in the rail cavity is 30 MPa, the time needed for actual rail pressure rises from 0 to 30 MPa is 1.27 seconds. Obviously, the requirement of the cold start of the 10 seconds for the marine diesel engine is satisfied.

Figure 4.73 shows the comparison between the simulation analysis results and the measured results of the fuel supply of the high-pressure pump and the fuel injection amount of the injector with the control of the pulse width of the injector. From the graph, it can be seen that the flow control device has a good follow-up characteristic, which means that, with the variation of the fuel injection volume of the injector, the fuel supply volume of the high-pressure pump is changed, so that the pressure fluctuation amplitude of the common rail chamber is controlled within a certain range. One point needed to be explained is that the difference between the fuel supply of the high-pressure pump and the injection amount of the injector is mainly caused by the larger principle fuel recovery of the electronically controlled injector. From the chart, it can be seen that the measured and simulated values of the fuel supply of the high-pressure pump, the measured values of the injector, and the simulation value are both basically matched. The maximum error of the fuel supply of the high-pressure pump is not more than 7% and the maximum error of the fuel injection of the injector fuel volume is no more than 5%. When the injection pulse width is 1500–2000 seconds, the error is small. The reason is that the response delay of the simulation parameter input of the solenoid valve is constant and the measured value is variable. In this pulse width interval, the values of the two are close.

The pressure fluctuation of the common rail cavity can be improved effectively by increasing the fuel supply frequency to the rail cavity and reducing the unit fuel supply. Therefore, the response time of the solenoid valve and one-way valve in the flow control

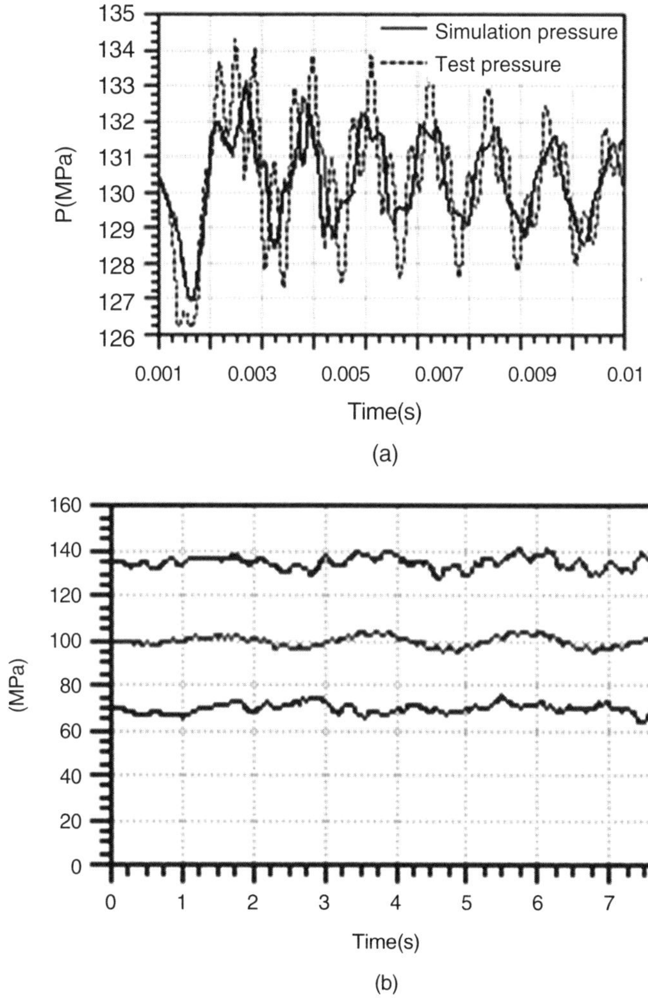

Figure 4.71 Pressure fluctuation of the common rail tube.

Figure 4.72 Test result of the starting condition.

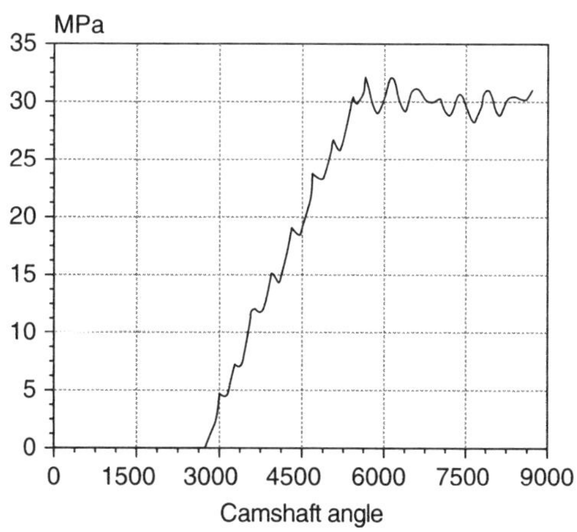

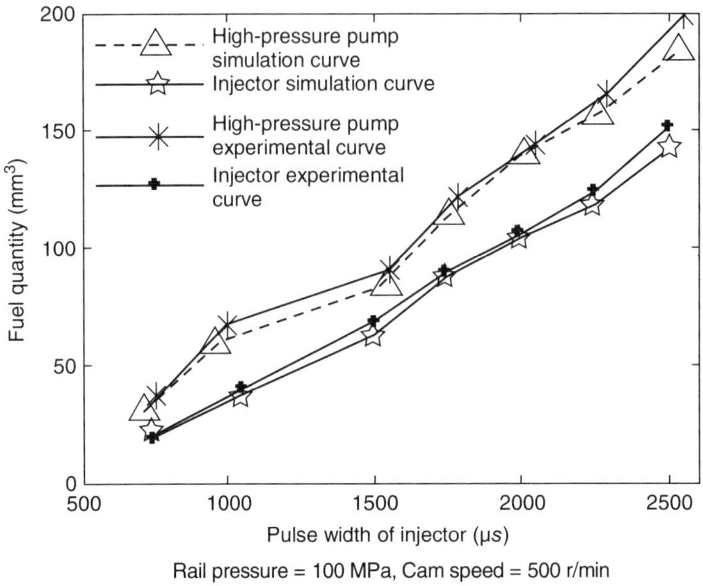

Figure 4.73 Comparison of the fuel supply of the high-pressure pump and fuel injection of the injector.

device should be improved, in order to make the solenoid valve keep a good accuracy under the small pulse width. Therefore the flow control device can precisely adjust the fuel supply to the rail chamber and the accuracy of the rail chamber pressure control is also improved.

4.3.4 Experimental Study on the Regulation of Common Rail Pressure by the Flow Control Device

The most effective and final test method used to determine whether the structure design of the flow control device of the high-pressure common rail system is reasonable, whether the parameter is chosen properly, or whether the rail pressure can be precisely controlled is to experimentally test it. At the same time, through experimental research and performance testing, the accuracy of theoretical design results and simulation mathematical models can be corrected and tested. This test is mainly a performance test for the fuel pressurization flow adjustment type flow control device. It mainly includes a test for one-way valve sealing performance, a test for dynamic response of the flow control device, a test for common rail pressure control, a test for the effect of the solenoid valve duty ratio in the flow control device to the rail pressure fluctuation, and other tests.

4.3.4.1 Test Device

All the tests were done on the high-pressure common rail system test bench. Figure 4.74 is a photo of the test bench, and Figure 4.75 is a composition diagram of the test device. This device mainly includes a combined high-pressure pump, a flow control device, a common rail tube, a fuel injector, a BY-120-6A type extra high-pressure portion overflow valve, a pump drive, and a speed regulation system.

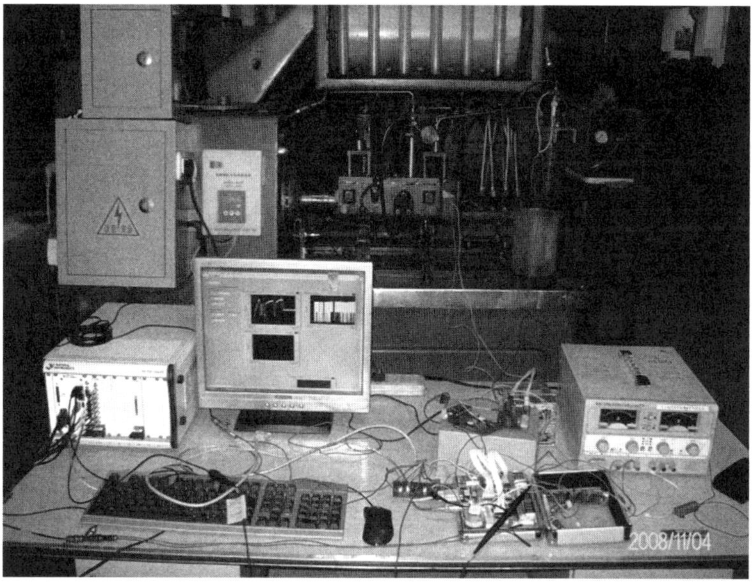

Figure 4.74 Test of common rail pressure control.

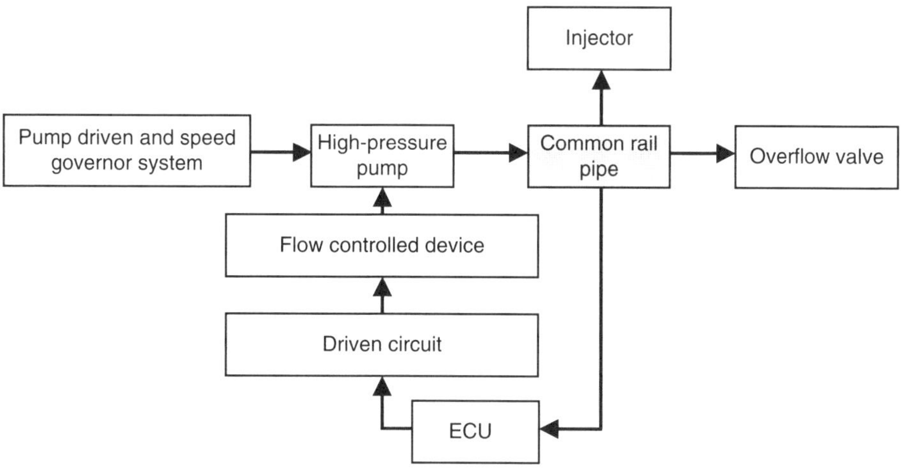

Figure 4.75 Sketch of a common rail system test bench.

4.3.4.2 Sealing Performance Test of the One-Way Valve

The fuel pressurization stroke adjustment flow control device is mainly composed of an electromagnet and two one-way valves, as shown in Figure 4.76. The main objective for checking the performance of the one-way valve is to test the sealing performance of the designed valve. The sealing performance not only affects the volumetric efficiency of the pump but also affects the control strategy of the rail chamber pressure.

When the control part of the test device is stopped, the flow control device works in basically the same way as the general high-pressure pump, the only difference being that a one-way valve is added to its suction port. When the rotation speed of the

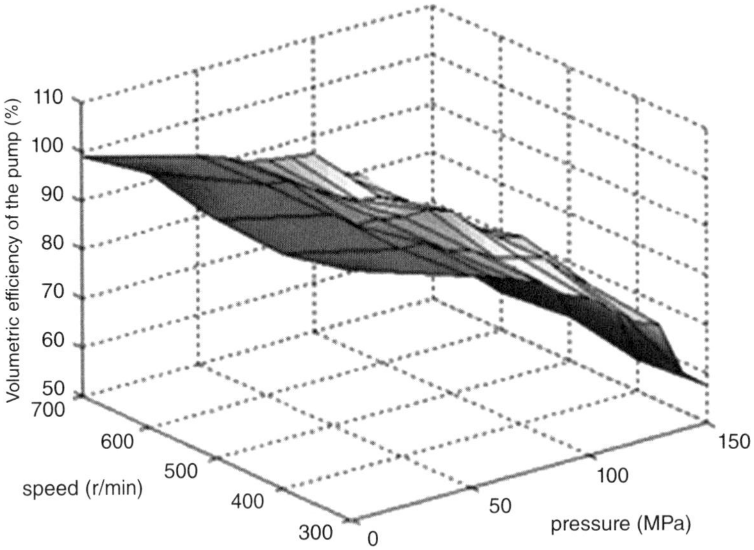

Figure 4.76 Volumetric efficiency of the pump.

camshaft of the high-pressure pump is fixed and the fuel in the common rail cavity is gradually pressurized by 0–150 MPa, the fuel flow rates under fuel pressure at all levels are measured in 60 seconds and the volume efficiencies at all levels are calculated. By changing the camshaft speed and repeating the above process, the fuel volumes and volumetric efficiencies of the high-pressure pump under different pressures and speeds can be obtained. Then the volume efficiencies of the pump before and after the installation are compared and the sealing performance of the one-way valve can be determined.

Tests were made on two kinds of unidirectional valves, spherical and conical types. It was found in the test that the sealing performance of the spherical one-way valve is poor and its volumetric efficiency is only 50% lower than the pressure of the 80 MPa common rail; it is difficult to establish the rail pressure above the 100 MPa rail. The main reason is that the steel ball used in the spherical valve is purchased on the market and the shape error can be large.

4.3.4.3 Experimental Study on the Dynamic Response Characteristics of the Electromagnet

The dynamic response characteristics of the electromagnet have a great influence on the response of the flow control device. Through the test, the opening and closing response times of the electromagnet in the design can be tested and the dynamic characteristics of the electromagnet can be analyzed to meet the requirements.

When the high-pressure pump is working, the action of the electromagnetic core and the action of the one-way valve are coupled and are difficult to distinguish. In the test, the experiment is carried out with the high-pressure pump on the test device stopped, which means that when the pump stops working (stationary state), a driving pulse signal is sent to the electromagnet and an acceleration sensor connected to the iron core is used to test the response of the core. Figure 4.77 is a photo of the high-pressure pump of the

Figure 4.77 High-pressure pump flow control device.

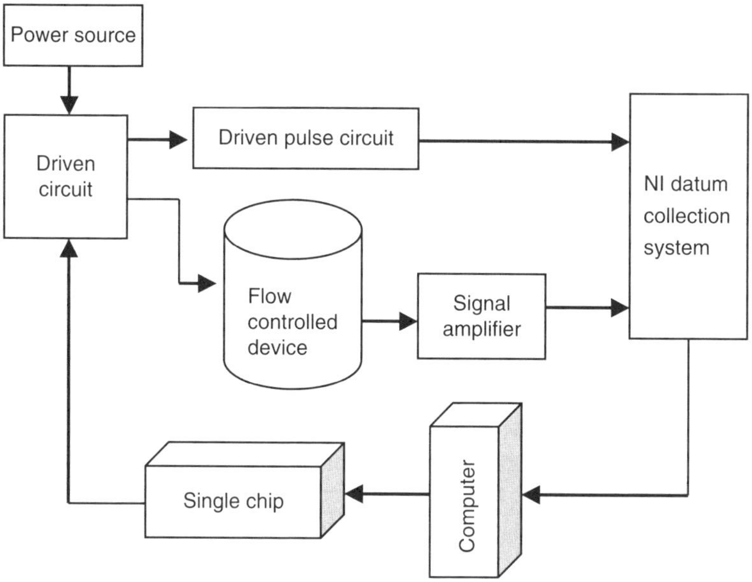

Figure 4.78 Sketch of a test device for response characteristics of a solenoid valve.

developed electromagnet flow control device. Figure 4.78 is a test device used to test the dynamic response characteristics of the electromagnet of a flow control device.

Figure 4.79 is the acceleration curve of the iron core of the electromagnet and the test result of the pulse signal when the pulse control signal is 50 HZ and the duty ratio is 50%. The test results are calculated and analyzed to show that, under this condition, the starting time of the electromagnetic iron core is 1.1 ms and the closing time is 0.9 ms. If the speed of the pump shaft is calculated as 750 r/min, the cam rotation angles are 4.95° and 4.05°, respectively. In the experiment, an eccentric cam is used as the high-pressure

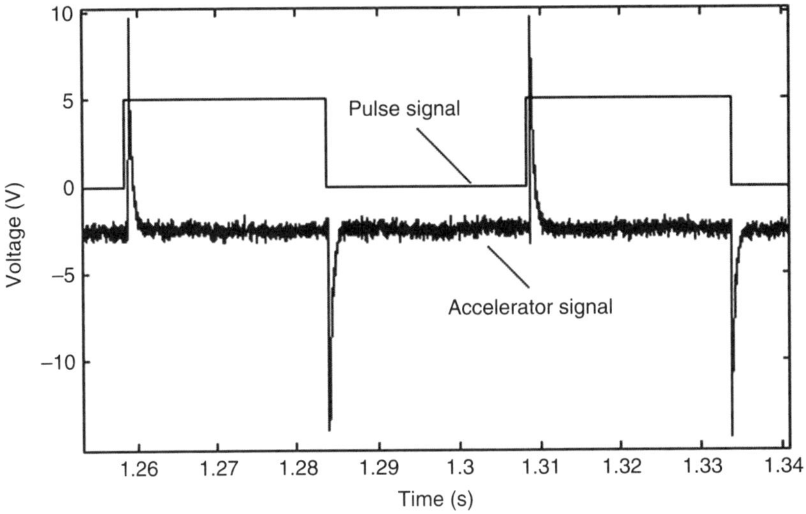

Figure 4.79 Response characteristic of the flow control device.

pump camshaft and the fuel pressure stroke continuous angle is 126°. Therefore, if the influence of plunger speed and other elements are not considered, the preliminary analysis shows that the response of the electromagnet core will account for 7% of the whole fuel stroke continuous angle, which is expected to meet the requirements of the electromagnet dynamic response characteristics.

4.3.4.4 Test of Pressure Control in the Common Rail Chamber

The flow control device is used to control the common rail pressure, but the performance of the device cannot be explained by the dynamic characteristics of the electromagnet in the flow control device. Therefore, the flow control device should be installed on the high-pressure pump to carry out the pressure control test of the whole common rail system.

Control of the common rail pressure is divided into four modes according to the working condition and state of the diesel engine. In the normal mode, the combination of an open loop and a closed loop is used as the compound control method. The open-loop control plays the role as pre-adjustment, which can improve the dynamic response characteristics of the fuel pressure control. The closed-loop control improves the control accuracy and enhances the adaptive performance of the system's fuel pressure control.

In the starting stage of the common rail system, in order to realize the rapid starting of a diesel engine, it is very important to establish fuel pressure quickly and achieve the target starting fuel pressure. However, when the diesel engine starts, its rotational speed is low, so the magnetoelectric sensor will fail to detect the signal of the down end port signal in time and drive the PCV valve to supply fuel as the normal cam phase. Therefore, the PCV valve can only be driven at a certain frequency. This control mode of the PCV valve is called the starting mode. Its purpose is to supply more high-pressure fuel to the common rail tube and to promote the rapid establishment of the rail fuel pressure. The starting mode of the rail fuel pressure is a type of open-loop control mode. The advantage of the fuel pressurization stroke adjustment flow control device is that the solenoid valve does not need to be energized under the starting condition, and the fuel supply under the full stroke of the plunger will help to improve the fuel pressure in the rail as soon as possible during the initial stage of starting the diesel engine.

After the solenoid sensor identifies the precise phase of the oil pump cam, the corresponding common rail pressure control modes, such as the pre-control mode, the closed-loop control mode, or the stop supply mode, can be selected according to the deviation between the target injection pressure and the actual rail pressure. If the deviation is less than 3 MPa, then the closed -loop control mode is chosen. If the actual common rail pressure is far below the target injection pressure, which means that the deviation is greater than 3 MPa, the pre-control mode is selected. If the actual common rail pressure is much higher than the target injection pressure, which means that the deviation is greater than 3 MPa, then the stopped supply mode is selected.

In the pre-control mode, the microcontroller drives the PCV valve with a larger fuel supply continuous angle T_{on}. Alternatively, when the solenoid valve is closed, more high-pressure fuel is supplied to the common rail to quickly reduce the deviation between the actual rail fuel pressure and the target injection pressure. The closed-loop control mode adopts the PID algorithm to control the fuel injection pressure, and proportional gain and integral gain are selected according to the range of deviation in order to achieve higher accuracy and a faster response. In the stopped supply mode, the ECU drives the full stroke of the PCV valve, and the high-pressure pump stops supplying high-pressure fuel to the common rail.

In the pressure control test of the common rail cavity, the rail target pressure is directly given by the microcontroller. The microcontroller calculates the fuel supply continuous angle T_{on} from the pressure difference. When the solenoid sensor detects the signal of the down end port, the microcontroller will give the flow control device the initial pulse, the electromagnetic valve will open, and the fuel in the plunger chamber of the high-pressure pump will return to the low-pressure circuit, so that the pressure cannot be established. After the microcontroller applies the pulse of T_{on}, the solenoid valve is closed and the remaining fuel in the plunger is supplied to the common rail cavity. Therefore, the amount of fuel supplied to the high-pressure rail tube is determined by the time of the power on and off of the control valve. Figure 4.80 is a sketch of the pressure control test device for a common rail cavity.

4.3.4.5 Test Results

Figure 4.81 shows the common rail pressure fluctuation with a cam speed of 600 r/min and a target pressure of 120 MPa, which is controlled by a closed-loop PID. From graph (a) and graph (b), it can be seen that the amplitude of the pressure fluctuation of the common rail chamber can be well controlled within 3% and the flow control device meets the design requirements.

The starting condition test mainly tests the time needed for the fuel pressure of the common rail chamber to reach the set value after the system is started. Figure 4.82 is the test result of a single cylinder pump with a six-cylinder engine rail tube. The target pressure value of the common rail is 30 MPa and the actual time needed for rail pressure to rise from 0 to 30 MPa is 1.3 seconds. Obviously, the requirement of a cold start of 10 seconds for a marine diesel engine is satisfied.

Figure 4.83 shows the test results of a target pressure from 90 to 60 MPa, then to 30 MPa, and after stability, from 30 to 60 MPa, and then to 80 MPa. From the graph,

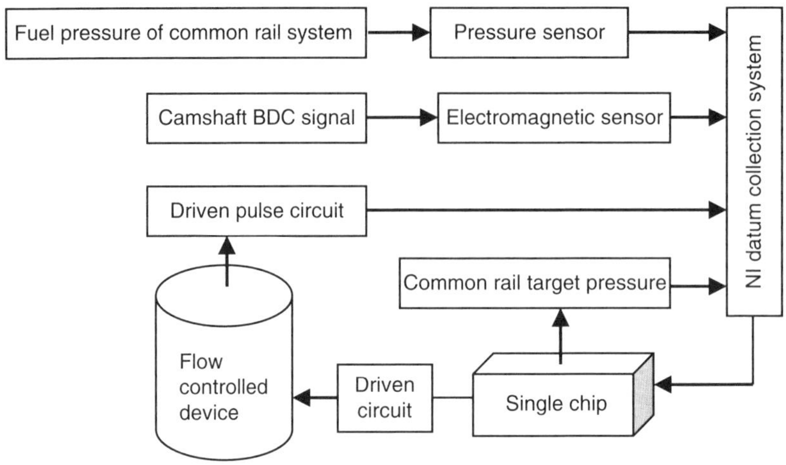

Figure 4.80 Sketch of the rail pressure control test device.

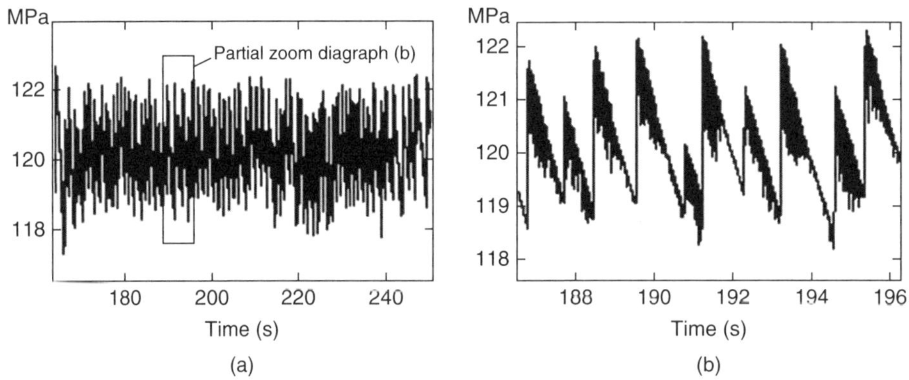

Figure 4.81 Pressure fluctuation of the common rail chamber.

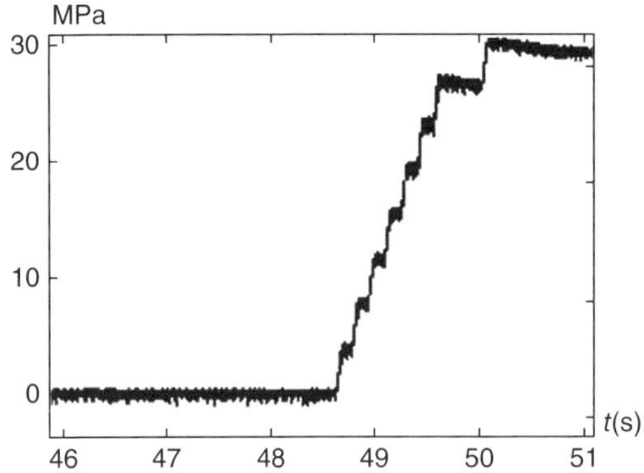

Figure 4.82 Test result of the starting condition.

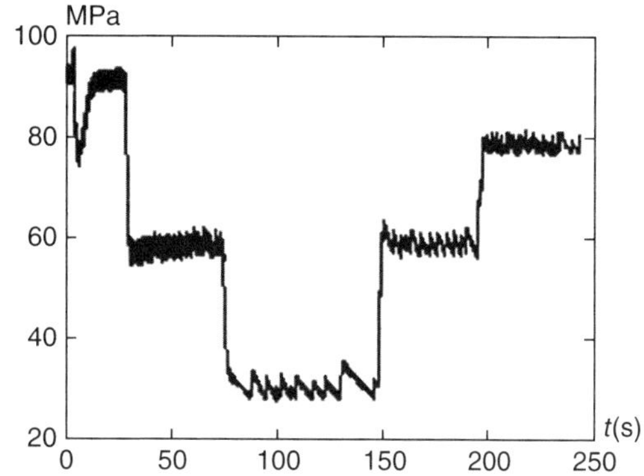

Figure 4.83 Test result of the pressure change control in the common rail cavity.

it can be seen that at the moment of 15 seconds, the measured fuel pressure decreases from 90 to 60 MPa in only 1.3 seconds, and the fuel pressure decreases in a timely manner to meet the need of the diesel engine target pressure drop control. The steady state control effect between 30 and 75 seconds is very good, while the overshoot is small. At the moment of 148.8 seconds, the measured fuel pressure increased from 30 to 60 MPa in 1.8 seconds. The pressure rises very fast and the fluctuation is not large, which meets the diesel engine target pressure rise control requirements, and the fuel pressure stability is between 151 and 195 seconds. The experimental results show that when the target pressure changes, the pressure of the common rail cavity can be quickly adjusted to the target value (the duration of the change process is no more than 1.8 seconds), and the pressure control state is stable after adjustment. The pressure fluctuation range is $\Delta P < 3\% \sim 5\%$.

Under transient conditions, the diesel engine rapidly accelerates and decelerates, or the load changes abruptly in a short time. In order to improve the dynamic response performance of the diesel engine, there must be a perfect control strategy and a precise control mode, and the flow control device must also have a good dynamic response. As for fuel pressure, a large change rate of the target fuel pressure at this time increases the difficulty of controlling the fuel pressure. When the rapid acceleration or load increases, it is necessary to accurately increase the injection pressure in time to meet the requirement of increasing the amount of fuel injection and increasing the rotation speed of the injection. It is necessary to reduce the fuel pressure in a timely manner. In fact, the actual following performance of the fuel pressure is needed so that overshoot and fuel pressure oscillation cannot appear. This experiment sets the target common rail pressure using a microcontroller, instead of the physical process of rapid acceleration, rapid deceleration, and sudden change of load. The advantage is that the target pressure value of any point can be set without the installation of an injector, which makes the test process simpler.

Figure 4.84 shows the pressure fluctuation of the common rail cavity when the target pressure is set to 60 MPa and the cam speed varies from 500 to 800 r/min. In the test, the common rail tube is not attached to the injector. The decrease of fuel pressure is mainly due to the leakage at various places and the leakage from the overflow valve. Therefore, if the duty ratio of the solenoid valve is not changed, the speed increases, the fuel supply increases, and the pressure of the rail chamber will be increased. It can be seen from the graph that the pressure of the common rail cavity can be well controlled at 60 MPa, which indicates that when the speed increases and the fuel supply

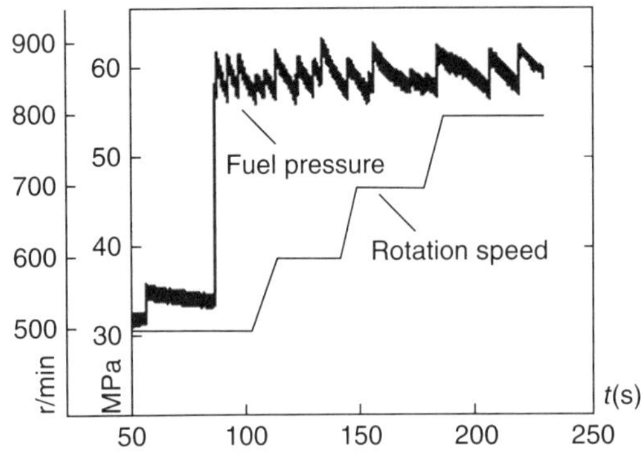

Figure 4.84 Test result of the varied speed.

increases, the microcontroller can change the duty cycle, the flow control device can change the fuel supply volume, and the pressure of the common rail chamber can be precisely controlled.

4.3.4.6 Experimental Study of the Influence of the Duty Ratio of the Solenoid Valve on the Pressure Fluctuation of the Common Rail

In the common rail pressure of the pressurization fuel stroke adjustment mode, the amounts of fuel and rail pressure are controlled by regulating the energized end time of the solenoid valve in the flow control device. Early closure can increase the fuel supply and late closure can reduce the amount of fuel supply, which means that the effective stroke of the plunger after the shutoff of the solenoid valve determines the amount of fuel supply or the pressure of the common rail tube. The opening and closing of the solenoid valve is controlled by a microcontroller pulse signal. Therefore, the fuel flow into the common rail tube can be controlled by controlling the duty cycle of the pulse signal, in order to achieve the purpose of controlling the common rail pressure.

Figure 4.85 is an experimental study of changing the influence of the duty ratio of the pulse signal to the rail pressure fluctuation on the common rail system test bench, when the cam speed is 372 r/min, without an injector, and the microcontroller set target pressure is 100 MPa. The starting point of the flow control device corresponds to the bottom end point of the plunger, the end point of the impulse signal corresponds to the starting point of the effective fuel stroke, and at point C fuel pressurization ends, corresponding to the top dead center of the plunger.

As shown in Figure 4.85a, the duty cycle $k = 17\%$ and the solenoid valve opening duration is 14 ms. During this time, although the plunger is going upwards, the fuel is not sent to the common rail tube, but passes through the low-pressure one-way valve and then flows out of the plunger cavity. At the end of the pulse signal, the solenoid valve is closed and the low-pressure one-way valve is closed. The plunger chamber establishes the pressure to pressurize fuel to the common rail tube and the high-pressure one-way valve is opened, until the plunger moves to point C to end the pressurization. The effective stroke $l = 9.48$ mm and the common rail pressure rise is 4.1 MPa. In Figure 4.85b, $k = 34\%$, the fuel supply is reduced, the effective travel $l = 7.89$ mm, and the rail pressure rise is 3.4 MPa. In Figure 4.85c, $k = 58\%$, the effective stroke $l = 4.36$ mm, and the pressure of the rail pressure rise is 2.7 MPa. In Figure 4.85d, $k = 73\%$, the effective stroke $l = 2.07$ mm, and the rail pressure rise is 1.2 MPa. The test results are slightly smaller

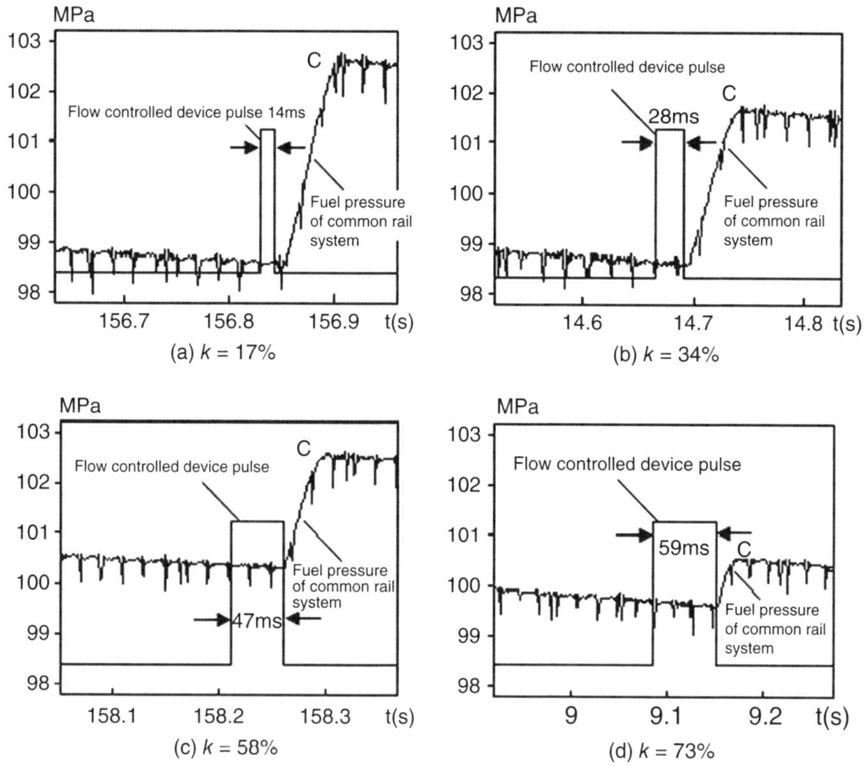

Figure 4.85 Influence of the duty ratio of the solenoid valve on the common rail pressure fluctuation.

than the theoretical results obtained by the formula (4.67), because the volumetric efficiency of the high-pressure pump equipped with a flow control device at 100 MPa is 80%. After correction, the test result is consistent with the theoretical result.

The smaller the duty cycle of the pulse signal, the larger is the effective fuel pressure stroke, the more is the fuel flow into the common rail tube, and the faster is the pressure rise of the common rail tube. The larger the duty cycle of the pulse signal, the smaller the fuel amount of the common rail tube will be, and the smaller the pressure fluctuation. In the tests with different speeds and pressures, it is found that the maximum achievable duty cycle of the solenoid valve is 86% and the rail pressure is increased by 0.7 MPa, which means that the resolution of the rail pressure controlled by the flow control device is 0.7 MPa. In a single supply cycle, the response characteristic of the solenoid valve and the maximum duty ratio are needed to activate the opening and determine the precision of the rail pressure control. In order to control the common rail pressure and improve the resolution of the flow control device at a high rotational speed, the response time of the solenoid valve must be improved and the good opening characteristics in a wide pulse area need to be maintained. In addition, the resolution of the flow control device can be improved by reducing the diameter of the plunger and increasing the plunger stroke.

5

ECU Design Technique

Normal operation of the high-pressure rail system of the diesel engine is realized by the ECU process whereby information is sent by various sensors after calculation of the scheduled program and then the parameters are limited to the allowable voltage level and are sent to the relevant executing agencies. Therefore, ECU is the core of the control of the high-voltage rail system.

5.1 An Overview of Diesel Engine Electronically Controlled Technology

The research of electronically controlled technology of the internal combustion engine started in the middle of the 1970s in the twentieth century, when the developed countries in the world began to apply it to automobiles, generator set, construction machinery, diesel locomotives, and ships. The core of electronically controlled technology is the electronically controlled injection system. Its main components include the control of oil quantity, injection timing control, injection pressure, and integrated control of the injection rate.

At present, the enforcement of environmental regulations, the application of diesel engine electronically controlled technology in the automobile industry in developed countries have been quite common, but the most advanced electronically controlled technology of the high-pressure common rail diesel engine is still only developed by a few companies, such as the German BOSCH company, Japan's Denso Company, Delphi Company of the USA, etc.

The research of diesel EFI (electronic fuel injection) technology started late in China. Since the late 1980s of the last century, the main work focused on research of the electro-hydraulic fuel advance device and the electronic governor. Since the early 1990s, many companies started to study EFI technology, and some progress has been made in control theory, system composition, and development of key components. At present, the main focus is on the actuator of the high-pressure rail fuel injection system. Tianjin University, Shanghai Jiao Tong University, and Wuhan University of Technology are all doing research on medium voltage rail accumulator EFI technology, in which Tianjin University developed the PAIRCUI system, which can better control the variable injection rate. There are many research companies studying the high-pressure rail system. Shanghai Jiao Tong University, Tsinghua University, and Guangxi Yuchai Group are all

Common Rail Fuel Injection Technology in Diesel Engines, First Edition.
Guangyao Ouyang, Shijie An, Zhenming Liu and Yuxue Li.

doing this part of the research. Yuchai Group and Delphi Company have developed the domestic high-pressure rail diesel engine EFI products, but the key to the high-pressure common rail injection system still relies on imports. According to the published data, the development of the high-pressure common rail system is still in the stage of system trial and bench test. For the control of the high-pressure rail system, due to reasons of technical security, there are few published data in China. The majority are using MAP.

5.1.1 The Development of ECU

In the development of the electronically controlled diesel engine, the research of the electronically controlled unit (ECU) has made rapid progress, mainly due to the rapid development of the electronically controlled industry and the application of various modern control technologies in the field of diesel engines. In order to meet the performance requirements of all aspects, electronically controlled technology has penetrated into various parts of the diesel engine, adopting many new techniques. The content of a diesel engine that is electronically controlled has been gradually developed from the original single control of the fuel system to the comprehensive management system including engine system control, fault diagnosis, real-time management, and other functions. Many kinds of advanced electronically controlled technology have been used more and more widely in the ECU.

5.1.1.1 The Application of Control Theory in the Research of an Electronically Controlled Unit

With the development of modern control technology, the electronically controlled mode of engine has evolved from the classical PID theory of the classical control theory to modern control methods, such as optimal control, adaptive control, fuzzy control, neural network control, and predictive control. The introduction of modern control theory makes the electronically controlled system more adaptive to a diesel engine, such as the complex multivariable system, the time-varying system, and the non-linear system. At present, there are several types of control methods and new control theories that are widely used in the engine electronically controlled system.

5.1.1.1.1 Adaptive Control and Robust Control

Adaptive control is based on theories such as state space analysis, system identification, parameter estimation, optimal control, stochastic control, and stability theory. The rapid development of electronic technology creates material conditions for complex adaptive control, which makes the adaptive control theory applicable to the ECU of the engine. Draper and Li put forward the adaptive control of the internal combustion engine in 1951. They proposed a control system that can make the internal combustion engine achieve the best performance with uncertain performance characteristics. In recent years, many researchers have done a lot of research work in the design of the adaptive control system, which have been successfully applied in engine control, thus giving the engine control system a good robustness.

In recent years, the application of model-based adaptive control technology in engine control has been developing rapidly. Adaptive control technology can greatly improve the control accuracy, which is mainly used for engine steady state control. Further research is still needed in the application of dynamic working conditions.

5.1.1.1.2 Neural Network and Fuzzy Control

The neural network is a multiple input and output system. It has strong capabilities in information integration, information complementarity, and redundancy. It has obvious advantages in the design of control schemes for multivariable, large-scale systems, and complex systems. These can meet the control requirements of the diesel engine system.

The fuzzy theory has something in common with the neural network, such as a parallel processing structure, and has the ability to use non-linear mapping. The fuzzy neural network has been a hot topic in the field of intelligent control in recent years. It not only has the characteristics of an easy understanding and is able to express human experience, but also has a parallel processing function, fault-tolerant ability, as well as the powerful self-organizing and self-learning ability of the neural network.

In foreign countries, the Japanese Toyota Corporation has applied the neural network to control earlier, and has put forward the concept of an "intelligent engine." At the same time, an experiment using neural network technology to control the engine was also carried out. In addition, M. Michael developed a neural controller to realize the real-time adaptive control of the air–fuel ratio of the automobile fuel injection system. Lenz U and others proposed a spark ignition engine control strategy with a high-precision control air-to-fuel ratio in a combustion chamber using the neural network method. G. V. Puskorius developed a feedback neural network-based controller to improve the serious non-linearity, variable delay, and dynamic characteristics of the idling process. Many universities and research institutes in China, such as The Academy of Armored Forces Engineering, Shanghai Marine Diesel Engine Research Institute and Huazhong University of Science and Technology, have also carried out research work of a neural network in engine control.

The essence of the neural network and fuzzy control is its parallel structure. At present, most controllers use a pipelining operation or superscalar structure system. The parallel computing chip is still in the research stage, far from the controller area. This makes the theory of the fast fuzzy neural network algorithm on the actual processor work slowly, and the resource consumption is large and there are still some problems to be solved when the real-time operation of dynamic process control is relatively high. In addition, the mathematical basis of both is not perfect, and there is no definite criterion and method for the performance analysis of the stability and dynamic characteristics of the system.

5.1.1.2 Function Expansion of the Engine Management System

With the rapid development of electronic computer technology and control technology, the electric control function of the vehicle engine is gradually being developed from the original fuel injection system to the engine integrated electronically controlled device with fuel injection control, turbocharging, electric control, idle speed control, intake swirl control, electric fan drive, idle speed stability control, EGR control, cold start electric control, diesel electric control hydraulic starting system, cooling control, and fault diagnosis. An electronically controlled engine has gradually formed an integrated engine management system including control, processing, fault diagnosis, and real-time management.

At present, the newly developed electronically controlled diesel engine has the functions of monitoring and diagnosis. This has become an important feature of modern engines. For example, COMMINS uses the CELECT and CENTRY electronic engine management system for comprehensive control and management of the engine. The two systems are fully compatible with INTELECT series software of COMMINS Company, and achieve continuous monitoring of environmental conditions and engine load, making it possible to control the air–fuel ratio accurately, in order to achieve the purpose of reducing emissions and improving efficiency.

5.1.1.2.1 Fault Diagnosis Function for an Electronically Controlled Engine

The complexity of a modern engine electronically controlled system is increasing. It requires the fault self-diagnosis system to be able to detect the working characteristics of the engine and monitor the working conditions of all parts of the electronically controlled system. For example, the fault diagnosis of the sensor, the fault diagnosis of the actuator, the fault diagnosis of the wiring circuit, and so on.

In 1988, the SAE, EPA, and California Air Resources Committee (CAR.B) put forward the OBD-I diagnostic system. The OBD-I system can monitor the control module, the sensors connected with the control module, the fuel meter system, the gas recirculation system, and so on. Subsequently, the standards of OBD-II were introduced. According to the regulations issued by the EPA in 1993, some light cars coming out of the factory in 1994 were equipped with the OBD-II diagnostic system. At the present time the OBD-III system is being used.

5.1.1.2.2 Field Bus Technology

The fieldbus control system is an open distributed control system, which makes the system fully digital computing and has a digital communication capability. Structurally, it uses a fully distributed scheme to put the control function completely on the site. It improves the reliability and flexibility of the system, so it has many advantages compared with the traditional distributed control system and is a field communication network where devices can communicate in a variety of ways: point to point, point to many points, or broadcast. The unified configuration and task download of the field enable simple control tasks such as PID, digital filtering, compensation processing, and other simple control tasks to be dynamically downloaded to the field equipment. It can reduce the amount of transmission lines and hardware equipment, and save the cost of system installation and maintenance. It also enhances the interoperability and interchangeability of the equipment of different manufacturers. At present, there are a variety of fieldbuses: Foundation bus, LonWorks, Profibus, HART, CAN bus, etc.

CAN is the control local area network, which is introduced by the German BOSCH Company, and is used for data communication between the automobile internal measurement and the execution parts. Since its introduction, CAN has been attached to major companies. Some famous automobile manufacturers in the world, such as BENZ, BMW, PORSCHE, and ROLLS.ROYCE, have begun to use the CAN bus to realize data communication between the vehicle internal control system and testing and executing agencies.

5.1.1.2.3 Sensor Technology

In recent years, with the rapid development of control technology, computer technology, and broadband network technology, more and more applications of multisensor systems have been applied. Higher requirements are needed for the sensor, namely, high accuracy, good reliability, easy networking, and primary intelligent functions such as information processing and self-inspection.

With the application of new materials and new technologies, electronically controlled systems require sensors to be integrated and intelligentized. The intelligent sensor has the functions of signal amplification, processing, various compensation functions, and a self-diagnosis function. Its intelligent part should also have the ability of coding and decoding. The multifunctional sensor has the Holzer effect injection starting point sensor, developed by the Wolf Company in the United States. The internal integrated circuit includes Holzer components, amplifying circuits, a comparison shaping circuit, a zero adjustment circuit, a temperature compensation circuit, a voltage regulator, an output driver, and a monolithic pressure sensor developed by Japan, which is made on a silicon chip. The non-linear compensation sensor and the sensor with a self-diagnostic function are also studied.

The network intelligent sensor is a new generation of sensors with embedded technology as the core and an integrated sensor unit, a signal processing unit, and a network interface unit. The development trend is of miniaturization, intellectualization, and networking in the related fields of industrial process automation, power system automation, and other related fields. With the introduction of embedded technology, the power consumption of the sensor is reduced, the volume is reduced, the anti-interference and reliability are improved, and the requirements of the engineering application are more able to be met. The application of software technology such as non-linear compensation, zero drift, and temperature compensation allow the sensor to have high linearity and precision, while the application of network interface technology enables the sensor to access the network conveniently. The point-to-point connection between the sensor and the specific measurement and control equipment reduces the complexity of the field wiring and the quality of the cable.

5.1.1.3 Development of Computer Hardware Technology

The ECU is the core of the diesel engine electronically controlled system, and its hardware core is the microprocessor. At present, in addition to the commonly used 8 and 16 bit microprocessors, the 32 bit, and especially the 64 bit, microprocessors have been gradually used in the engine ECU. In recent years, in the diesel engine control system developed recently in the world, the CPU is basically using the 32 bit microprocessor. For example, the latest development of the Japanese Isuzu Motor Ltd. 4JxlTc diesel engine electronically controlled system ECU adopts the 32 bit microprocessor. The ECU of the MTU 2000 and 4000 series diesel engine electronically controlled device uses the Motorola 683XX Series Microprocessor.

5.1.2 Development of Electronically Controlled System Development Tools and Design Methods

5.1.2.1 Application of Computer Simulation Technology

In the development of the engine electronically controlled system, the development of the control algorithm is very important. The on-line adjustment and adjustment of the control algorithm is very expensive and time-consuming. In the design process, the system simulation technology can be used to reduce the cost and speed up the development speed. The system simulation technology runs through the whole development, debugging, and improvement process of the electronically controlled system. For example, at the beginning of the design, the simulation technology is used to demonstrate the scheme, and the technical comparison is carried out to optimize the reasonable scheme. In the debugging stage, the relationship between the system response and the parameters is analyzed by the simulation technology.

In recent years, with the rapid development of computer hardware and software technology, more commercial simulation software systems are appearing. Among them, MATLAB from American Math Works is quite popular. One of the components it carries, Simulink, is a software package for modeling, simulation, and analysis of a dynamic system. It supports continuous, discrete, and mixed linear and non-linear systems. It also supports systems with multiple sampling frequencies. It is a universal simulation software. At the same time, some special simulation software systems have emerged in the field of engines. For example, the en-DYNA software of the German Resis Company is a model base of engine dynamics analysis based on the MATLAB platform. It can build a real-time engine model from the model base. The output torque of each cylinder can be calculated by a semi-physical model. The pre-processing module can produce the general engine model and combustion data, so that the latest ECU can be tested.

In addition, the EASY_5 software of Boeing Company is used for engineering analysis, and the AMESim software of France's IMAGINE has successfully designed in fuel injection, braking system, power transmission, cooling system, and so on. These are powerful examples that the simulation technology has played a more and more important role in the design of the engine and its electronically controlled system. The application of advanced simulation software can greatly reduce the time spent by simulation workers in modeling and programming, in order to facilitate the improvement of system performance.

5.1.2.2 Computer-Aided Control System Design Technology

Computer-aided control system design (CACSD) can not only control the design and offline simulation but can also establish a real-time rapid control prototype (RCP) and promote the transformation from the verified design to the product controller and hardware-in-the-loop simulation (HILS).

The RCP is the rapid establishment model of control objects and the controller, which can perform multiple times for off-line and on-line experiments for the whole control system to verify the feasibility of the software and hardware schemes of the control system. After the designed controller is put into production, it must be tested in detail before it is put into the market. Now the HILS is widely used. When the system is tested, the controller is real and the rest should use the actual products if possible. If the real product is not available, then we should use the real-time digital model to simulate the external environment of the controller and test the whole system.

Many special hardware and software products have been launched abroad. One of the typical products used for the development of ECU is the dSPACE real-time simulation system developed by the German dSPACE Company. It is a software and hardware platform based on MATLAB/Simulink/RTW control system development and hardware/software in the loop simulation. It integrates the development, programming, and testing of ECU in a unified environment, which can accelerate and simplify the development process and quickly implement the modification of control parameters and can link with the OSEK standard real time operating system.

Audi's engine HILS uses this working platform. In order to test the engine ECU in an interactive and automatic way, all ECU operation modes can be completed in one night, so the development cycle of ECU is greatly reduced.

5.2 Overall Design of the Controller

The design of a diesel engine ECU can be divided into two parts: hardware design and embedded software design. There is a strong correlation between the hardware design process and the software design process. Strong hardware support is the premise of implementing the core control strategy of ECU, and excellent software programming is the guarantee of giving full play to the advantage of hardware performance. Only when software and hardware cooperate well can we design ECU with excellent performance.

In order to achieve the best combination of software and hardware, the hierarchical design method of top-down design is used in this paper to design a layered ECU. To determine the structure of the underlying system according to the requirements of the upper level ensures that the performance of the system meets the design requirements.

5.2.1 Controller Development Process

The design object of this section is to examine the electronic controller of the oil supply system of the high-pressure rail diesel engine. It is a kind of electronic equipment with a certain control ability and the development process of the conventional electronically controlled equipment is as follows:

(1) The requirements and design goals are defined in a word description way according to the survey.
(2) Put forward the structure of the system according to past experience.
(3) Hardware design and processing of hardware circuits should be done by the hardware engineer.
(4) The control scheme is designed by the control engineer and the control law is described in the form of an equation.
(5) The control law is realized by manual programming by a software engineer.
(6) Integration of the code into a hardware circuit is peformed by a system engineer or an electronic expert.
(7) The system is tested using real control objects or test stations.

The flow chart of the design is shown in Figure 5.1.

This section gives a preliminary research on the principle of fuel supply and electronically control of the high-pressure rail diesel engine, which is far removed from the actual development process of a diesel electronically controlled system. In the design of hardware and software of the controller, there are relatively few considerations given about the adaptability of the working environment and the electromagnetic compatibility testing of the equipment. Therefore, the research work in this section does not include an environmental test, electromagnetic compatibility testing, an ECU engine bench test, and any subsequent development process.

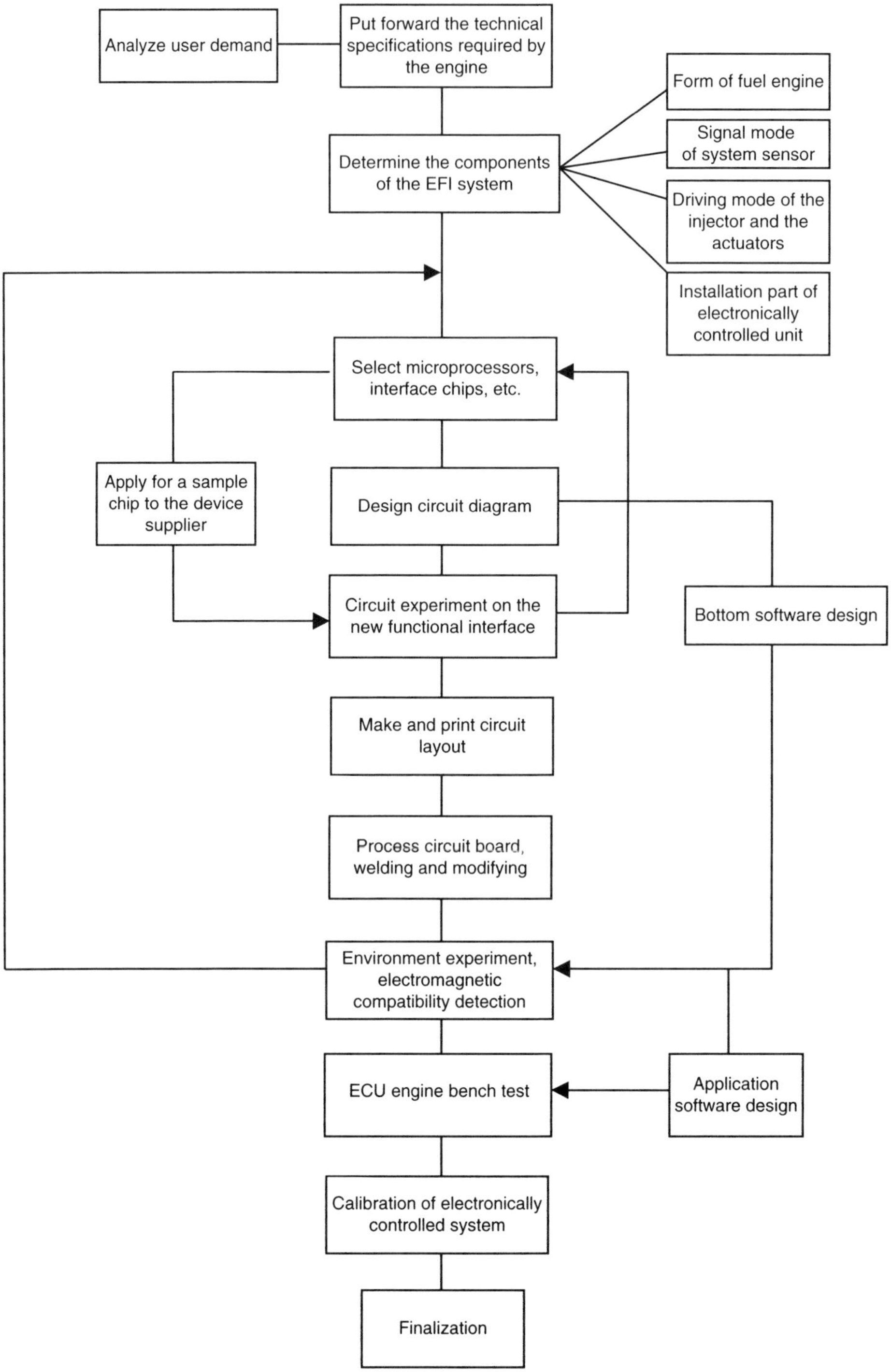

Figure 5.1 ECU design flow chart.

5.2.2 Hierarchical Function Design and Technical Indicators of the Controller

The working principle of the high-pressure rail system is completely different from that of the traditional mechanical fuel system, and has a higher flexibility. On the one hand, this flexibility provides a possibility for optimization of the diesel engine operation index and intelligent development of the diesel engine. On the other hand, this flexibility gives higher requirements for its electronically controlled components. If the electronic controller does not synchronize with the mechanical part of the diesel engine according to the requirements, the whole diesel engine will not run. In order to ensure normal operation of the high-pressure rail diesel engine, use is made of its advantages of a wide control range and high flexibility. The electronic controller of the high-pressure common rail system must first be able to control the basic power output function of the diesel engine, such as starting, speed regulation, steady speed, and so on. On the basis of realizing the basic functions, we also need to realize the functions of increasing the output power, reducing operation pollution, improving work economy, and improving diesel engine performance and life. Finally, in order to ensure reliability and maintainability, additional functions such as fault self-diagnosis, bus communication, and other functions need to be added.

According to the above requirements of the controller function, combined with current research practice of high-pressure rail technology, the electronic controller designed in this section has the function structure shown in Figure 5.2.

(1) Conversion of the diesel engine operative condition control module.
The top layer of the hierarchy is conversion of the diesel engine operative condition control module. The module determines the state of the diesel engine based on the information input from the man–machine interface, the communication interface, and the sensor group. Then the lower layer modules, including start, warming, acceleration and deceleration control, and speed stabilizing control modules, are invoked to control the operation of the diesel engine.

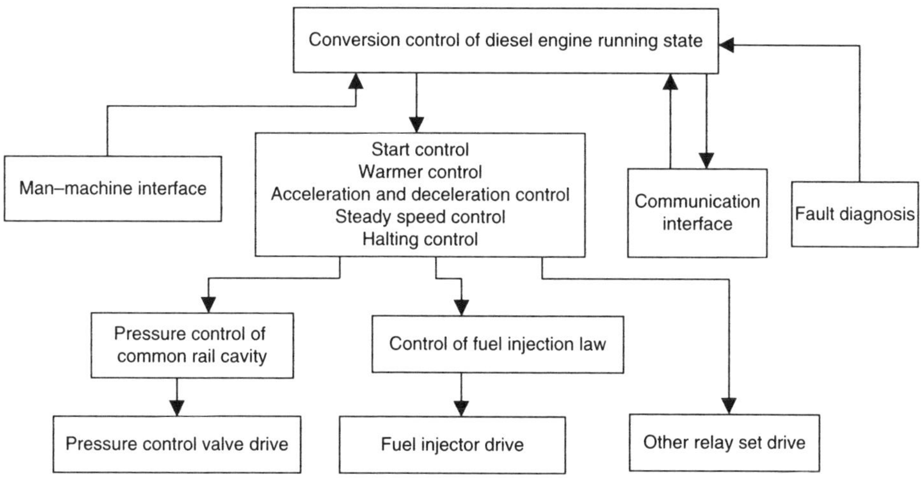

Figure 5.2 Function structure diagram of the controller.

(2) Start control module.
The starting process in this paper refers to the process of the diesel engine from static acceleration to the minimum stable speed process. The control contents of the process include the control and timing protection of the starting motor, the establishment of the oil pressure in the rail chamber, and the oil supply control in the starting process of three items. For starting control, it is required to start reliably, stably, and rapidly.

(3) Warming control module.
The warm machine process in this section is when the diesel engine is started by the freezing state, in order to ensure stable operation for the operation life of the diesel engine, and the automatic running stage of the diesel engine is carried out. This stage is based on the temperature of the cooling water and the oil. When the cooling water temperature or lube oil temperature is below a set temperature, ECU will enter the warming process automatically until the cooling water temperature reaches the set minimum steady working temperature. The process can also be manually terminated to meet the requirements of a special case.

(4) Acceleration and deceleration control module.
The acceleration and deceleration control module is the main process of the diesel engine control system. Since the ultimate control target of the designed ECU is revolution speed, the quality of the two control processes is directly related to the running quality of the diesel engine.
The aim of a steady speed control is to optimize operation parameters under the premise of a stable speed. The goal of acceleration and deceleration control is to require an acceleration and deceleration process to be quick with a small overshoot. Because the system does not carry out the actual calibration and the matching test, these two specific indicators cannot be directly verified, so the specific indicators are not given.

(5) Shutdown control module.
The shutdown process is the process of the diesel engine stopping the output power to the complete stationary process of the diesel engine. The control of the process is simpler than the other processes. However, in order to ensure the next successful start and extend the service life of the diesel engine, the normal shutdown process is not just to stop the oil supply but to do it in a series of steps. The downtime control program is set up for the automatic execution of the process.

(6) Man–machine interface module. The operation of the diesel engine needs manual control, so a good man–machine interaction interface is essential. Strictly speaking, the man–machine interaction interface is divided into a machine side control interface and a remote control interface. A telecontrol interface transmits information to a diesel engine controller through a communication interface. The side control interface is located directly near the diesel engine, usually the switch, the knob, the indicator, and the display screen, etc.

(7) Communication interface module. In order to realize the information transmission of the diesel engine and its upper control system, the communication function is essential. ECU receives the instructions from the superior control system and sends the data to the superior system through the communication interface module. The communication interface module can also be used to calibrate the diesel engine.

(8) Fault diagnosis module. Fault diagnosis technology involves multidisciplines and is a difficult function to realize. At present, the actual application of self-diagnosis technology is still at a low level. According to the actual situation, the controller designed in this section only realizes the basic functions of sensor fault self-diagnosis, diesel engine running status alarm, and so on.

Based on the requirements of the above control targets, the specific functions of the rail chamber pressure control module and the fuel injection law control module at the bottom of the functional structure are as follows:

(1) Pressure flexible control of the common rail chamber. The pressure of the fuel in the rail can be controlled by ECU. Generally speaking, the high common rail chamber pressure will bring a better atomization effect, create a better combustion environment, improve the system's economy, and optimize the emission performance of the system. In fact, considering the various factors in the operation of the diesel engine, the pressure of the rail chamber is not always high enough. The higher the pressure of the rail chamber, the higher is the energy consumption of the high-pressure pump, reducing the service life of the high-pressure fuel system, making it more difficult to establish the pressure of the common rail chamber of the diesel engine, and increasing the fuel consumption of the slack and low-speed processes. Therefore, in order to improve the overall performance of the diesel engine, it is necessary to optimize the pressure of the common rail chamber for different working conditions. According to the working capacity of the high-pressure oil pump, the adjustment range of the pressure of the high-pressure common rail chamber is 30–150 MPa. In order to ensure the stability of the working process, the fluctuation of pressure in the working process should be no more than 3% of the maximum pressure.
(2) Flexible control of the fuel injection law. The flexible and accurate control of the oil supply law can accurately achieve multiple injection of oil using the time control of the oil supply and fuel supply. Realization of this function is the greatest embodiment of the flexibility of the high-voltage common rail system and the basis for the optimization of the working process of a high-voltage common rail diesel engine.

According to the time series analysis of the combustion process of the diesel engine, the index of the flexible control of the fuel injection law is determined as follows:

(1) The two-stage fuel injection with a pre-injection and a main injection can be realized.
(2) A flexible and adjustable injection advance angle in the range of −58° to 20° is achieved.
(3) The oil supply control pulse width is up to 0–25 ms and the control precision is 0.2 μs. The actual single cycle fuel supply can be adjusted within the range of 25–200 mg and the error of the supply is less than 3% of the maximum oil supply.

5.2.3 Input Signal

According to the character of the input signal, the input signal of the controller can be divided into two types: the man–machine interface input signal and the sensor input signal.

5.2.3.1 Man–Machine Interactive Interface Input Signal

The input signal of the human–machine interaction interface includes the start switch signal, the stop switch signal, the automatic/manual switch signal, the operation state selection switch, the speed setting knob signal, and the common rail pressure manual control knob signal.

According to the properties of these signals, it can be divided into two types: switching signal and continuous signal.

5.2.3.1.1 Switching Signal

The switch signal includes the starting switch signal, the stop switch signal, and the automatic/manual switch signal. The start switch signal occurs by the starting switch. The start switch is a non-self-locked button switch used to control the start of a diesel engine. The signal is only valid when the diesel engine is waiting to start, as the signal is invalid during operation of the diesel engine. The stop switch signal occurs by the stop switch, and the stop switch is a non-self-locked button switch used to control the parking of the diesel engine. An automatic/manual switching signal, a total of two gears, is used to control the operation of the diesel engine. After selecting the automatic gear, the process of starting the diesel engine is automatically controlled. The controller will choose the warming process according to the working condition. After selecting the manual gear, the diesel engine can control the revolution speed manually after the engine is started.

These signals are all input through a corresponding switch on the outside. Through the switching signal input circuit, the external open and off signals are converted to 0 and 5 V voltage signals. In general, the switching process is accompanied by vibration. In order to ensure the stability of the signal received by the control core, the hardware anti-chattering circuit is added.

5.2.3.1.2 Continuous Signal

The continuous signal includes the speed setting knob signal and the manual control knob signal of the common rail pressure. The speed setting knob is used to set the target speed of the diesel engine and the common rail pressure manual control knob is used to manually set the pressure control of the common rail chamber. The continuous signal is a 0–5 V voltage signal generated by partial pressure of a potentiometer.

5.2.3.2 Sensor Input Signal

The sensor input signal includes the inlet temperature sensor signal, the cooling water temperature sensor signal, the lubricating oil temperature sensor signal, the inlet pressure sensor signal, the common rail chamber pressure signal, the relief valve drive current signal, the crankshaft position signal, and the TDC signal.

5.2.3.2.1 Temperature Input Signal

The temperature input signal includes the inlet temperature signal, the cooling water temperature signal, and the lubricating oil temperature signal. The controller receives these temperature signals through a sensor mounted on the components of a diesel engine. The input sensor signal is set to a 0–5 V DC voltage signal after a conditioning circuit. Through the A/D conversion, the controller can obtain the temperature parameters of the corresponding point.

5.2.3.2.2 Pressure Input Signal

The pressure input signal includes the inlet pressure signal and the common rail chamber pressure signal. Through the pressure sensor installed in the intake pipe, the controller can obtain the pressure data of the inlet pipe. Combined with the intake temperature data, the controller can calculate the air volume in the cylinder in order to optimize the target oil supply. Through the pressure sensor installed on the common rail chamber, the controller can obtain its pressure parameters. On the one hand, the pressure parameter is used to provide feedback for the pressure control of the common rail chamber and, on the other hand, it is used to determine the pulse width of the control signal of the fuel supply.

Because the pressure data of the common rail chamber is very important to the operation of the diesel engine, once the sensor fails, it will affect the basic operation of the diesel engine. Therefore it is very important to judge the state of the common rail pressure sensor. Once the sensor fault is determined, the control strategy of the common rail chamber pressure is converted into an open loop strategy at the same time as an automatic alarm is turned on. According to the current of the solenoid valve through the pressure of the common rail chamber, checking the MAP diagram of the open loop of the current/pressure and determining the pressure of the common rail chamber should ensure that the diesel engine can still work at this time.

5.2.3.2.3 Pulse Input Signal

The pulse input signal includes the crankshaft position signal and the top dead center signal, both of which are the key signals and are essential to the normal operation of the diesel engine. The crankshaft position signal is output from the magnetoelectric sensor mounted on the crankshaft body. The structure of the magnetoelectric sensor is shown in Figure 5.3.

The way it is installed is shown in Figure 5.4. When the iron gear rotates, the teeth change across the front of the magnetoelectric sensor probe, which changes the magnetic resistance of the sensor, thus changing the flux of the flow through the coil so that the induced electromotive force is generated on the coil.

As each tooth is crossed over the sensor, the sensor will output a periodic sine wave. Therefore, each turn of the crankshaft will produce a sine signal of 60 cycles. The top dead center signal is also produced by the magnetoelectric sensor mounted on the body above the protruding shaft. When the crankshaft is transferred to the first 70° position of the top dead center point of the first cylinder, the sensor will output a sinusoidal signal of a cycle. Through a signal conditioning circuit, the output sinusoidal signal is converted to a pulse signal with a steep rising edge and a descending edge. These two signals are shown in Figure 5.5.

The pulse generation circuit of the controller determines the time when the oil control pulse occurs according to the top dead center signal and the crankshaft position signal; the specific process will be introduced in the next section.

5.2.4 Output Signal

According to the power of the output signal, the output signal can be divided into two kinds: power output signal and data interface signal. The starting motor control switch signal, the driving signal of the injector solenoid valve, and the common rail pressure

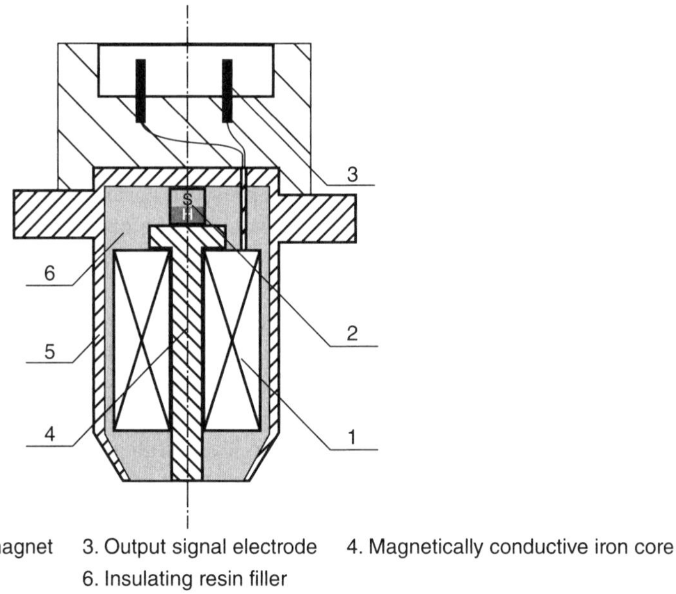

Figure 5.3 Magnetoelectric sensor structure: 1. Coil, 2. Permanent magnet, 3. Output signal electrode, 4. Magnetically conductive iron core, 5. Magnetic shell, and 6. Insulating resin filler.

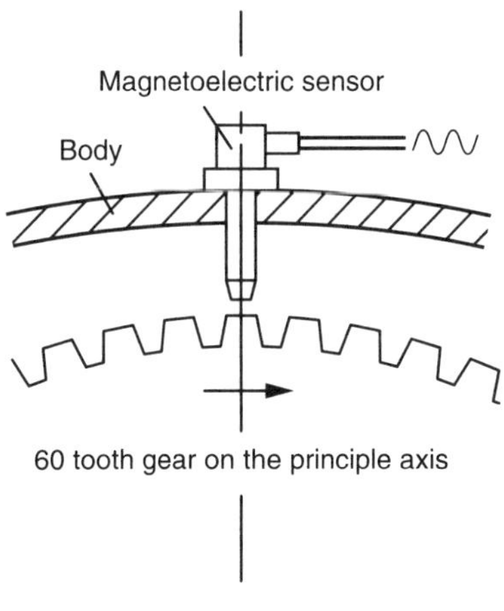

Figure 5.4 A schematic diagram for measuring the position of a crankshaft.

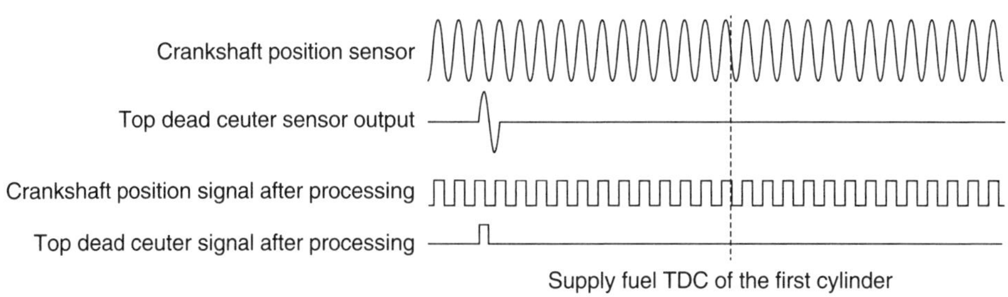

Figure 5.5 Pulse signal.

control solenoid valve drive signal all belong to the power output signal. The display output interface signal and the communication interface signal belong to the data output signal.

5.2.4.1 Starting Motor Control Switch Signal

The starting motor control switch signal is produced by the starting motor relay. When the diesel engine is waiting to start, the user presses the start button and the controller turns to the start state, outputting the starting signal of the starting motor, driving the starting motor relay coil, and connecting the starting motor. The controller will disconnect the starting motor relay when the user loosens the start button or the starting motor will run more than its single maximum working time. In the running state of the diesel engine, the control switch signal of the starting motor will remain closed.

5.2.4.2 Drive Signal of the Electronically Controlled Injector

The controller designed in this section can output the driving signal of the six-way electronically controlled injector at the same time. These signals will drive the solenoid valve of the electronically controlled injector to control the injection of the injector. In order to meet the needs of the electronically controlled injector, the demand for the driving signal is very high.

5.2.4.2.1 Time Precision Requirements

The purpose of the time precision of the driving signal of the electronically controlled injector is to achieve accurate injection timing and injection control. Therefore the time precision requirements can be divided into requirements of the driving current start time and the requirement of the pulse width of the driving current.

The injection timing is generally described by the injection advance angle. In this example, the precision index of the injection timing is a 0.5° crankshaft angle. In the case of the rated speed of 1500 r/min of the diesel engine, the time of the 0.5° crankshaft angle corresponding to the diesel engine is 56.5 μs. In the design of the controller, the crankshaft angle and the injection time are connected by the crankshaft position signal. The crankshaft position signal has the character of every 6° for a cycle; to achieve the 0.5° fuel supply timing accuracy, in this example the crankshaft position signal cycle is equally divided into 64 parts. The control target resolution of the timing of the oil supply is $6/64 = 0.09375°$ and meets the requirement for the 0.5° crankshaft rotation angle.

The electronically controlled injector can be regarded as a solenoid valve and a hydraulic valve in series, and the opening of the solenoid valve and the opening of the hydraulic valve are both delayed. Even in the same injector, the delay of the two parts is not fixed and the opening delay of the solenoid valve is influenced by the size of the driving current. The delay of the hydraulic valve is influenced by the pressure of the common rail chamber. The effect of these delays must be fully considered when the controller outputs the starting signal of the electronically controlled injector and is compensated automatically. In this way, the correct timing of the oil supply is guaranteed. Using the calibration test, the injection delay time of the electronically controlled injector is 1.38–1.78 ms when the pressure of the orbit is within the range of 80–150 MPa and the driving voltage is 120 V. In order to achieve accurate timing of the oil supply, the load time of the drive current must be accurately controlled. The delay compensation timing clock frequency is 5 MHz and the accuracy of time compensation is 0.2 μs, which can meet the requirements.

In addition to the time delay of the injection, the electronically controlled injector also has a stop-injecting delay. When the driving current is cut off, the injector will be delayed for a period of time to stop the injection. The delay time is closely related to the size of the driving current of the solenoid valve and the pressure of the injection. The stop-injecting delay is the main factor affecting the pre-injection control and is also very important for determination of the main injection pulse width.

5.2.4.2.2 Current Waveform Requirements

In order to improve the opening speed of the solenoid valve and reduce its calorific value, there are certain requirements for the driving current waveform of the injector solenoid valve. The ideal drive current waveform is shown in Figure 5.6. It is required that the starting current can be quickly loaded and the solenoid valve can be quickly actuated. The maximum starting current should meet the requirement of the starting of the solenoid valve. It is necessary to reduce the current and reduce the impact between the armature and the limit plate before the solenoid valve actuation. When the solenoid valve is completely actuated, keep the current in a stable state. Moreover, under the condition of reliable actuation, the current should be kept as small as possible, which will improve the closing response speed of the solenoid valve, reduce the calorific value of the solenoid valve, and increase the life of the solenoid valve.

5.2.4.2.3 Power Requirements

In order to ensure that the solenoid valve can be quickly sucked up, the armature must be able to get enough energy in a short time. According to the elastic coefficient K of the spring of the solenoid valve, the initial compacting force F_0, and the displacement l of the armature of the solenoid valve, the minimum energy W required for each absorption of the solenoid valve can be calculated:

$$W = F_0 \times l + \frac{K \times l^2}{2} \tag{5.1}$$

To release the energy in a very short time (about 0.3 ms), the instantaneous power is very large. Considering the energy loss caused by heating the solenoid valve coil, the actual driving energy is larger than W. Therefore, the actual instantaneous power

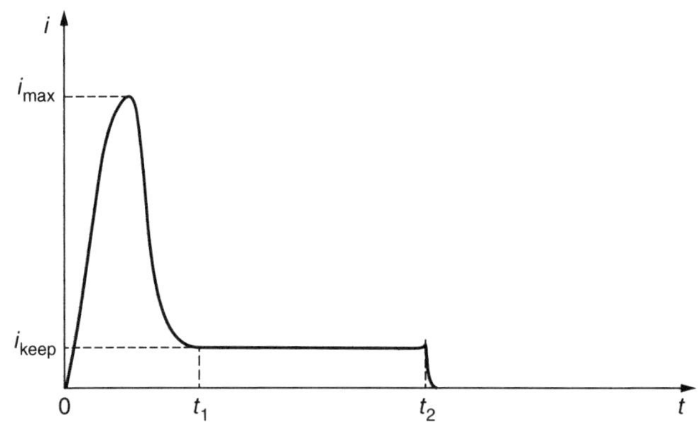

Figure 5.6 Ideal driving current waveform.

is much bigger. Due to intermittent work, the average power that the solenoid valve needs is actually quite small. Obviously, the average power and its matching ordinary power cannot meet the requirement of this special large current pulse load, so we need to design the drive circuit separately so that it can keep the current running smoothly.

5.2.4.3 The Driving Signal of the Solenoid Valve Controlled by the Common Rail Chamber Pressure

The proportional electromagnetic relief valve produced by the improved proportional valve factory of a Ningbo electrohydraulic proportional valve is shown in Figure 5.7.

The electromagnetic force is proportional to the driving current. In order to ensure the stability of the electromagnetic force, the valve needs to provide a constant driving current. According to the calibration test of the common rail chamber pressure control solenoid valve, the relationship between the driving current and the control pressure of the solenoid valve of the proportional solenoid valve is shown in Figure 5.8 when the oil temperature is 20°.

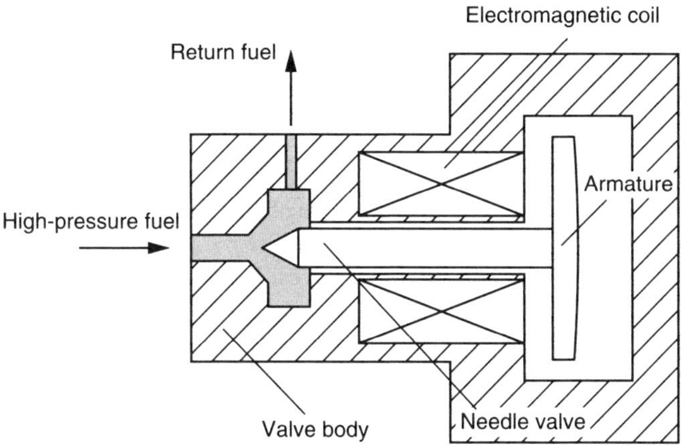

Figure 5.7 Electromagnetic relief valve.

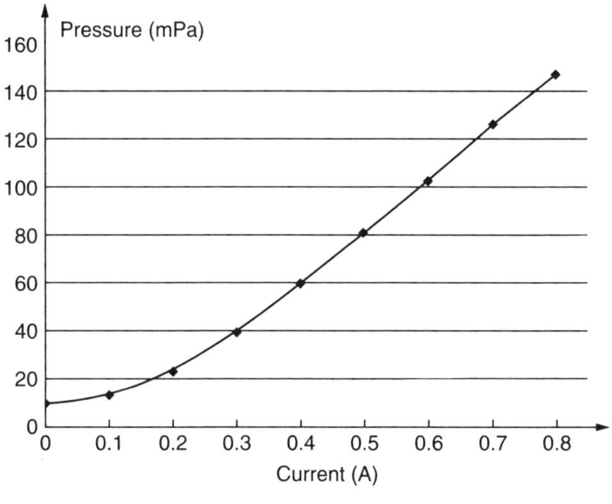

Figure 5.8 The relationship between the driving current and pressure.

To ensure that the pressure of the common rail chamber can be controlled in the range of 30–150 MPa, its driving circuit should provide the adjustable driving current of 0–0.9 A. In order to ensure that the pressure fluctuation of the common rail chamber is less than 3 MPa, the fluctuation of its driving current should be less than 0.013 MPa.

5.3 Design of the Diesel Engine Control Strategy Based on the Finite State Machine

The control strategy research of the diesel engine has been a hot topic in the diesel engine industry. Because the diesel engine is a complex, multivariable and time-varying non-linear system, its operation and management control processes are quite complex. This section describes the control process of a marine common rail diesel engine using the hierarchical finite state machine (FSM) method and designs a control strategy for a high-pressure common rail system suitable for a marine diesel engine.

5.3.1 Brief Introduction of the Finite State Machine

5.3.1.1 Finite State Machine Definition

The FSM is also known as the finite state automation (FSA). It is a tool for modeling the behavior of an object. Its role is to describe the state sequence of an object in its life cycle and how to respond to various events from outside.

An FSM is an abstract machine composed of a set of states, a set of inputs, a set of outputs, a set of state transition functions, a set of events, and a set of action.

State. A condition in which the object is in its life cycle. The object in a specific state will meet certain conditions, perform certain actions, or wait for certain events.

Event. It takes a certain position in time and space and is a meaningful thing for the state machine. Events usually cause changes in the state, prompting the state machine to switch from one state to another.

Transition. A relation between two states indicates that an object will perform certain actions in the first state and will enter second states when an event occurs while a specific condition satisfies.

Action. These are atomic operations performed in state machines. The so-called atomic operation means that they cannot be interrupted by other messages while they are running, and they must be carried on continuously.

An FSM can be defined as a five-tuple of $M = (Q, \text{sigma}, \text{delta}, Q_0, F)$:

(1) $Q = \{Q_0, Q_L, \ldots, Q_n\}$ is a set of finite states. At any given time, the FSM can only be in a determined state Q_i.
(2) $\Sigma = \{\sigma_1, \sigma_2, \ldots, \sigma_m\}$ is a set of finite input. At any given time, the FSM can only receive a determined input σ_j.
(3) $\delta{:}Q \times \Sigma \rightarrow Q$ is state transfer function. If, in a certain time, FSM is in a state of Q_i in Q and receives an input σ_j, then $\sigma_j \in \Sigma$. Then the next moment will be in a state of determination $q' = \delta(q_i, \delta_j) \in Q$. The $q = \delta(q, \varepsilon)$ is defined here. For any state q, the FSM does not have any state transfer when an empty input is received.

(4) $q_0 \in Q$ is the initial state of the FSM, which receives the input state. The FSM starts receiving input from this state.
(5) $F \in Q$ is the set of the end state. An FSM no longer receives input after reaching the final state.

5.3.1.2 State Transition Diagram

A state transition diagram is a form of describing the working process of an FSM. For the transfer function δ for a finite state machine $M = (Q, \Sigma, \delta, q_0, F)$, we use a circle (node) to represent the state. The state of the existence of the transfer relationship is connected by a directed arc and the corresponding input is tagged by the side arc. The initial state of a node is marked with an arrow. The directed graph, established by the above rules, is called a state transfer graph. In addition, the state conversion table and the state transition matrix are also used.

5.3.2 Design of the Operation State Conversion Module

As a power device, the working process of a diesel engine is complex, but in general the operation process can be described by a limited numbered state transfer. The design of the system divides the engine working state into seven kinds: the shutdown waiting state, the startup state, the warming state, the acceleration and deceleration state, the steady speed state, and the fault state. In the steady state, it is divided into three modes: fuel economy model, low emission mode, and normal mode. When the oil machine is in each of these states, the operation of the state is realized by the corresponding control module.

The working process of the controller can be represented by the state transition diagram, as shown in Figure 5.9.

State 1. The state of self-examination, interface initialization, and waiting for startup
State 2. Startup state
State 3. Warming state
State 4. Acceleration and deceleration state
State 5. Steady speed state
State 6. Normal shutdown state
State 7. Fault processing state
State 8. Emergency shutdown state

According to the state transfer relationship shown in Figure 5.9, the state transfer process of a normal diesel engine running process is as follows.

First, when the controller is energized, the system automatically enters "State 1." In this state, the system will perform a self-checking program. First check the control core program memory and MAP diagram memory and then check the supply voltage and the state of each sensor. When all the devices are confirmed, the communication interface is initialized. When these processes are finished, the controller is in a state waiting to start.

When the user presses the start button, the controller enters the start state. According to the requirements of the diesel engine starting process, the driving relay coil is driven, the starting motor is connected, the common rail chamber pressure is established and the oil supply is supplied according to the start of the MAP diagram.

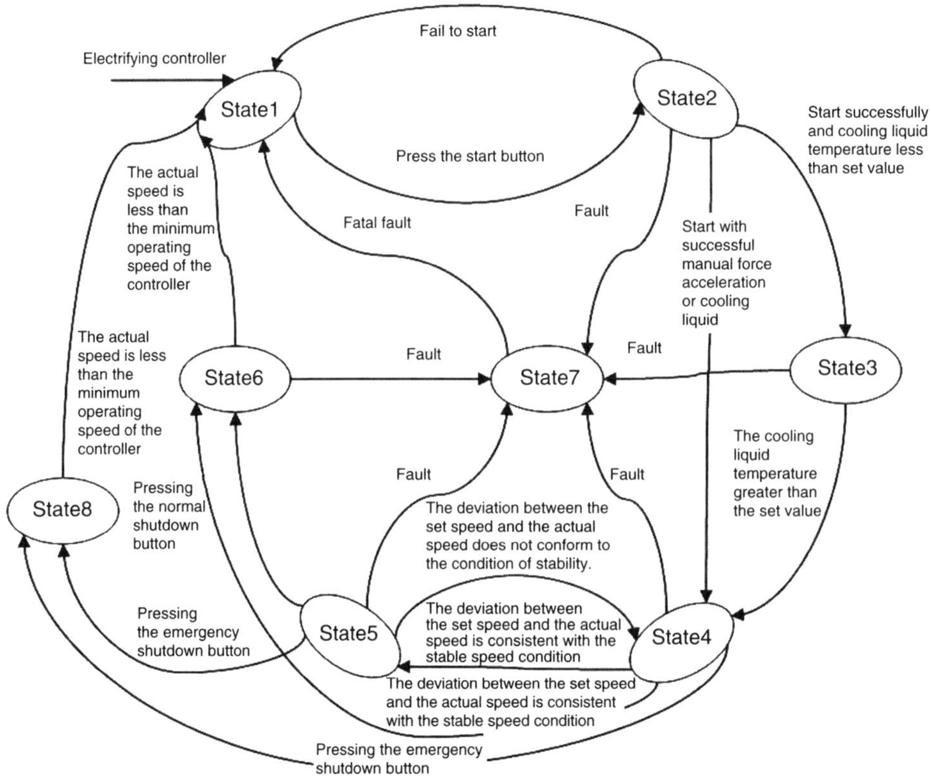

Figure 5.9 State transfer diagram of the top layer of the controller.

When the startup is successful, the controller determines whether the diesel engine is in the warming state according to the temperature of the cooling water. When the cooling water temperature is low, the system will automatically turn into the warming state. At this time, the system will control the diesel engine according to the requirements of the warming machine. When the cooling water temperature is in accordance with the normal working temperature, the system is transferred to the acceleration and deceleration state, adjusting the speed to an appropriate speed set by user. If the user is set to manual control, the system will skip the warming process, directly enter the acceleration and deceleration state, and adjust the speed to the set speed of the user.

When the speed adjustment is stable, the deviation between the actual speed and the set speed meets the condition of stability. The system automatically moves into the steady speed state and maintains the stability of the diesel engine speed according to the steady speed MAP diagram. In this process, if the deviation between the actual speed and the set speed exceeds the steady speed range, the system will automatically enter the state of acceleration and deceleration. Adjust the oil supply rule by adjusting the MAP diagram according to the dynamic condition until the speed is adjusted to the range of the steady speed control.

When the diesel engine has finished working and the user presses the shutdown button, the controller turns to the shutdown state. The diesel engine will run for a period of time before shutdown to prevent a bubble forming in the oil supply system, the lubrication system, and the cooling system, where a bubble is caused by the sudden breaking of oil. If an emergency shutdown is needed, press the emergency shutdown button; the system then enters the emergency shutdown state and the fuel supply is stopped quickly.

When the system speed is less than 100 r/min, the system turns to the waiting state. At this time, the fuel supply has stopped and the controller is waiting to start again.

When the diesel engine fails in the working process, the controller will enter the fault processing state, send and display the fault information, and evaluate the effect of the fault. If the fault is not fatal, it will run with a fault and the performance quality will decline, but the diesel engine can only complete the basic speed control function.

The steady-state mode discriminant module accepts the speed signal and the target speed signal (the accelerator pedal signal). By comparing the actual speed and the speed of the target, the system is judged to be in steady state or dynamic, and the discriminant signal is output to the working mode coding module for processing. When the system has a known speed of the target, the criterion for determining the engine speed stability or not needs a quantitative criterion.

The steady-state speed is a stable value near the target speed. It may fluctuate, and its fluctuation range depends on the steady-state error index of the dynamic module. Therefore, once the difference between the target speed and the actual speed is beyond the critical value, that is, entering the dynamic condition, then

$$| N_{mea} - N_{set} | < E_{max} \tag{5.2}$$

where

N_{mea} = measured speed

N_{set} = set speed

E_{max} = permissible maximum speed deviation in the state of stability

Once the conditions in this formula are not set up, the system enters a state of acceleration and deceleration. This decision method is correct in judging the transition from the steady state to the acceleration and deceleration state. However, it is not sufficient to determine whether the diesel engine is stable or not. The common speed regulation is shown in Figure 5.10. It shows that there is an upward and down oscillation of speed in the process of speed regulation. When the actual speed oscillates to the speed of the target speed, the speed deviation satisfies the formula (5.2) criterion when the time is between the $t_1 \sim t_2$ interval. It is obviously inappropriate to enter the steady state control at this time. Therefore the criteria for the steady-state process need to be strengthened.

Generally speaking, the speed stability also includes the small change rate of the rotation speed; that is, the angular acceleration is small. In practical applications, it can be satisfied at a certain speed sampling frequency:

$$| \Delta N_k - (\Delta N_k - 1) | < \Delta N_{max} \tag{5.3}$$

where

ΔN_k = kth speed sampling data

$\Delta N_k - 1$ = $(k - 1)$th speed sampling data

ΔN_{max} = maximum speed fluctuation allowed by the steady speed process

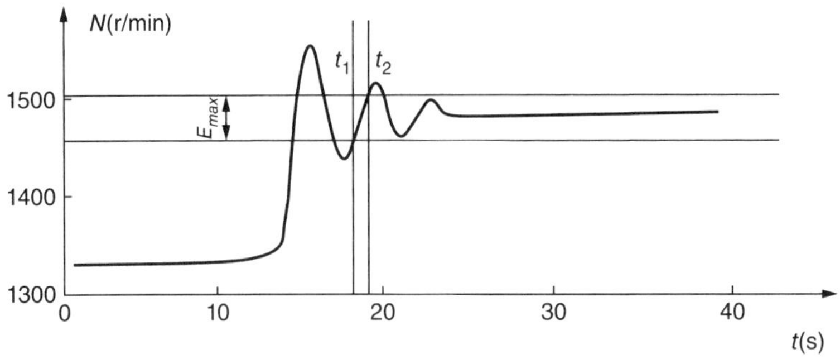

Figure 5.10 Speed curve of the speed regulating process.

When the two conditions are all satisfied, it can be considered that the system enters the steady state. Otherwise the controller is always in the state of acceleration and deceleration.

5.3.3 Design of the Self-Inspection State Control Strategy

The self-inspection state is the initial state after the system is energized. In this state, the working process of the controller can be represented by the state transition diagram, as shown in Figure 5.11.

State 1. Program self-examination state
State 2. Power self-check state
State 3. Self-detection state of the sensor
State 4. Communication interface, timer initialization
State 5. Startup waiting state
State 6. Fault display state

5.3.4 Design of the Starting State Control Strategy

The starting state control includes three tasks: starting motor control, starting oil amount, and starting time common rail chamber pressure control. The three control tasks are independent of each other at the same time, but there is a strong correlation between them. The state transfer diagram of the starting motor control is shown in Figure 5.12.

State 1. The control signal of the starting motor is high and the timing value is added.
State 2. The control signal of the starting motor is low and the timing value is zero.

The common rail pressure at the startup time is controlled by an open-loop control. By gradually setting up the duty ratio of the PWM signal, the pressure of the common rail chamber is gradually increased in order to reduce the load on the starting motor. At the same time, the impact is reduced. When the pressure reaches the starting pressure standard, the pressure of the common rail chamber enters the closed-loop control and maintains the starting pressure.

The oil supply of the starting process is controlled by a fixed supply of oil. The size of the specific oil supply is 1.5 times that of the oil supply under the lowest speed of the empty car.

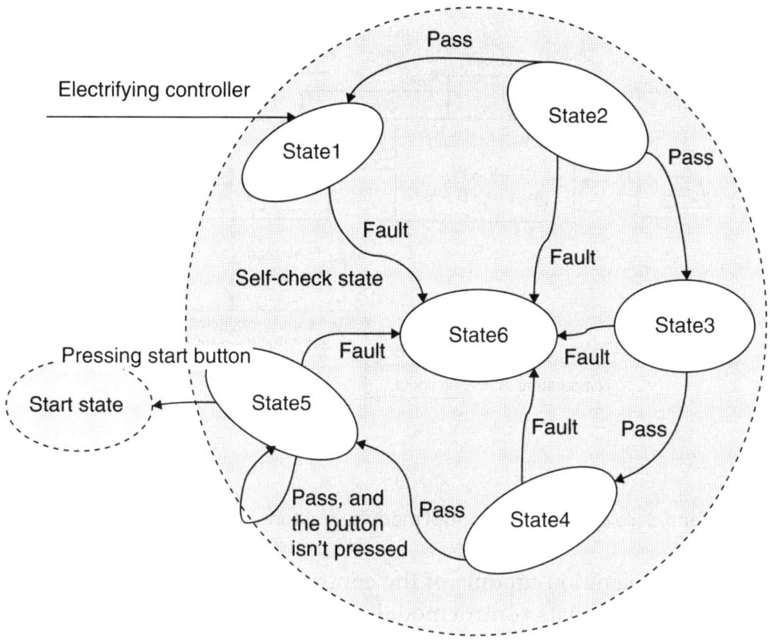

Figure 5.11 Self-checking state transfer diagram.

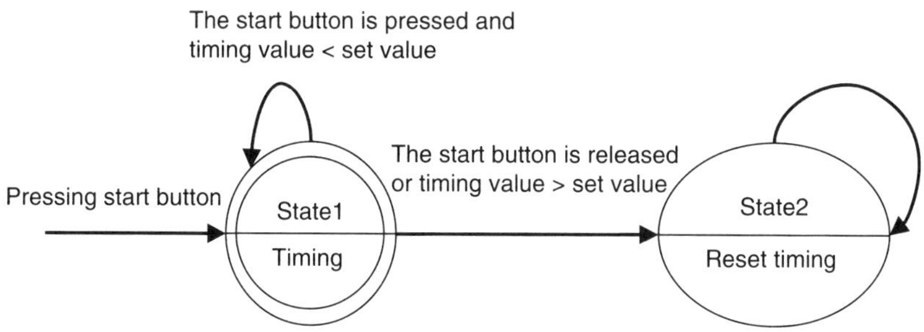

Figure 5.12 Control state transfer diagram for the starting motor.

5.3.5 Design of a State Control Strategy for Acceleration and Deceleration

The acceleration and deceleration state control module plays the role of a rapid adjustment of speed when the diesel engine is in the state of speed transition. The principle of making the control strategy is to make the adjustment of the dynamic condition fast and accurate, so that the engine can quickly enter the steady state and optimize the injection parameters. The goal of the control strategy optimization is the dynamic response time and the speed overshoot, so the target of the acceleration and deceleration state control is the diesel engine speed. The way to control the oil supply is to change it, while the other parameters of the oil supply are controlled by a simple MAP diagram. The realization of the parameters of the fuel supply can be a numerical equation or a table based on the specific values solved by the equation, so as to facilitate real-time control.

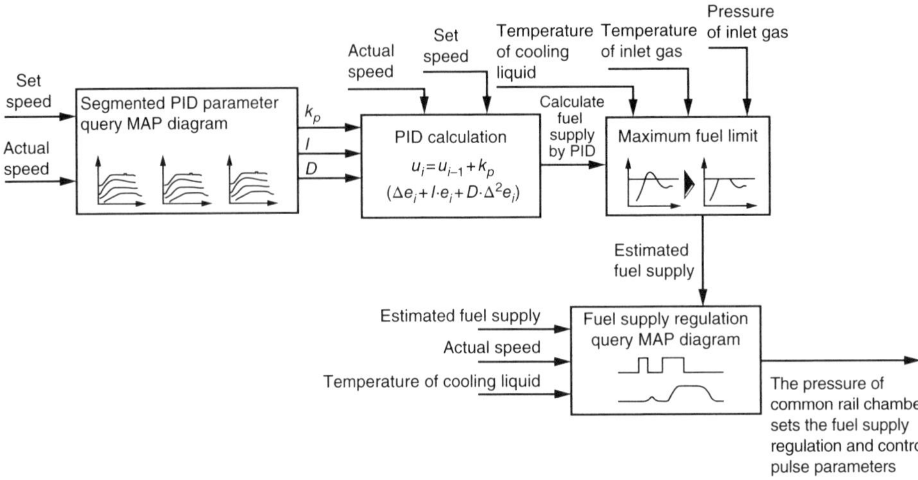

Figure 5.13 Acceleration and deceleration state control module structure.

The acceleration and deceleration module of the controller designed in this section takes a piecewise PID control, and its control module structure is shown in Figure 5.13.

5.3.6 Design of a Stable Speed Control Strategy

The goal of the stable speed control strategy is to optimize the combustion process of the diesel engine under the condition of the diesel engine speed stability. At the end of the acceleration and deceleration process, the output power of the diesel engine is balanced with the external load. The fuel injection parameters of the diesel engine cannot be optimized at this time. In the steady-state module, there is an optimized MAP diagram of the diesel engine speed–fuel injection quantity–fuel injection time relationship. Through the current injection quantity and speed control, MAP can be used to adjust the fuel supply law and then optimize combustion.

5.3.7 Principle of the Oil Supply Pulse

The control signal generator module of the injector solenoid valve is used to determine the starting point of the injector solenoid valve by combining the pulse counting with the timing and using the timing method to determine the end point of the control pulse. When working, first determine the starting point of the oil supply pulse and the angle difference α of the crankshaft on the rising edge of the upper stop point signal and then calculate the complete crankshaft position and number of pulses contained in the crankshaft angle difference, n_1:

$$n_1 = (\text{INT})(\alpha/6) \tag{5.4}$$

where (INT) represents the integer part.

The remaining angle value can be determined using the timing method. First measure the number of clock pulses corresponding to the position pulse of a crankshaft, N. Because the crankshaft inertia of the diesel engine is large and the short time speed

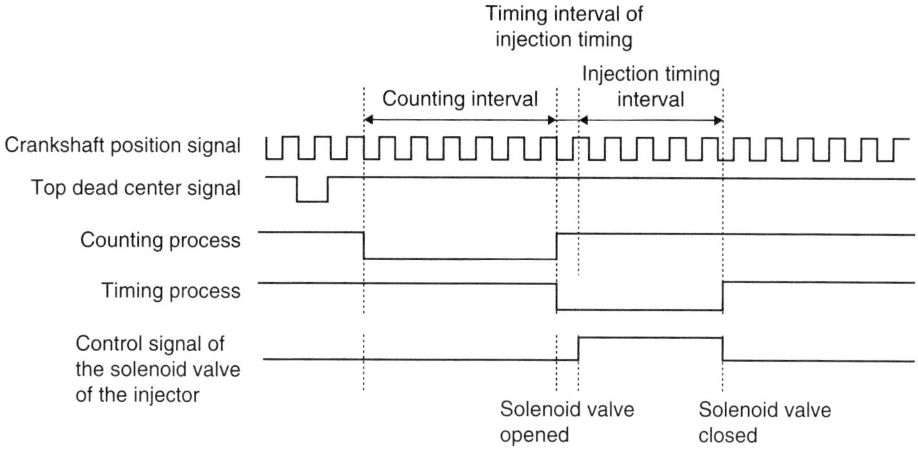

Figure 5.14 Time sequence diagram of the fuel injection control signal generator module.

change can be ignored, the cycle of the next crankshaft position can be predicted according to N:

$$n_2 = (\text{INT})\{N[\alpha(n_1 \times 6)]/6\} \tag{5.5}$$

$$n_3 = n_2 + N_{pw} \tag{5.6}$$

where N_{pw} is the number of time clock cycles corresponding to the pulse width of the injection control pulse.

When the count of the signal pulse of the crankshaft position is finished, the trigger timer starts. When the value of the timer is at the same time as the N_2 phase, the fuel supply control signal is set to "1." When the value of the timer is at the same time as N_3, the fuel supply control signal is set to "0." The process of the oil supply control signal ends. Its sequence diagram is shown in Figure 5.14.

The principle of the oil supply control pulse in a single cylinder injector is as described above. In the actual controller design, we also need to consider the following aspects: matching of the control pulse and the driving circuit, pre-injection pulse generation, the oil supply law control data interface, timing of the multicylinder oil supply pulse, process optimization, etc. The solutions to these problems will be described in detail later in Section 5.5 where field programmable gate array (FPGA) development is discussed.

5.4 Design of the ECU Hardware Circuit

After determining the working principle and control strategy of the system, the hardware circuit can be designed. This section will introduce the design process of the ECU hardware circuit.

5.4.1 Selection of Core Controller Parts

CPU and ASIC (specialized integrated circuits) are the cornerstones of contemporary hardware technology. CPU is a logical device to perform binary operations by instruction. A task is decomposed into a sequence of instructions expressed as a code, which is executed by a CPU to complete the operation required by the task. ASIC is a logical device designed to complete a task. A task is decomposed and expressed as a logical circuit. As the required input signal arrives, the circuit executes the task and sends the result to the output.

Flexible generality is the greatest advantage of CPU. In theory, all digital-based tasks can be expressed as an instruction sequence by CPU through an algorithm. However, the way CPU is doing a task is often not the best and the efficiency is not high. For example, if $a = bc + d$, CPU must be divided into two steps: first, $e = bc$ and then $a = e + d$. If ASIC is used, by properly configuring the gate circuit, it can be completed in one step. The internal gate circuit in CPU is an enormous number, but only a few gates work in each operation. The rest is in the idle state, but the electricity is still needed and the heat is generated. In order to improve performance, CPU can only adopt the strategy of increasing the main frequency and increasing the scale of integration. As a result, high cost, heat dissipation, and power consumption become the cost of a flexible and commonly used system.

ASIC is a special device that can only perform one or a few tasks, but the structure is optimized. It has high efficiency, no excess gate circuit, low power consumption, low heat generated, low cost, and large volume. It is widely used in various kinds of embedded systems and consumer electronics products, but its function has less variety. The design process of the mature ASIC chip is complex and the design cost is very expensive, so it is not suitable for small batch production.

One device currently has both CPU flexibility and ASIC efficiency, which is a programmable logic device. A programmable logic device is a kind of universal device that can configure its internal logic function through a program. It has the features of a flexible function and a high-efficiency structure. A field programmable gate array (FPGA) is a large-scale programmable logic device. The reconfigurable computing scheme based on the FPGA can have both the versatile flexibility of CPU and the efficient cheapness of ASIC. It is a technical scheme that combines the advantages of both. Considering the characteristics of the controller, the FPGA is used as the core component of the control system.

5.4.1.1 Characteristics of FPGA

FPGA is the product of the development of ultra-large-scale integrated circuits and computer aided design technology. Through the configuration of the user's internal gate array connection, FPGA can implement a variety of different digital logic functions, such as simple logic gate operation, high-speed CPU, etc. The application of FPGA can integrate circuit board level products into chip level ones, reduce power consumption, and improve system reliability. It is widely used in space, automatic control, communication, and other fields. Compared with CPU, FPGA has the following advantages.

A single FPGA can be configured to be a number of parallel digital logic processing units. These units can process data in parallel without interference with each other, significantly improve the real-time performance of the processing, and increase the utilization of the system resources.

FPGA uses all hardware operation methods, so the reliability and anti-interference ability of the work process is much stronger than that of the CPU using software operation methods.

FPGA -based design has no relevance to the specific devices used. That is, the same design can be applied directly to a variety of FPGA devices and the designers will not be restricted by the devices, as the design has independent intellectual property rights. There is a direct relationship between the instruction system of the single chip and the single chip computer. A series of single chip programs cannot be applied directly to another series of single chip. Once the hardware circuit is upgraded and the single chip model is replaced, the programmer needs to learn the programming method of another single chip computer.

The device pin functions of the FPGA can be defined freely by the designer in accordance with the needs. Therefore, the circuit board can be designed first and then the pin is defined according to the circuit board connection mode, which brings great convenience to the circuit board. By configuring different contents of the FPGA device, the identical circuit board may have completely different functions.

The cost of FPGA's system hardware development is low. A computer with an electronic design automation (EDA) software installed with a corresponding FPGA development and an FPGA download cable can be used for the professional development of FPGA. FPGA development software is fully functional and the result of logical timing simulation is correct and reliable, which can greatly reduce the cost of development. The user can make a logical design first; after the logical sequence simulation, the user can determine the size of the chip resource required by the system and can then choose the model of the FPGA chip, making it tailored, which will make the hardware design of the system more optimized.

Using system on programmable chip technology, by configuring the CPU soft core, the FPGA can become a monolithic hardware system embedded with high-speed CPU and realize the function that a single CPU cannot perform.

5.4.1.2 Selection of Core Auxiliary Devices

According to the advantages of FPGA devices summarized in the Section 5.4.1.1, it is reasonable to choose the FPGA device as the control core. After the initial design of the system on the EDA software QUARTUSII, we select the Cyclone series FPGA chip EP1C3 of ALTERA Company. The FPGA chip provides 104 programmable input/output pins, 2910 programmable logic units, embedded 7 Kbyte programmable storage units, and a PLL clock module.

Because FPGA adopts volatile storage technology, it needs to configure every time after powerup and input the functional design data of FPGA, so that the FPGA chip has the functions that designers need. The configuration process is actually a serial communication process. In this paper, ALTERA's special configuration chip EPCS1 is selected. The chip uses FLASH storage technology and the capacity is 1 Mbit. When the system is powered up, the device can automatically configure the FPGA. The configuration time varies according to the content, but the longest time does not exceed 4 ms. This relatively short time will not affect the work of the diesel engine.

Using FPGA only as the core of the controller will deliver the best performance by building the SOPC system, but it will also lead to a very large amount of work. It is clear that it is difficult to achieve at present, so a PIC series of an 8 bit single chip is selected as the auxiliary device of FPGA to assist in the work of signal processing.

A PIC microcontroller is the product of Microchip Company of America. The production history of the PIC 8 bit MCU is only 11 years, but now its output has ranked second in the world (after Motorola Company). Now there are more than 120 kinds of PIC single chip. It can be divided into 5 series: PIC12's 8 pin series, PIC16C5 basic series, PIC16C6 intermediate series, PIC17 advanced series, and PIC18C expansion (high and new product) series. The PIC MCU pin also comes in many varieties; there is (PIN) 8/14/18/20/28/40/44/64/68/80/84, and so on. It can be seen that the specification of the PIC 8 bit single chip variety is very complete.

The PIC chip microprocessor uses the Harvard bus structure to separate the data bus from the instruction bus in the chip and comes in different widths.

The selected PIC16F877A is a higher configuration of the single chip microcomputer (SCM) based on the PIC16F87X series of single chip. The internal resources of this SCM are integrated with the following resources:

8 K × 14 bit Flash program memory erase duplication
368 general memory RAM units
8 channel 10 bit AD converter
two 8 bit timer/counters, with a 16 bit programmable timer counter
2 capture/comparison/pulse width modulation CCP modules, combined with a timing counting module, can realize input capture, output comparison, and pulse width modulation output function
1 main synchronous serial port module can realize the two working modes of SPI and I2C
1 universal synchronous/asynchronous transceiver USART module, which can be used for two-wire serial communication, can be defined as full duplex asynchronous mode, so as to achieve communication with PC or other SCM systems; it can also be defined as a half-duplex synchronization mode, which can communicate with the serial EEPROM system
1 parallel slave port PSP module is used to connect high-speed data transmission and exchange with other parallel data buses with open bus, DSP, or microprocessor

One internal watchdog timer module that can reset the program in time when a program runaway occurs. These two devices are composed of a high and a low. FPGA gives full play to its internal logic configuration, the advantages of parallel processing, and the main function of the controller, realizing the functions of an MAP query, oil supply control pulse occurrence, data display, and so on. The PIC microcontroller has the advantage of many internal modules, completes the function of peripheral data processing, realizes an input analog signal A/D conversion, uses a common rail chamber pressure PWM control, and has external communication functions.

5.4.2 Control Core Circuit Design

After determining the control core device, this section will introduce the circuit connection design process of the two chips.

5.4.2.1 FPGA Circuit Design

Because the pin of the device can be programmed, the peripheral circuit of the FPGA device is relatively easy to design. Besides the logic output and input pin, the peripheral circuit of the FPGA also includes the FPGA power supply, the FPGA configuration circuit and the logic level conversion interface between the FPGA and the SCM. This section introduces the design of these FPGA peripherals.

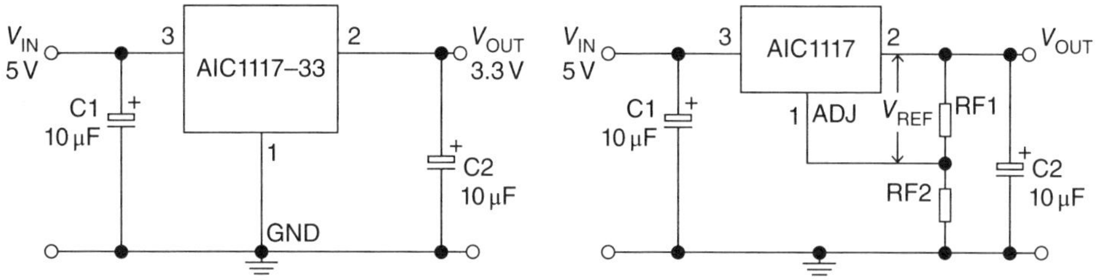

Figure 5.15 FPGA power supply circuit.

5.4.2.1.1 *Power Supply Design*

The work of the FPGA device selected in this article requires two different kinds of voltage and its logical interface voltage standard is 3.3 V, so it is necessary to supply a driving power of 3.3 V. The voltage range of the power supply is between 3.0 and 3.6 V, the working voltage of the core part is 1.5 V, and the voltage range of the power supply is between 1.425 and 1.575 V. For this purpose, two power sources are needed for FPGA, and the power supply of the control core of the power module is 5 V. In this model, the power circuit of FPGA is designed by using the low dropout voltage regulator chip AIC1117 series. The device can provide the driving current of 800 mA, which is enough FPGA to work properly. The designed power supply circuit is shown in Figure 5.15.

The power supply for 3.3 V can be directly converted from the AIC1117-33 chip and the power supply of 1.5 V needs to be converted to a voltage regulator chip. The relationship between the input and output voltages is shown by

$$V_{OUT} = V_{REF} \times (1 = RF_2/RF_1) + I_{ADJ} \times R_{F2} \tag{5.7}$$

$$V_{REF} = V_{OUT} - V_{ADJ} = 1.25\text{V}$$

$$I_{ADJ} = 55\ \mu A$$

After calculation, the selection of RF_1 is 125 Ω and RF_2 is 25 Ω.

5.4.2.1.2 *Configuration Circuit Design*

FPGA devices need to be configured before work. This section selects a dedicated configuration chip, EPCS1, to configure it. Meanwhile, the data configuration of EPCS1 adopts the way of "AS (active serial) Programming," and is configured by a PC interface through a 10 pin interface through ByteBlaster II. Its configuration circuit is shown in Figure 5.16.

After the design is completed, the designed data is downloaded through the cable of ByteBlaster II into the special configuration device EPCS1. Since the EPCS1 device adopts FLASH technology, the information inside it will not be lost after a power failure. Therefore, during the next powerup process, the EPCS1 device will automatically configure the FPGA so that it has the design function.

5.4.2.1.3 *Logic Voltage Matching Circuit*

In order to reduce the power consumption of the system, the interface of the FPGA device adopts the 3.3 V standard voltage and the interface standard of the PIC single chip is the 5 V TTL standard. In order to ensure the normal two-way communication

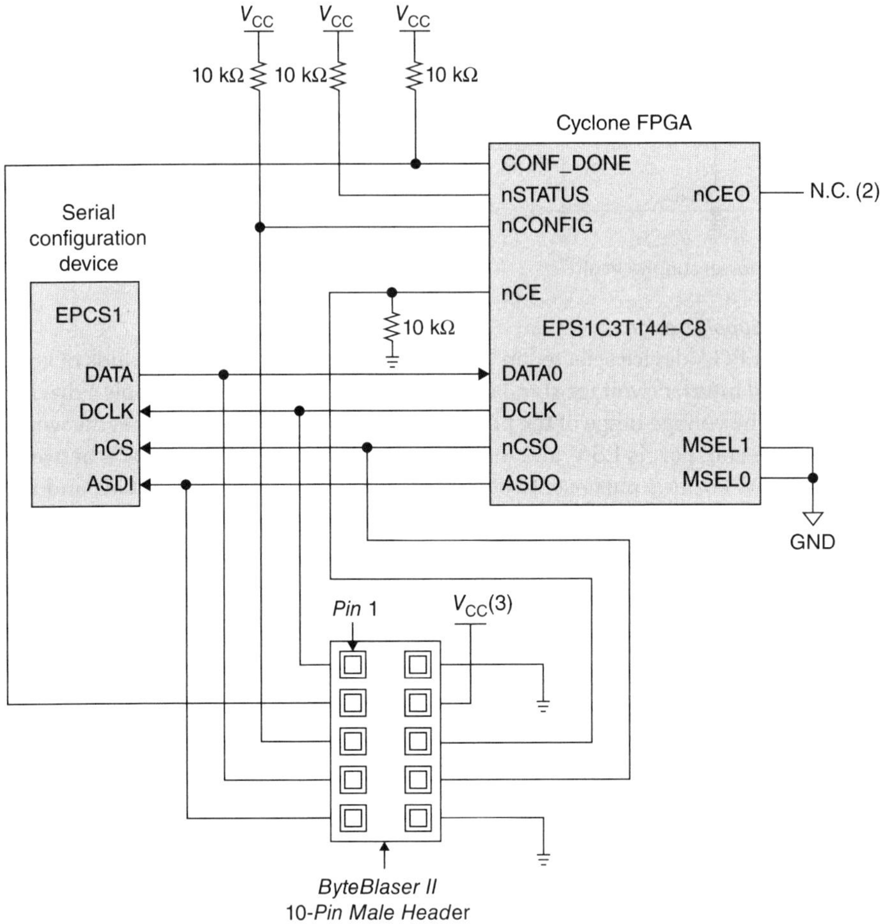

Figure 5.16 FPGA configuration circuit.

between the two different standards, a logical voltage matching circuit needs to be provided. The bidirectional logic conversion chip 74lvcx4245 is chosen here, which can provide eight channels and the logic conversion of different standard voltages. Its structure is shown in Figure 5.17.

The logic structure in Figure 5.17 shows that A and B channels can use two different logical levels. By setting the voltage of T/R pins, we can determine the direction of data transmission. The OE pin can make the data path of the level conversion chip in a high resistance state and cut off the communication channel. Its true table is shown in Table 5.1.

Since the communication between the FPGA and PIC is controlled by the FPGA, the T/R tube is controlled by the FPGA and the OE pin is directly grounded.

5.4.2.2 Circuit Design of SCM

Because the integration degree of the PIC single chip is very high, the program memory, the AD conversion circuit, and the PWM pulse generator are all integrated inside

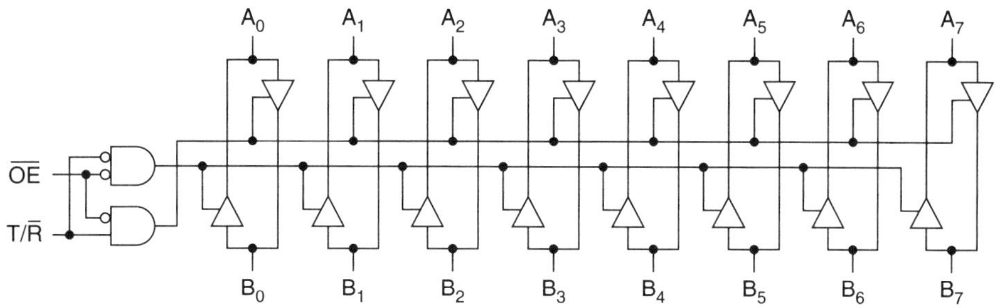

Figure 5.17 Structure of 74lvcx4245.

Table 5.1 Truth table for the 74lvcx4245 chip.

Inputs		**Outputs**
OE	***T/R̄***	
L	L	Bus B data to bus A
L	H	Bus B data to bus B
H	X	HIGH-Z state

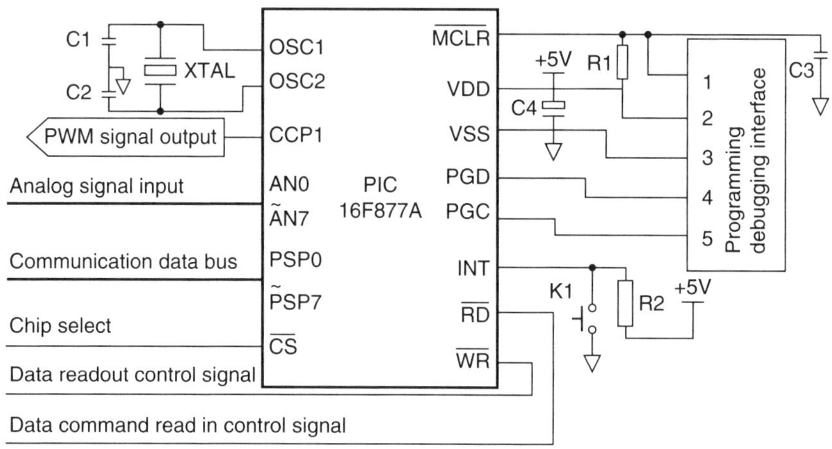

Figure 5.18 Circuit diagram of the MCP system.

the chip. The MCU circuit is therefore composed of a single chip, a reset circuit, a clock circuit, and a programming debugging interface. The actual circuit is shown in Figure 5.18.

A clock circuit composed of crystal XTAL and two capacitors, C1 and C2, are shown in Figure 5.18. In this circuit we use a 20 M crystal oscillator, the clock frequency of the system is 20 M, and C1 and C2 use 18 pF ceramic capacitors in order to ensure a stable clock circuit.

Because the system does not need manual reset during the working process, the reset pin is directly connected to a high-voltage source in order to prevent a reset caused by external interference or voltage fluctuation, and a capacitor of 1 μF is added to C3.

K2 is the emergency shutdown button of the system. When the common rail system fails, we press the button, which triggers the interrupt of the single chip computer and starts the corresponding interrupt program. The pressure of the common rail chamber is set to a minimum and the system can be resumed only after resetting.

The programming debugging interface is used to write the program memory to the internal program memory of the SCM for debugging.

The pressure feedback signal and the current feedback signal are analog voltage signals with a range of 0–5 V. The pressure feedback signal is provided by a pressure transmitter installed on the common rail chamber. The current feedback signal is generated by the current sampling and output signal in the power drive circuit of the electromagnetic overflow valve.

The communication data bus is an 8 bit data bus. The ECU core control unit writes the control data to the PIC SCM system through the data bus or reads the A/D measurement signal. The data are written or read by the ECU core control unit through a chip select signal, data write instruction, and data read signal control signal.

5.4.3 Design of the Sensor Signal Conditioning Circuit

Between the output of the sensitive element and the input end of the amplifier, it is possible to have power interference, electrostatic interference, electromagnetic coupling interference, and common mode interference. Such signals inevitably have noise, which will be overwhelmed by noise in serious cases. Therefore, it is very important to reduce the noise and improve the signal-to-noise ratio. For the sensory units, the signal should be amplified due to the very small amount of charge in the output. When the signal is amplified, the noise pollution caused by the amplifying circuit should be minimized. Because the resistance of the piezoresistive sensitive element is caused not only by the strain, it is also affected by the temperature change. Therefore, the temperature compensation must be considered.

5.4.3.1 Design of the Signal Conditioning Circuit for the Temperature Sensor

In order to improve combustion and achieve monitoring of the running state of the diesel engine, the controller designed here measures the intake temperature, cooling water temperature, and lubricating oil temperature in real time and designs a thermal resistance sensor, which changes the temperature signal to a change in the resistance value. The sensor is a thermistor induction device with a negative temperature coefficient (NTC). The resistance of this thermistor decreases with the increase of temperature. Its temperature and resistance characteristics are non-linear. This characteristic can be defined by the Steinhart-Hart equation. The calibration results of common thermistors are shown in Figure 5.19. The resistance value is given in a ratio form (R/R25). The ratio indicates the ratio of the current temperature resistance to the resistance at 25 °C. Usually the same series of thermistors have similar characteristics and the same resistance/temperature curve.

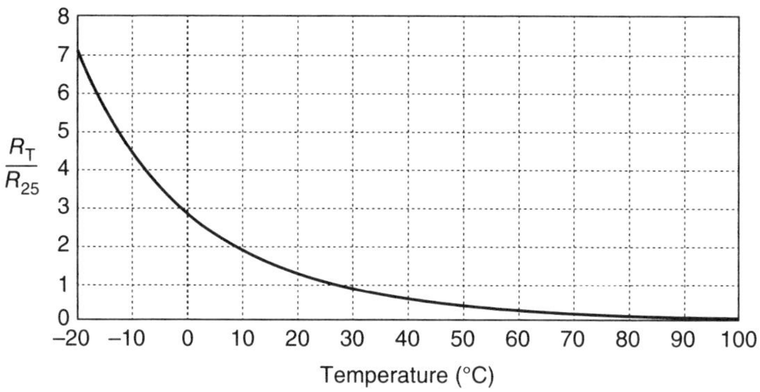

Figure 5.19 Calibration curve of the thermistor.

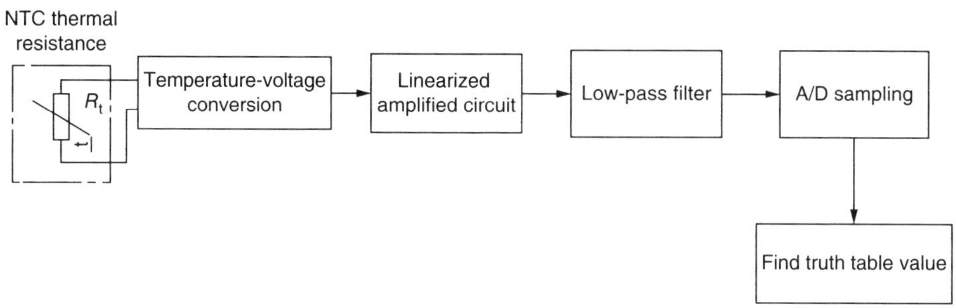

Figure 5.20 Principle of the temperature measurement.

Because of the non-linear characteristics of the thermistor, the measured data need to be linearized by software. In this paper, a look-up table method is used. The principle of its measurement is shown in Figure 5.20.

First, through the temperature voltage change, the circuit changes the resistance change of the thermistor to the change of the voltage. Using a linear amplifier to amplify a weak voltage signal to a voltage of 0–5 V can work in an A/D sampling device. Then the partial noise interference is filtered through the low pass filter, thus preventing the frequency aliasing in the A/D sampling. After the processed signal is sampled by A/D, the core device of the controller is input and the temperature value of the thermistor is obtained by checking the table.

Because the thermal resistor is essentially a resistance element, when the current flows through it, some heat will be generated. Therefore, in the circuit design, we should prevent self-heating of the thermistor. It allows the system to measure the heat of the thermistor, not the ambient temperature. Then a current switch is added to the measuring process circuit of the temperature signal. During the measurement process, the thermistor is energized. After the AD sampling is finished, the power supply of the thermistor is stopped. The temperature measurement sensor circuit is shown in Figure 5.21.

The electric bridge is composed of four resistors, R1, R2, Rt, and Rp1, where the change of Rt's resistance converts to the voltage change of Ux. By constructing a linear amplification circuit with negative feedback by N1, the amplified signal is passed through the low-pass filter constructed by N2 and the last output signal is directly connected to the

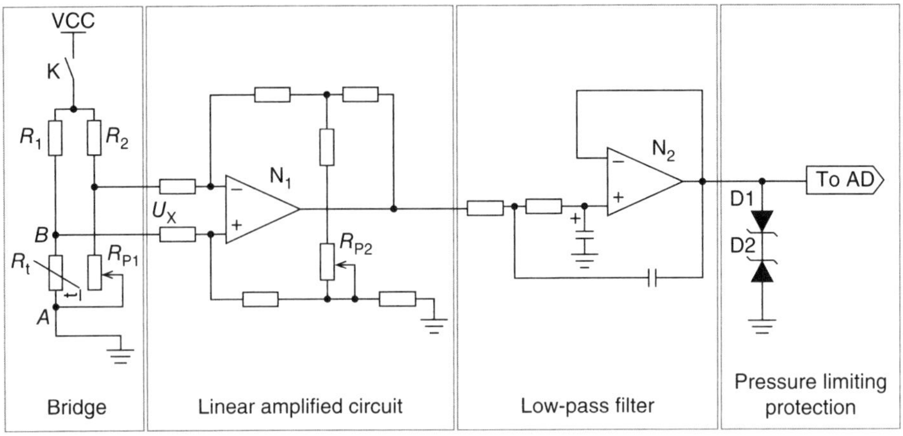

Figure 5.21 Temperature measurement circuit.

A/D conversion interface. The adjustable resistance Rp1 is used to adjust the zero point of the temperature measurement circuit and Rp2 is used to adjust the amplification gain of the temperature measurement circuit. D1 and D2 are used to limit the output voltage and protect the safety of the AD acquisition circuit. K is a measuring switch, which is disconnected when we are not sampling. It closes before sampling and then disconnects after the sampling is finished.

5.4.3.2 Design of the Signal Conditioning Circuit for the Pressure Sensor

Similar to the temperature sensor, the pressure sensor uses a semiconductor piezoresistive sensor. The resistivity changes with the intensity of the stress, that is, the "piezoresistive effect." This sensor has a high sensitivity and can directly reflect small stress changes. An electric bridge made of a semiconductor material on the inner pressure diaphragm of the sensor is shown in the equivalent circuit diagram in Figure 5.22.

When the bridge is measured, a constant current source is connected to the two ends of the bridge. Under the action of pressure, the resistance of the bridge arm will change. The voltage at the other ends of the bridge changes with the change of the bridge arm resistance. In this way, the pressure on the diaphragm of the sensor is transformed into the voltage output of the bridge. After the voltage is amplified and filtered, it can be converted to A/D; that is, the pressure value of the measured point is obtained. According to this principle, the schematic diagram of the pressure measurement circuit is illustrated by using the operational amplifier circuit shown in Figure 5.23.

In the circuit shown in Figure 5.23, the operational amplifiers N1 and N2 constitute a constant current source with current feedback. R5 is a current sampling resistor. N2 acts as a voltage follower in the feedback loop, which increases the internal resistance of the feedback loop, reduces the effect of feedback current, and reduces the current loss due to feedback. The operational amplifiers N3 and N4 constitute the first level common mode amplifier, in which the adjustable resistance Rp1 is used to adjust the gain of the amplifier. The amplifiers N5 and N6 constitute a comparative amplification circuit and the adjustable resistance Rp2 is used to adjust the zero point of the pressure measurement. N6 is used as a voltage follower to set the zero point reference voltage.

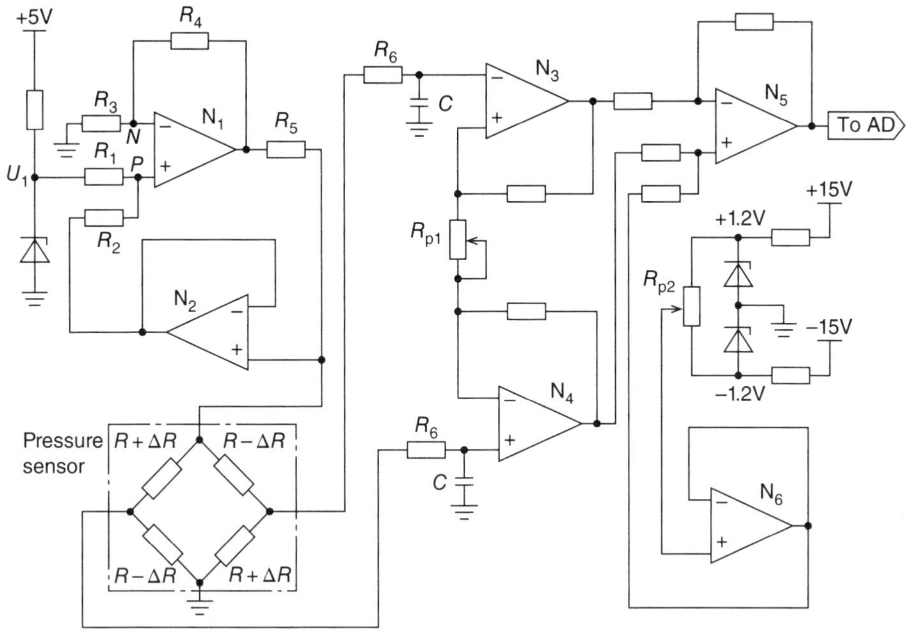

Figure 5.22 Equivalent circuit of the pressure sensor.

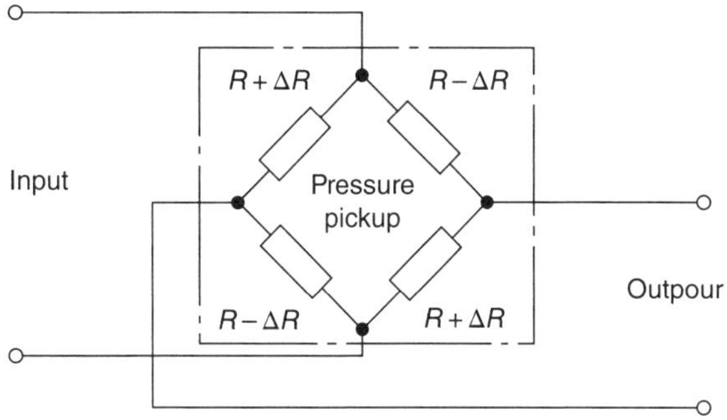

Figure 5.23 Pressure measurement circuit.

5.4.3.3 Design of the Pulse Signal Conditioning Circuit

There are two sets of pulse signal in the design. They are the crankshaft position signal and the top dead center signal. These two signals are key signals. The fuel supply control pulse processes the circuits of the crankshaft position and the TDC signal, and the two signals are used as scales. Therefore, their conditioning circuits have high requirements. The specific requirements are as follows:

(1) Requirement of the voltage amplitude.
Due to the use of the solenoid sensor, when the rotational speeds are not the same, the output signal amplitude will have a large difference. However, the voltage required by the controller should be in the voltage range of the logic level standard, so the signal conditioning circuit should limit the voltage amplitude to the standard range and prevent leakage pulse or high voltage.

Table 5.2 Definition of the TTL level standard and the CMOS level standard.

Logic standard	GND	VCC	VOH (Min)	VOH (Max)	VIL (Min)	VIL (Max)
3.3 V COMS	0.0 V	3.3 V	Vcc−0.1 V (3.2 V)	0.4 V	0.8 Vcc (2.64 V)	0.2 Vcc (0.66 V)
3.3 V TTL	0.0 V	3.3 V	2.4 V	0.4 V	2.0 V	0.8 V
5.0 V COMS	0.0 V	5.0 V	3.5 V	0.4 V	0.7 Vcc (3.5 V)	0.3 Vcc (1.5 V)
5.0 V TTL	0.0 V	5.0 V	2.4 V	0.4 V	2.0 V	0.8 V

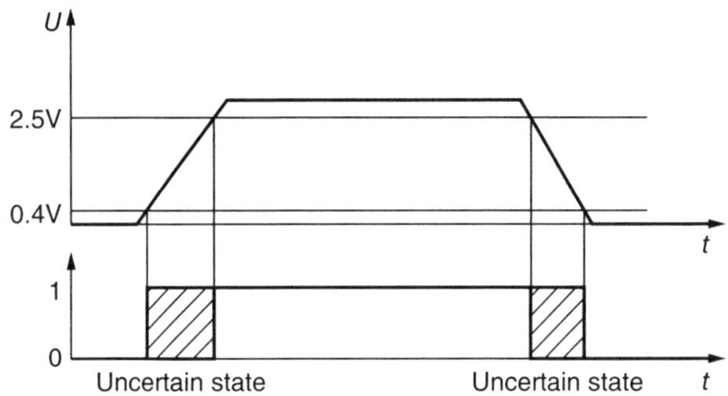

Figure 5.24 The logic uncertainty caused by the rise time and the fall time.

(2) Requirement of rise time.
In the crankshaft position, the TDC signal are pulse signals. In the working process, the rising edge is used as the working points. Because the voltage in the actual circuit is not likely to mutate, there is a rise time of the signal. In the digital circuit, the definition of TTL level standard and CMOS level standard are shown in Table 5.2. From Table 5.2, it can be seen that the voltage of the 3 V TTL standard with logical zero is 0–0.4 V and the voltage of logical one is 2.5 V. When the voltage is between 0.4 and 2.5 V, the state of signal logic will be unstable, as shown in Figure 5.24, at 0–0.4 V. In order to avoid the timing error caused by this unstable state, the rise time of the pulse signal is required to be as short as possible.

(3) Requirement of signal isolation.
The crankshaft position signal and the TDC signal occur through the solenoid sensor. Because the electromagnetic interference due to the diesel engine operation environment is very serious, in order to prevent the introduction of external interference into the controller it is necessary to use a high-speed photoelectric isolation device to isolate the two signals and reduce the disturbance to the controller.

(4) Requirement of phase.
Because the original signal passes through the multistage treatment before it reaches the control core, there must be a phase difference between the input and output of a signal conditioning circuit. However, in order to ensure the stability of the fuel supply control pulse, the phase difference of the signal conditioning circuit must be stable and compensatable.

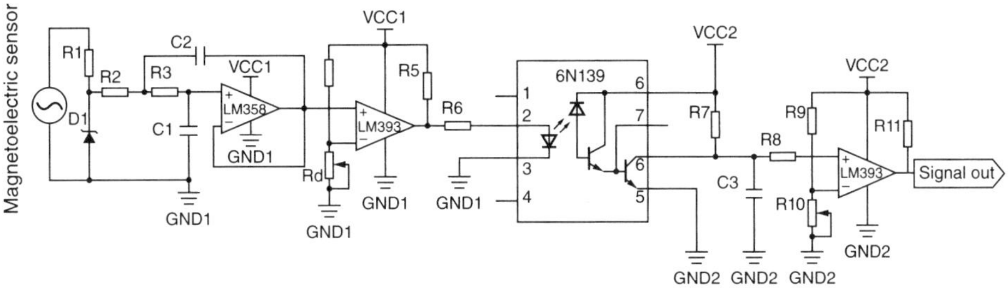

Figure 5.25 Pulse signal conditioning circuit.

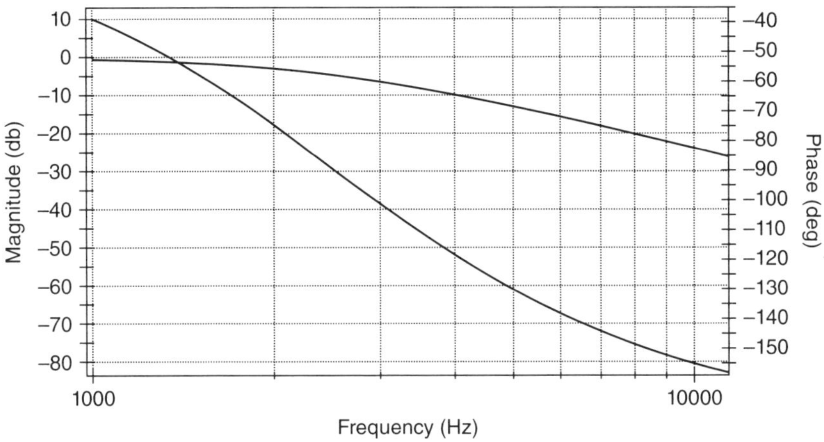

Figure 5.26 Frequency domain response and phase delay of the filter.

Based on the above requirements of the four aspects, the design of the present pulse signal conditioning circuit is shown in Figure 5.25.

The circuit is divided into four parts. The first part adopts a single supply amplifier LM358 and capacitors C1, C2, R2, and R3 to constitute a second-order Bessel filter, wherein the capacitance of C1 is 0.0047 μF, capacitance of C2 is 0.0068 μF, resistance of R2 is 8.25 K, and resistance of R3 is 14.7 K. The frequency domain response and phase delay of the filter are shown in Figure 5.26. The second part is a voltage comparator composed of a single voltage comparator LM393. LM393 can work normally under the condition of a 2 □ 18 V single power supply, and transforms the sine wave into a square wave.

The third part is an isolating circuit composed of the optocoupler 6N139. The two power supplies, VCC1-GND1 and VCC2-GND2, at both ends of the optocoupler are electrically isolated. A voltage signal of both ends isolated by the optical signal, the optical isolation device 6N139 is a high-frequency optocoupler. Its frequency response characteristics is better than that of the common optocoupler and the signal output passes through the optocoupler and the fourth part of the circuit, the voltage comparator, to reduce the rise time of the output signal.

5.4.4 Design of the Power Drive Circuit

Divided by the function, the power driving module in the ECU designed in this section is made up of two independent circuits. One is the power driving circuit of the common rail chamber pressure control solenoid overflow valve; the other is the power drive circuit of the electronically controlled injector solenoid valve. This section will introduce the circuit design process of these two parts.

5.4.4.1 Design of the Power Drive Circuit of the Pressure Controlled Solenoid Overflow Valve in the Common Rail Chamber

The pressure of the common rail chamber in the high-pressure common rail system is required to be in the range of 30 □ 150 MPa with flexible control. The pressure of the common rail chamber is controlled by the proportional electromagnetic valve produced by the improved Ningbo electrohydraulic proportional valve factory. The valve can adjust the fuel pressure of the common rail cavity flexibly in the range of 0 □ 170 MPa.

By measurement, the proportional control coil resistance of the proportional solenoid valve is 10.4 Ω and the inductance is 91 mH.

In order to realize the adjustable flexibility of the common rail pressure cavity, a current source that can provide an accurately adjustable driving power is needed. A manual adjustment drive of the original configuration is an adjustable constant current source with a range of 0 □ 2 A, where the input voltage of the constant current source is 220 V (AC) and the output frequency is 100 Hz of a PWM drive signal. In the constant current source, the current feedback is set up by operation and an amplifier, and the constant driving current can be guaranteed to a certain extent. Because the electromagnetic relief valve was modified, the drive current needed to reach 150 MPa decreased from 1.5 to 0.82 A, and the original driving circuit is a 220 V power supply circuit. The circuit is purely analog, so the control accuracy is not high and the volume is huge. The electronic control interface is the 0 □ 5 V analog interface, which does not match the digital interface of the controller core, so redesigning the new power drive circuit is needed.

In this model, a PWM circuit with a double closed-loop feedback of current and pressure is used to drive the proportional solenoid valve. The PWM signal is generated by the control core. Because the maximum driving current of the PWM signal output by the control core is 20 mA, it is not enough to drive the proportional solenoid valve, and therefore more switches are needed in the driving circuit to amplify the control signal.

4N35 is selected as the optocoupler U1, and the current of the optocoupler though the input end only needs 10 mA. The maximum PWM driving current of the core output is 20 mA, which can directly drive the optocoupler. The role of the resistor R1 is to limit the current flowing through the optocoupler input end. When the PWM signal is high, the current flows through the emitting diode of the optocoupler and breakovers in the photoelectric diode. The 12 V power on the MOSFET switch diode as the gate voltage rises drives the MOSFET to the breakover. Then there is current through the solenoid valve coil. When the PWM signal is low, no current flows through the emitting diode, resulting in a cut-off in the photoelectric receiving diode and a gate voltage drop in the MOSFET. When the MOSFET is cut off, the solenoid coil is formed through a freewheeling diode D1 to form a freewheeling loop, which ensures a stable current fluctuation and protects the MOSFET switch device from the impact voltage damage of the solenoid coil.

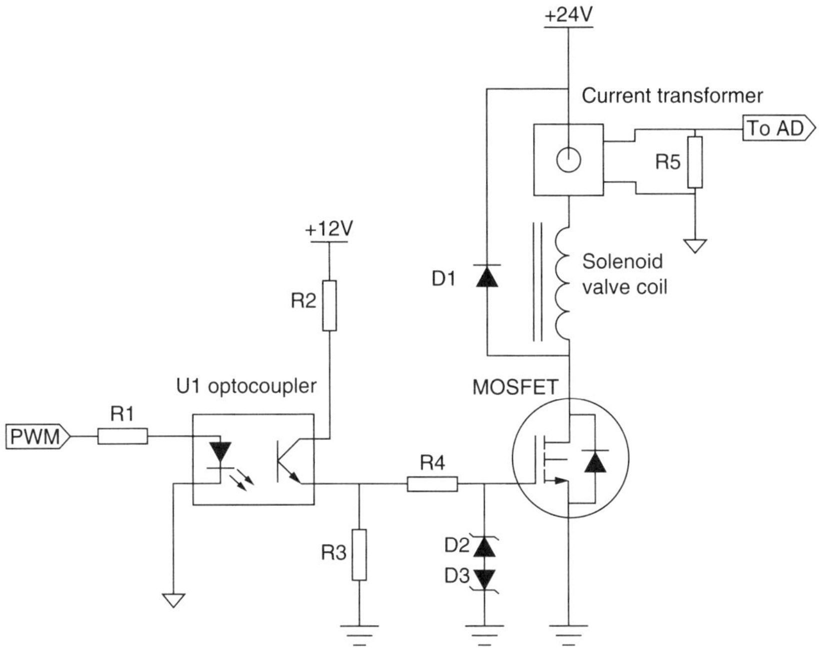

Figure 5.27 Drive circuit of the proportional overflow valve.

According to the driving current and power of the electromagnetic overflow valve, a drive circuit, as shown in Figure 5.27, is chosen.

5.4.4.2 Design of the Power Drive Circuit for the Solenoid Valve of the Injector

The function of the power drive circuit for the solenoid valve of the injector is to amplify a weak current control signal of 5 V to a strong current signal that can drive the solenoid valve of the injector. In order to make the injector work normally, the power drive circuit of the injector must ensure that the waveform of the output current satisfies the requirements of the injector solenoid valve driving current and that the current change response meets the specified requirements.

The high-voltage input of the power drive part is 130 V and the driving object is the solenoid valve coil of the injector. The internal resistance of the solenoid valve coil is 2.7 Ω, the inductance is 4.7 mH, the minimum open current is 15 A, and the minimum holding current is 4 A.

According to the above requirements, three different drive circuit structures are presented: a single power single switch drive circuit, a dual power dual switch drive circuit, and a single power dual switch drive circuit.

The circuit diagram of the single power single switch drive circuit is shown in Figure 5.28. In the circuit of Figure 5.28, the driving power is from only one VCC1. The current on each coil of the solenoid valve is controlled by a power MOSFET device, so it is called the single power single switch circuit. The current is kept by the PWM signal.

The advantages of the single power supply single switch circuit are having a simple structure and having a common position for the six drive signals, so that only a MOSFET

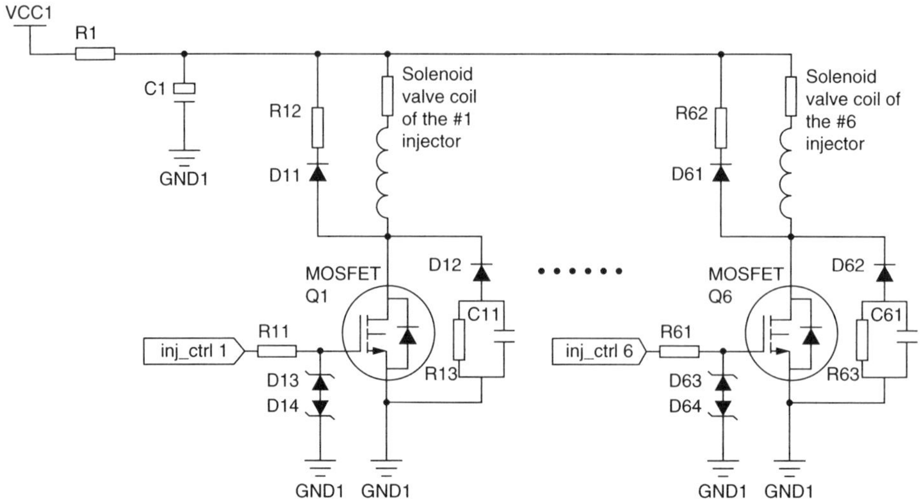

Figure 5.28 Single power single switch drive circuit.

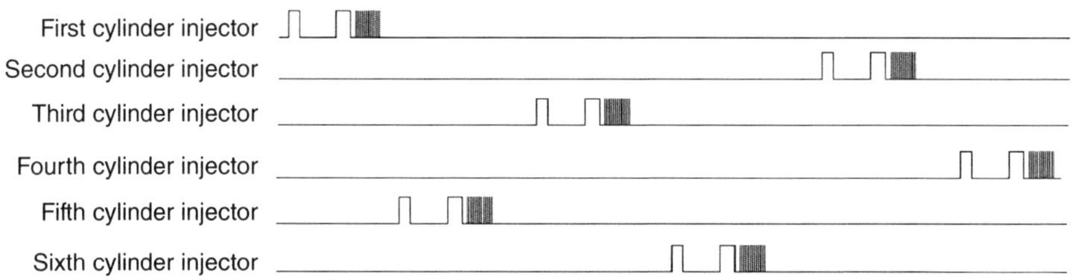

Figure 5.29 Input control signal of the single power single switch drive circuit.

switch power supply is needed. A disadvantage is that the power MOSFET switches in each channel are subjected to the impact of voltage or current during the switching process, so the RCD absorption circuit is required. In addition, the adjusting precision of the operation time of the driving current of the injector solenoid valve is influenced by the frequency of the depressurization PWM signal. The driving signal of the circuit is shown in Figure 5.29.

The circuit diagram of the dual power dual switch circuit is shown in Figure 5.30. According to the working principle of the circuit, the high-voltage switch of the six channels can be combined into one, so that the circuit is simplified as shown in Figure 5.31.

The advantages of the driving circuit of a dual power and dual switch are a stable driving current, smooth conversion between a high and low drive voltage, a tiny switching loss of switching devices, a small impact of current and voltage, and the fact that the driving pulse width regulation is not affected by the frequency of the PWM signal. The disadvantages are the use of a two-power supply and an increased number of switch tubes by one, which increases the complexity of the system. All power MOSFETs will be impacted by the switching voltage and current. It is still necessary to configure the RCD absorption circuit in order to account for this, and the circuit that drives the high-voltage switch tube needs two separate driving power sources.

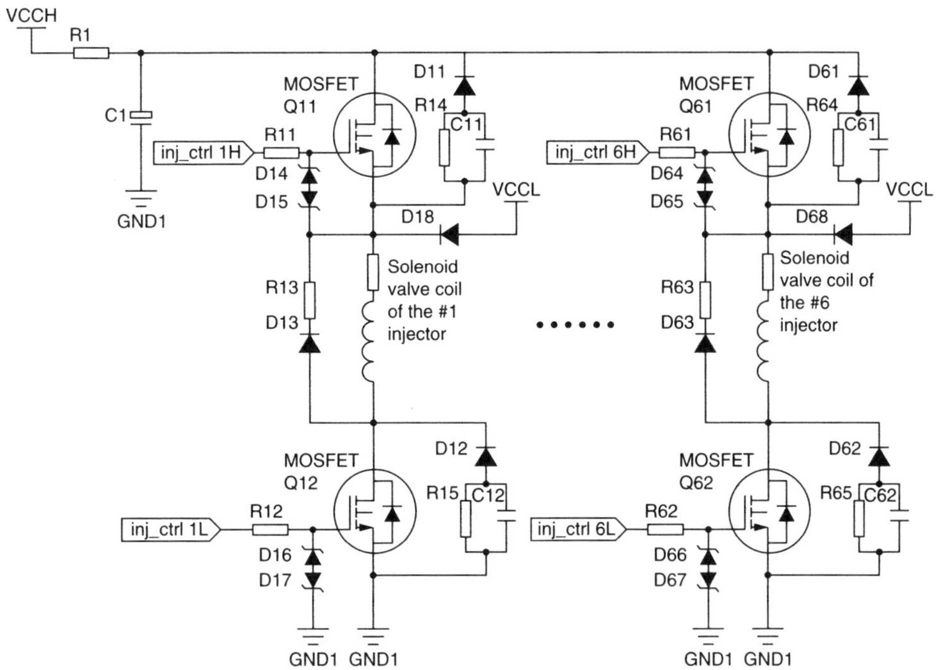

Figure 5.30 Dual voltage and dual switch driving circuit.

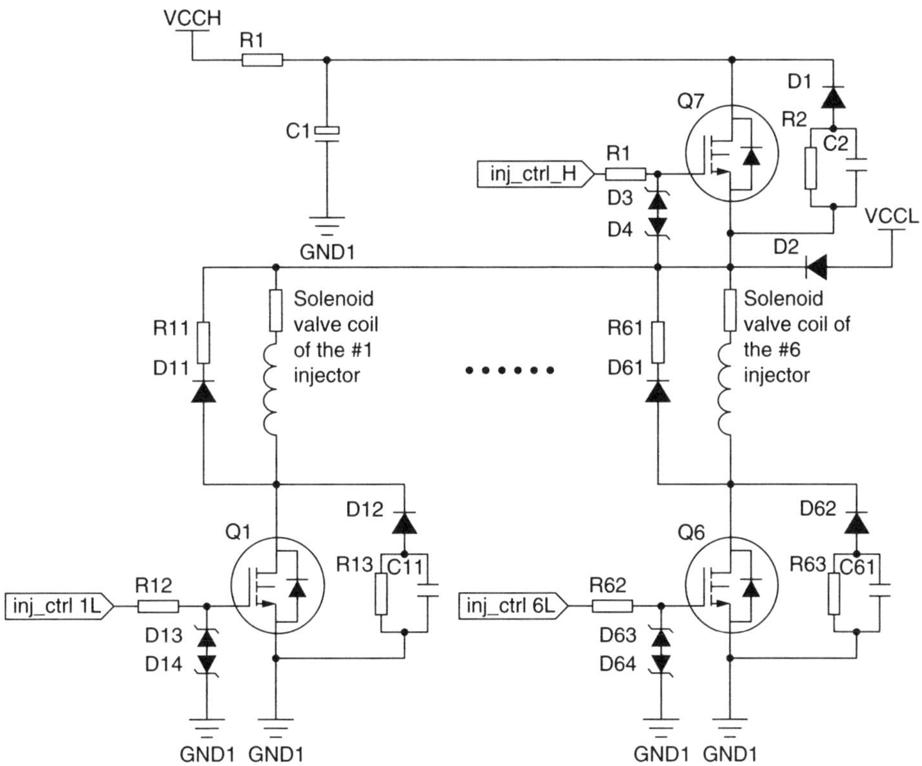

Figure 5.31 A simplified dual voltage drive circuit.

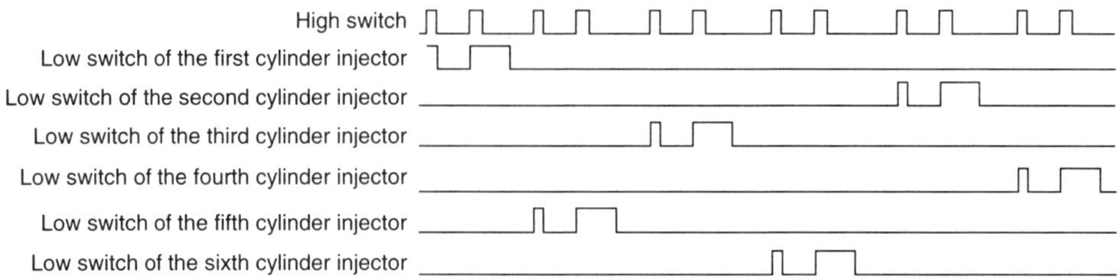

Figure 5.32 Double power dual switch drive control signal.

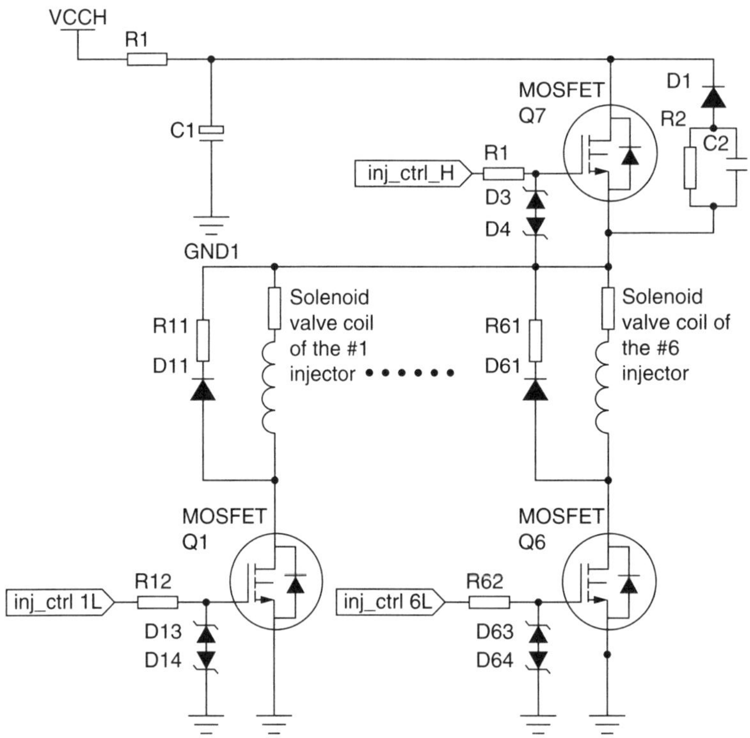

Figure 5.33 Single voltage and dual switch driving circuit.

The driving signal required for the simplified circuit is shown in Figure 5.32. In the driving control signal shown in Figure 5.32, in order to reduce the impact of the high-voltage switch opening on the other switch devices, the opening time of the low-voltage switch should be earlier than that of the high-voltage switch.

Combining the characteristics of the single power single switch circuit with the dual power dual switch circuit, a single power dual switch circuit has been designed. The circuit diagram is shown in Figure 5.33.

In the driving signal in Figure 5.34, there is a time difference between low-voltage and high-voltage switch control signals, which makes the opening time of the low-voltage switch MOSFET earlier and the cut-off time later than those of the high-voltage switch MOSFET. Then, a soft switch of the low-voltage switch MOSFET is realized. When

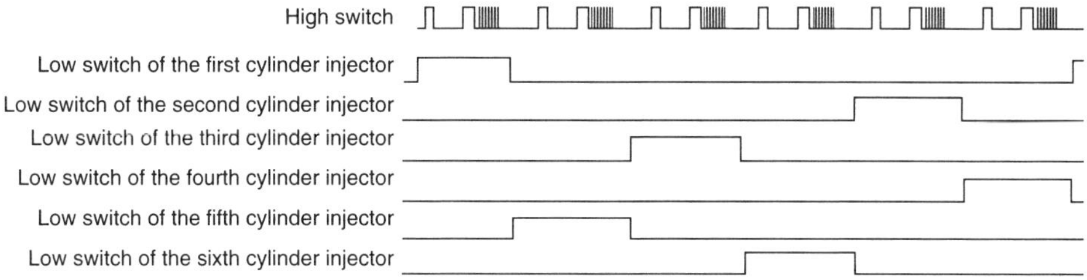

Figure 5.34 Single power dual switch drive control signal.

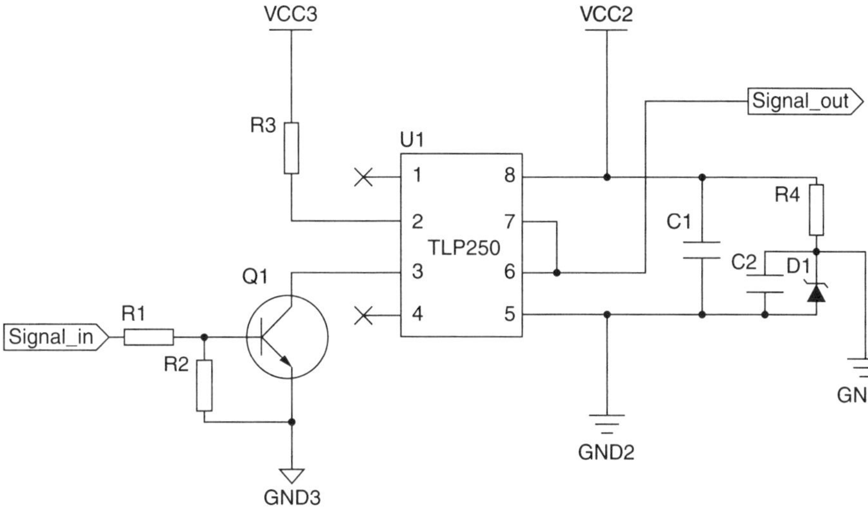

Figure 5.35 MOSFET switch drive circuit.

the low-voltage switch MOSFET turns on, because the high-voltage MOSFET switch device is in the off state, the low-voltage switch MOSFET turn-on current is 0. When the low-voltage switch MOSFET is cut off, because the high-voltage switch MOSFET is in the off state, the voltage of the low-voltage switch MOSFET is zero when it is cut off. Therefore, the absorption circuit low-voltage MOSFET switch device can be omitted, which greatly reduces the complexity of the circuit and improves the reliability.

Because the driving voltage of the power MOSFET is 15 V, in order to improve the switching speed of MOSFET devices, the gate of the MOSFET device not only needs drive voltage but also needs a larger driving current to establish the grid voltage. The output voltage of the EDA core device is 3.3 V and the maximum drive current is 20 mA. Namely, it cannot drive the MOSFET directly, and in order to reduce the interference, the circuit of the EDA core devices must be isolated using a power drive circuit. Therefore, an isolated drive circuit has been designed for a power MOSFET, as shown in Figure 5.35.

As shown in Figure 5.35, the MOSFET switch drive circuit adopts the MOSFET switch isolation driver chip TLP250. The chip is embedded with a photoelectric isolator, which can ensure electrical isolation between the input and output signals. The drive voltage waveform is shown in Figure 5.36. The instantaneous power of the injector solenoid

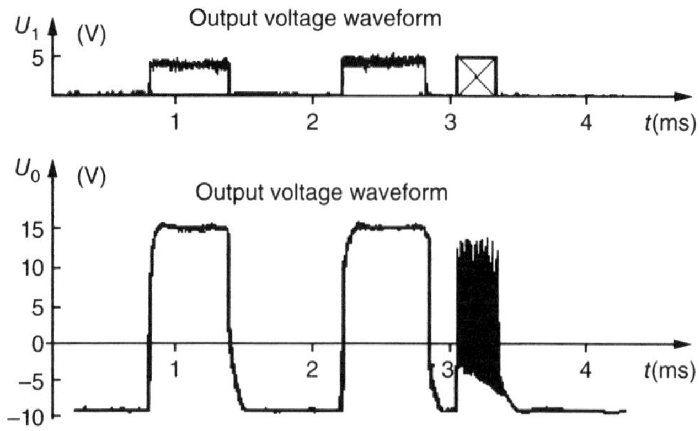

Figure 5.36 Voltage waveform of the MOSFET drive circuit input and output.

valve is relatively large and the drive current is up to 20 A. If 130 V of power is directly used to supply a 20 A current and the required output power is more than 2600 W, the actual test result is that the average driving single injector power consumption is less than 20 W. After adding the current limiting energy storage circuit composed of R1 and C1, when the injector is not in operation, the power supply charges C1 through L1. With the increasing voltage of C1, the current of the power supply decreases gradually. When the drive signal inj_ctrl of the solenoid valve of the injector drives the power MOSFET to turn on, the energy stored in capacitor C1 is released and the current through the solenoid valve coil rapidly increases from 0 to nearly 20 A. Meanwhile, R1 limits the mutation of the power current, limiting the current provided by the power to a smaller range. When the drive signal inj_ctrl of the solenoid valve is low, the power MOSFET cut-off and the power is charged to C1 through R1 to prepare for the next drive.

In order to ensure that every time the energy stored in C1 is enough to drive the injector solenoid valve, the capacity of C1 is large, but the increase of the C1 capacity will lead to a reduction of the charging speed, which will eventually lead to the decline of the driving voltage of the actual injector solenoid valve.

In order to ensure the current limiting performance of R1, R1 needs a certain resistance. If the resistance of R1 is too large, it will cause the charging speed to decrease. The energy consumption on R1 will increase and the efficiency will be reduced, so the parameters of R1 and C1 need to be carefully chosen.

According to the test of the diesel engine and the fuel injector, when the diesel engine is at the highest speed and full load, the fuel supply of a single injector in a single cycle is 286 mm^3. When the pulse width of the electronically controlled injector is 4 ms, the amount of fuel supply is 300 mm^3. The maximum design speed of the controller is 2000 r/min and the calculated minimum interval between two fuel supply processes is 6 ms.

Using the professional circuit simulation software Multsim2001, the circuit model, as shown in Figure 5.37, is set up, and the circuits under various R1 and C1 parameters are simulated.

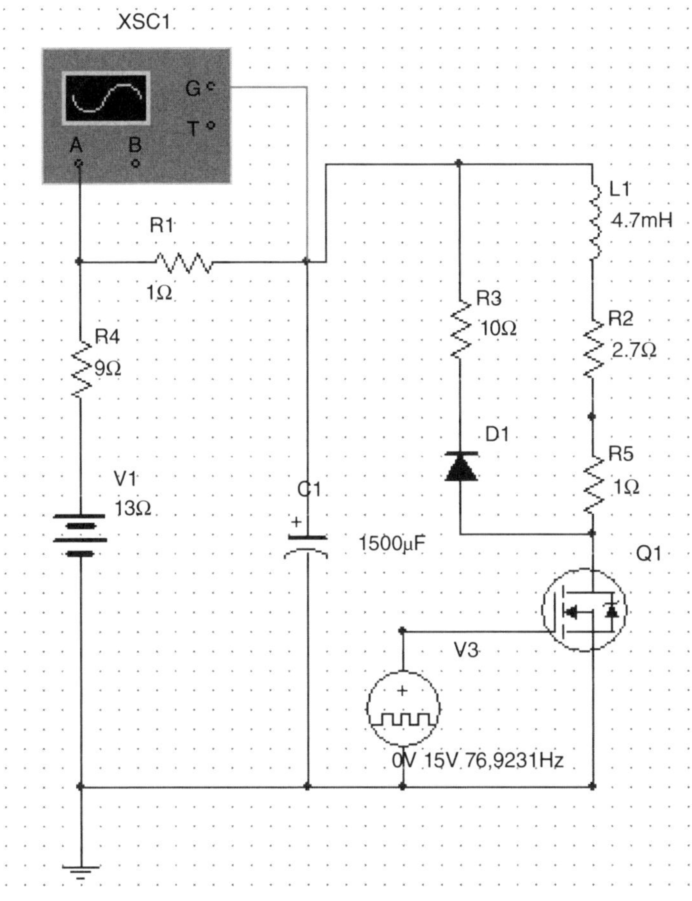

Figure 5.37 Simulated circuit.

By measuring the voltage waveform on the 1 Ω resistor in the circuit, the waveform of the current of the power supply can be obtained. Figure 5.38 is a simulation of the power source current waveform.

After calculating the energy and voltage drop consumed on the R1, and considering the size of the C1 device, R1 = 5 Ω and C1 = 1500 µF are selected.

5.5 Soft Core Development of the Field Programmable Gate Array (FPGA)

After designing the hardware circuit of the FPGA, it does not work immediately. It also needs to use EDA technology and the hardware description language VHDL to develop an FPGA soft core and give it certain functions. This section will introduce the process of developing a soft core for the FPGA.

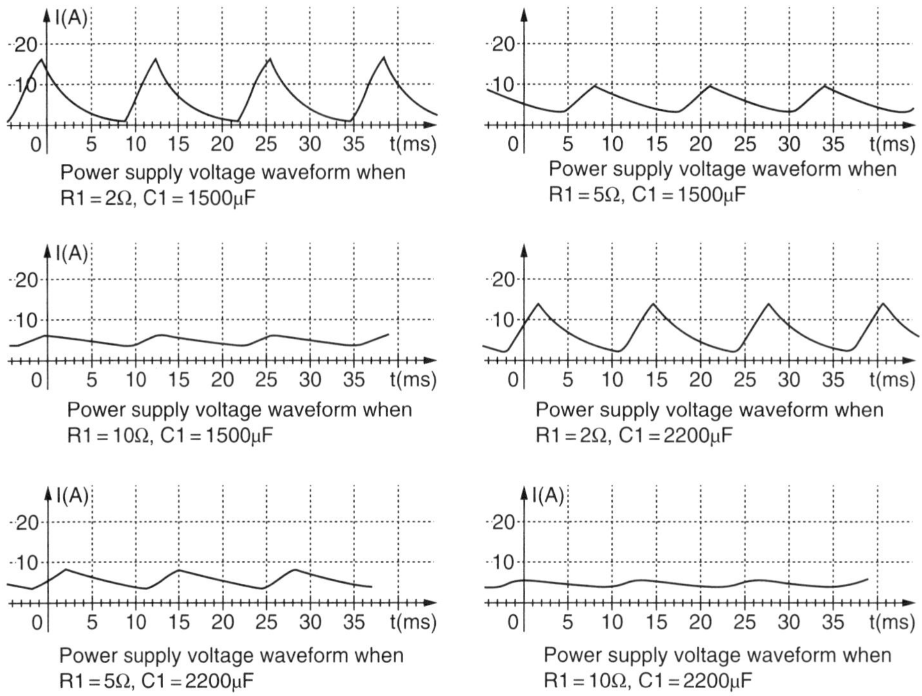

Figure 5.38 Simulation of the power source current waveform diagram.

5.5.1 EDA Technology and VHDL Language

5.5.1.1 Introduction of EDA Technology and VHDL Language

EDA is the abbreviation of electronic design automation. Namely, electronic design automation refers to the use of a computer to automatically design the electronic system. EDA technology is guided by computer and microelectronics technology. It brings together advanced technologies: computer graphics, topology, logistics, microelectronics technology and structural science, computational mathematics, and so on.

EDA technology uses the computer to replace artificial elements and completes the logic synthesis, layout, wiring, and design simulation of the digital system. The designer only needs to complete the description of the function of the system, which can then be processed by the computer software to get the design results. In addition, modifying the design is as convenient as modifying the software, which can greatly improve the design efficiency.

The HDL (hardware description language), which is widely used at present, is mainly divided into two kinds, VHDL and Verilog HDL. The most widely used one is VHDL. In 1982, under the planning of a branch case of the United States Department of Defense, IBM, TI, and some other companies began developing the VHDL (VHSIC HDL) language in 1983, where VHSIC stands for very high speed integrated circuit. In 1987, the language was standardized by the IEEE organization and was defined as the IEEE 1076–1987 standard. In 1993, it was further revised to the ANSI/IEEE 1076–1993 standard. Later, a VHDL program package, which can match the integration tool, was specifically named IEEE 1076.3 and became part of the 1076 IEEE standard. Recently, the new standard package IEEE 1076.4 (VITAL) was developed to become a model function library for the establishment of ASIC and FPGA.

5.5.1.2 Introduction of EDA Tools

The EDA design tool software that uses EDA occupies an important position in the application of EDA technology. EDA technology uses computers to complete automation of the electronic design, while EDA software based on the computer environment is the foundation of EDA technology. The special EDA design platform, QUARTUS II provided by ALTERA Company, is used in this section, and its boot and working interfaces are illustrated in Figures 5.39 and 5.40, respectively.

The structure of tool software QUARTUS II can be divided into five modules according to the main software involved in the EDA design process: the design input module, HDL synthesizer, timing and function emulator, adapter module, and download module.

Among them, the design input module is used for electronic design input and supports multiple expressions of electronic design input, such as the schematic input mode, the state diagram input mode, the waveform input mode, and the HDL language text input mode.

The HDL synthesizer is used to integrate the designs described by the HDL and transforms the design into the interconnect relationship of the chip. There are two kinds of commonly used HDL languages, VHDL and Verilog HDL. The HDL synthesizer introduced here is mainly for the two languages.

The timing and function emulator is used for sequential and logical simulations of design. Because of its high reliability, the emulator of EDA software plays a significant role in the design process of the FPGA.

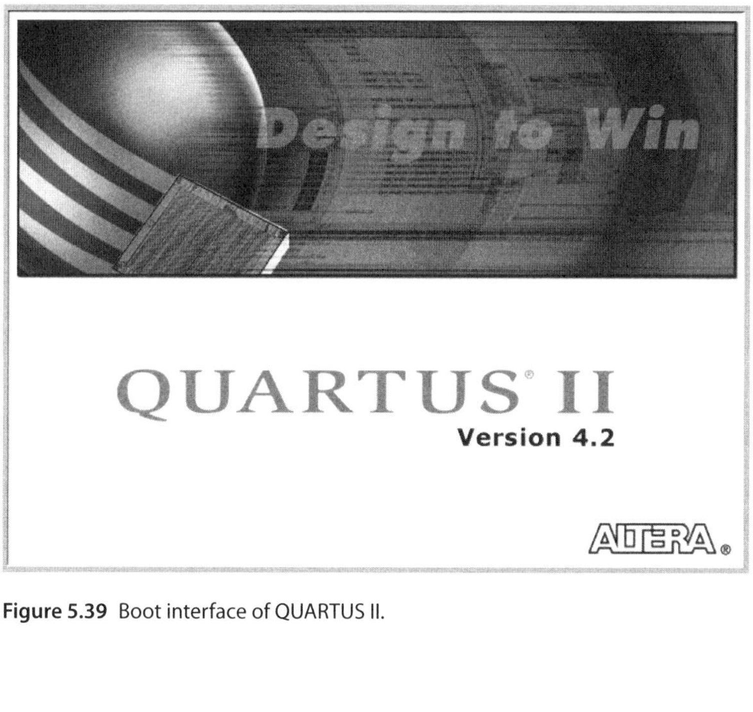

Figure 5.39 Boot interface of QUARTUS II.

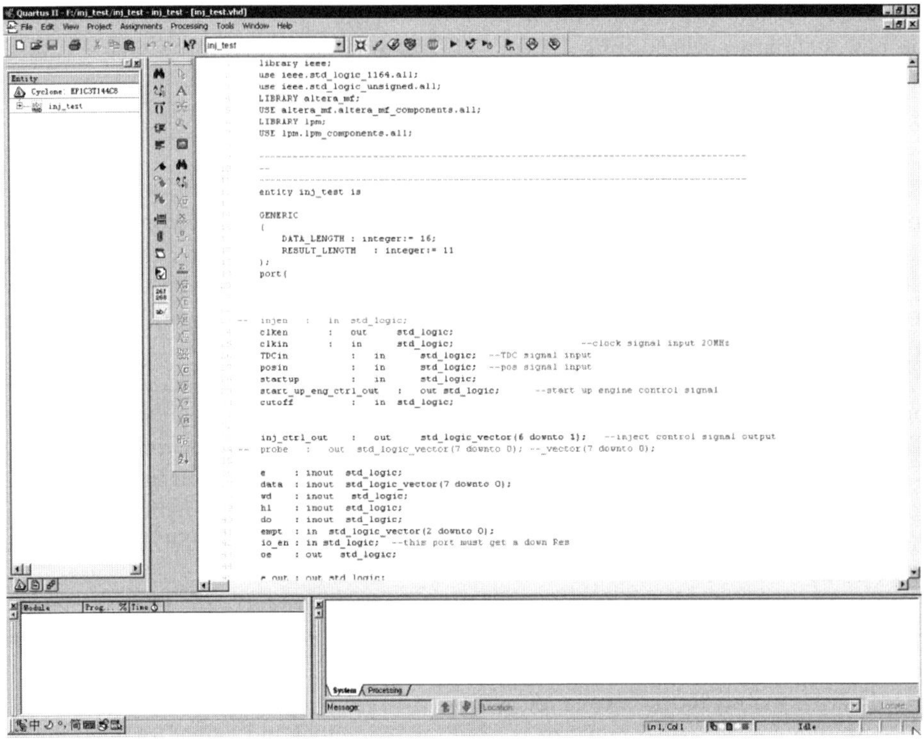

Figure 5.40 Working interface of QUARTUS II.

The task of the adapter module is to complete the layout and wiring of the target system on the device. An adaption is structure synthesis. The final output of the adapter is the download files that each manufacturer defines as its own, and is used for downloading to the device to implement the design.

The download module downloads the design to the corresponding actual device to realize the hardware design. Software parts are usually done by software that are specifically designed for device downloads or programming by the manufacture of programmable logic devices.

The complete FPGA development process is shown in Figure 5.41. In Figure 5.41, as shown in the development process, all steps before production can be completed by computer simulation, which greatly saves the design cost and improves the transparency and reliability of the device design.

5.5.2 Module Division of the FPGA Internal Function

In the design in this model, the EP1C3 chip integrates a number of functional modules, including the control strategy MAP query module, the pulse/switch signal acquisition module, the instantaneous rotational speed measurement module, the fuel injector control pulse generation module, the running state monitoring module, the display control module, the relay control signal generation module, and the core control

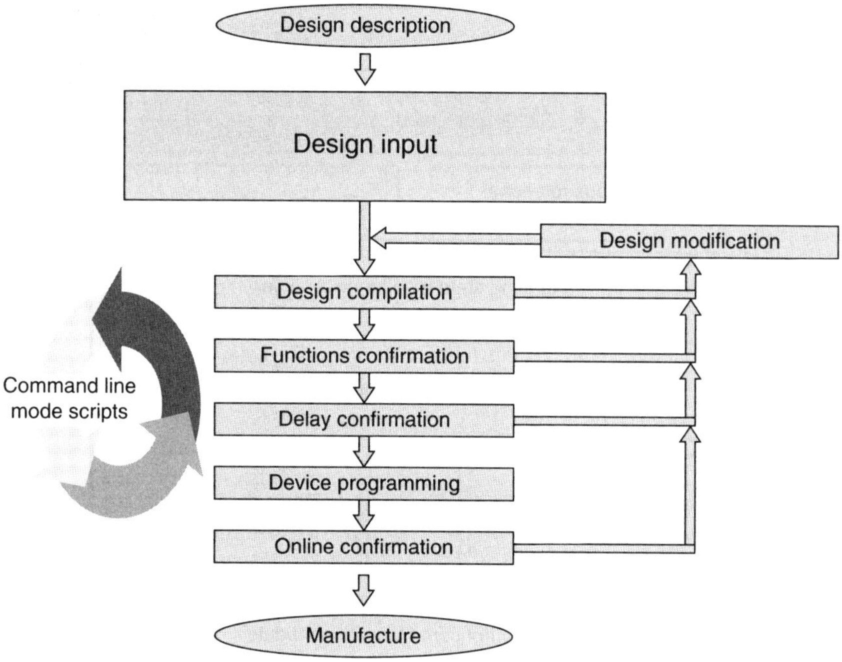

Figure 5.41 Development process of the FPGA.

communication module. The structure is shown in Figure 5.42. These modules are independent and can run in parallel.

The control signal MAP query module is the core of ECU control. This module queries the MAP stored in the FlashROM based on the acquired state information, obtains the final control parameters, and writes these control parameters to the corresponding address in the dual port RAM, which is read by the corresponding control module. The control signal MAP query module and peripheral module communicate through the dual port RAM. A 7 K programmable memory unit embedded in the FPGA can be configured as the multiple dual RAM memorizers different sizes and the same width. The memorizers can be read or written at the same time, and the read and write priority logic control is used to prevent the emergence of a read–write conflict, which enables the control strategy MAP query module to deal with multiple MAP query tasks at the same time, and greatly improves the response speed of the system.

In order to improve the reliability of the working system, a global clock control is used in all internal FPGA. A 20 MHz clock source is connected externally. In the timing module, multiple synchronous clock frequency signals are obtained through register frequency division, where CLK_1 is 10 MHz, CLK_2 is 5 MHz, CLK_3 is 2.5 MHz, and CLK_4 is 1.25 MHz. The waveforms of these clocks are shown in Figure 5.43.

All the modules in the FPGA are synchronized through these clocks, which greatly reduces the probability of the occurrence of the circuit glitches and improves the reliability.

The function of the instantaneous rotational speed measurement module is to measure the instantaneous rotational speed of the diesel engine. The module can output

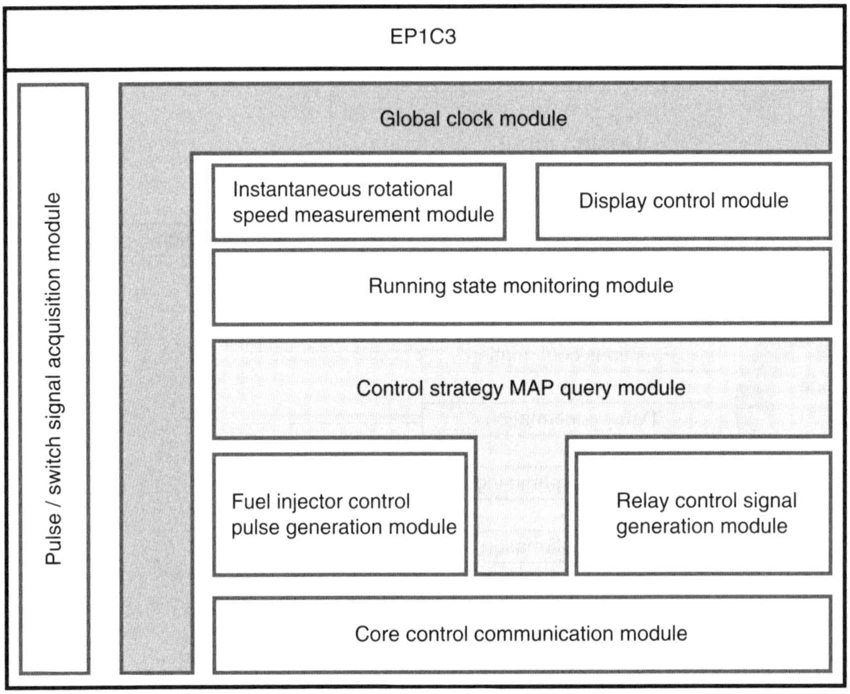

Figure 5.42 FPGA internal module structure.

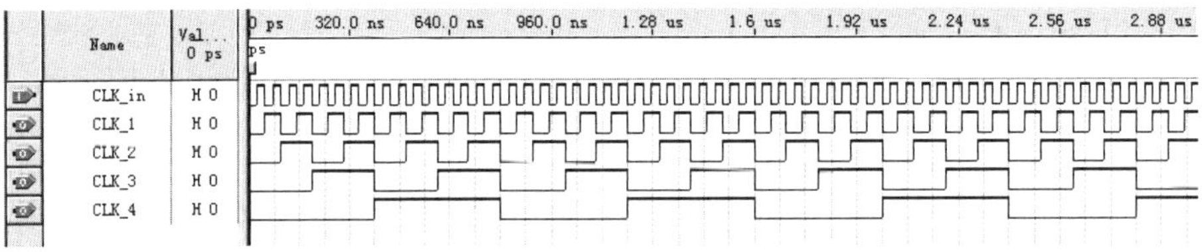

Figure 5.43 Waveforms of global clocks.

diesel engine instantaneous speed data of 16 bits, diesel instantaneous angular velocity difference data of 16 bits, and crankshaft position signal periodic data of 16 bits.

The measured instantaneous speed is used to guide the control strategy MAP query of the diesel engine. The measured periodic data of the pulse signal of the position of the crankshaft is used to assist the injector control pulse generation module to accurately determine the timing of the fuel supply. The instantaneous angular velocity difference data of the diesel engine can be used to determine the combustion state in the cylinder, as an assistant parameter for controlling the fuel supply.

The operation state monitoring module monitors all kinds of operation data, compares the measured data with the permissible data, immediately sends the alarm when the operation data exceeds the permitted range, and adopts emergency processing steps to limit or stop the operation of the diesel engine. In order to prevent a false alarm or a late alarm, the module uses two levels of redundancy design to ensure the reliability of the module.

The display control module is used to control the operation parameters of the LCD display, including the current operation parameters, the measurement data of the sensor group, and so on. When the failure occurs, the display module displays the fault information, which is convenient for the maintenance personnel to find the fault point.

The injector control pulse generation module sends out the injector control pulse according to the output parameters of the control strategy MAP diagram query module. The procedure to determine the timing of the fuel supply is divided into two steps. First, according to the current speed, the crank angle corresponding to the inherent delay of the injector is calculated. The angle difference between the fuel supply time and crankshaft position reference signal is deducted using the calculated delay angle; according to the new angle difference, the position of the crankshaft is counted. The part of the less than one crankshaft position signal period is determined by timing, and the timing value can be calculated by using the periodic data of the pulse signal of the crankshaft position.

When the fuel supply timing is determined, the width of the fuel supply pulse is determined using the timing method, and the functions of pre-injection and multisection injection can be realized.

Due to the machining accuracy, the timing frequency of the sensor and the injector electromagnetic response, mechanical hydraulic response, and other factors, the actual fuel supply starting point will have a certain delay compared to the ideal value. There is a relationship between the delay time and the driving voltages of the individual injector, the common rail pressure, and the fuel injector solenoid valve. In order to ensure the actual fuel supply starting point consistent with the set value, to generate a control signal there must be an advance value. To achieve the time compensation in advance, the delay time can be measured by experiment and making a compensation table. Then, the delay time can be obtained by checking a specific working table and then compensating for determination of the timing of the fuel supply.

The core control communication module controls and is responsible for the communication of parallel data communication between the single-chip microcomputer and FlashROM of an internal core. Communication with the MCU is by an 8 bit parallel communication mode, occupying a total of 12 pin resources, and the actual data is 16 bits, through a 3 bit transmission. The first byte represents the attribute data, the second bytes of data represents the high 8 bits of transmitted data, and the third bytes represents the low 8 bits of transmitted data. A parallel parity bit is also provided on the transmission interface to ensure the correctness of the internal data transmission.

FlashROM uses a 16 bit parallel communication interface and takes 39 pins, of which 16 bits are data communication pins, 20 bits are data pins, and 3 bits are reading and writing control pins. According to the limit of the device, the design of the interface has the highest data reading speed of 5 Mbytes/s, which can fully meet the requirement of the lookup table.

5.5.3 Design of the Rotational Speed Measurement Module

5.5.3.1 Measuring Principle

The principle of the rotational speed measurement module is to measure the cycle of the crankshaft position signal and to get the instantaneous speed of the crankshaft after calculation. When the crankshaft turns 6°, a crankshaft position signal pulse is generated. By measuring the period of the pulse signal, the crankshaft's instantaneous speed can be obtained. The reference frequency of the cycle is 5 MHz, so the resolution of the module to the pulse period is 200 ns. The speed measurement error dn caused by the resolution of the measuring clock can be calculated by

$$d_n = |n_m - n_r| = |1/T_m - 1/T_r| = |T_r - T_m|/(T_r \times T_m) \approx n^2 \times |T_r - T_m| \quad (5.8)$$

where

$$|T_r - T_m| = 2 \times 10^{-7}$$

$$n < 2000\,\text{r/min}, d_n < 0.8\,\text{r/min}$$

Because the rotational speed of the diesel engine runs much faster than 10 r/min, the measurement error of 0.8 r/min can meet the operation requirements.

Because the diesel engine is a power machine working periodically, the speed is not constant during the working process. Even in the same cycle, the speed fluctuation is obvious. Figure 5.44 shows a speed change curve of a four-cylinder diesel engine during its working process.

From Figure 5.44, it can be seen that the speed change of the diesel engine is obvious before and after the fuel supply. With this obvious change, it can be judged whether the combustion is carried out inside the cylinder. If the ignition fails, there will be no change in the instantaneous speed. Therefore, the speed measurement module is needed to measure the fluctuation of the speed to determine whether the cylinder combustion is successful. Therefore, to add a subtraction device with an output speed difference in the speed measurement module is needed.

In the realization of the speed control strategy, the controller is supposed to find the average speed of the diesel engine. Therefore, it is necessary to average the multiple speed measurements and add an accumulator to find the average value in the speed measurement module.

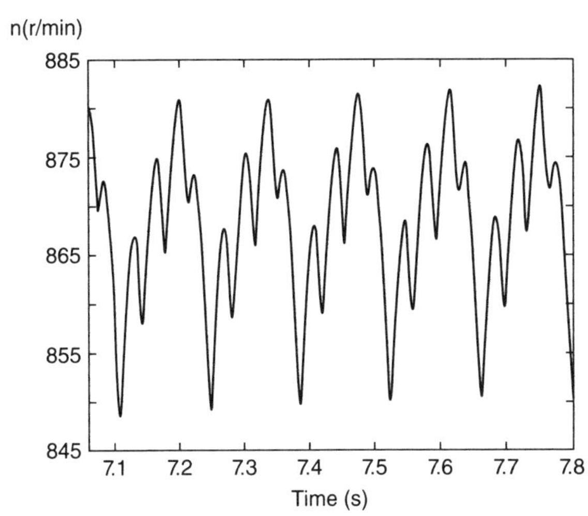

Figure 5.44 Rotational speed fluctuation curve of the diesel engine.

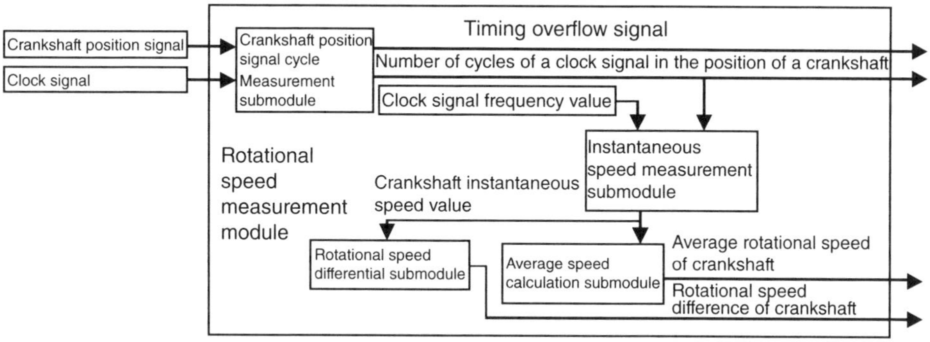

Figure 5.45 Internal structure of the speed measurement module.

5.5.3.2 Structure Design

According to the measurement principle described in Section 5.5.3.1, the rotational speed measurement module should include the crankshaft position signal cycle measurement submodule, the instantaneous speed calculation submodule, the average speed calculation submodule, and the rotational speed differential submodule. The structural relationship between these submodules is shown in Figure 5.45.

The crankshaft position signal cycle measurement submodule is the core of the speed measurement module, and its working sequence is shown in Figure 5.46. In the figure:

clk_2 = 5 MHz clock signal (input signal)
pos = crankshaft position signal (input signal)
mea_value = cycle number of clock signals in a cycle signal of a crankshaft position (output data)
dob_pos = signal after 1/2 frequency division of the crankshaft position signal (internal signal)
mea_timer = timer accumulating signals (internal signal)

When the crankshaft position signal periodic measurement submodule works, first the frequency of the crankshaft position signal POS is divided to generate the dob_pos signal, which can ensure the elimination of the measurement error caused by the change of the duty ratio of the crankshaft position. When the value of the dob_pos signal is "0," the value of timer accumulated signal mea_timer is set to 1; when the value of the dob_pos signal is "1," the timer accumulated signal mea_timer added "1" to the rising edge of each clock signal clk_2; when the value of the dob_pos signal changes from "1" to "0," mea_value reads the value of mea_timer, because the dob_pos signal is only synchronized with the pos signal and is not synchronized with the clock signal; before the value of mea_timer is 1, there is always a period of time that can make mea_value able to read the value of the mea_timer, so there will be no error reading.

When the crankshaft speed is lower than the minimum measured speed of 76.3 r/min, mea_timer will overflow. Therefore, to prevent the measurement error caused by mea_timer overflow, when mea_timer is going to overflow, the overflow signal will be set, and mea_timer value will be limited to prevent it from spillover again.

The crankshaft position signal cycle measurement submodule is divided into three processes. The first process is mea_p0, which is a timing process. The second process is

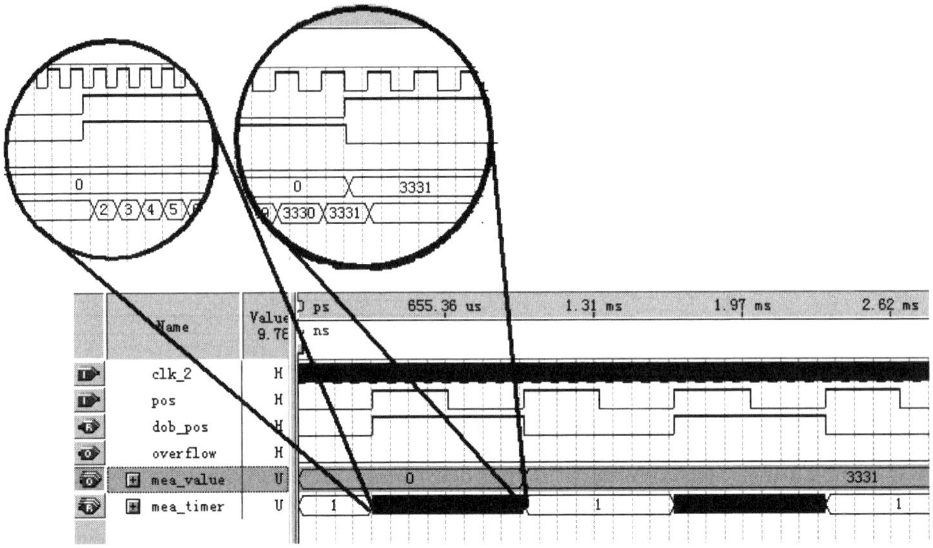

Figure 5.46 Working sequence of the crankshaft position signal cycle measurement submodule.

mea_p1, which is the frequency division process of the crankshaft position signal. The third process is mea_p2, which is the replication process for the output data mea_value.

The instantaneous speed calculation module is actually a division module and the calculation formula of the instantaneous speed is shown by

$$\text{Nmea} = 60/[(\text{mea\_value}/\text{fclk}) \times 360/6]\ (\text{r/min}) \tag{5.9}$$

where

Nmea = instantaneous speed of measurement
mea_value = cycle number of the clock signals in a cycle signal of a crankshaft position
fclk = frequency of the clock signal, which is 5 × 106 Hz

Substitute fclk = 5 × 106 into formula (5.8). Then

$$\text{Nmea} = 5 \times 106/\text{mea\_value}\ (\text{r/min}) \tag{5.10}$$

The 5 × 106 is translated into binary with 23 bits. If a divider IP provided by the software is used and a parallel divider with a 23 bit dividend and a 16 bit divisor is allocated, 543 logic units and 18.7% of the chip sources will be taken, and the actually used divisor of the divider is in the range (2500–65 536), and is not a constant. Therefore, to specially allocate a parallel divider is a waste of system resources, so a serial divider with a constant dividend is designed, and the quotient is limited to 11 bits, which is less than 2023. The working principle is similar to a manual process of the division operation. After 7–23 clock cycles, the results can be obtained. After compilation, 101 logic units are in use, occupying 3% of the system resources, and the working sequence of the instantaneous speed calculation module is shown in Figure 5.47, where

clk_2 = clock signal
mea_value = output value of the crankshaft position signal cycle measurement module (input)

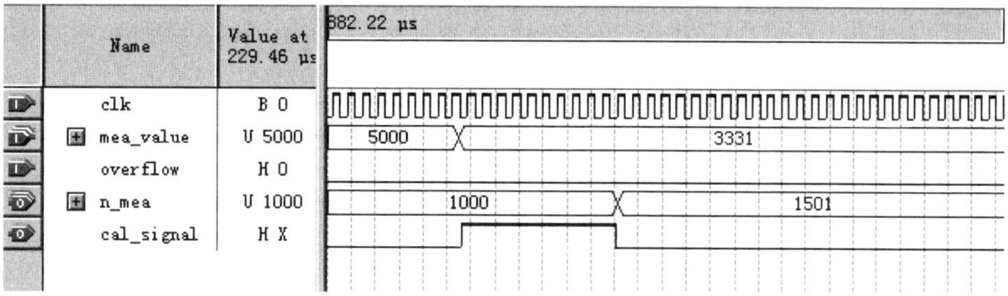

Figure 5.47 Operation time sequence of the speed measurement module.

overflow = module overflow sign of the crankshaft position signal cycle measurement module (input)
N_mea = calculated instantaneous speed value (output)
cal_signal = calculating process sign (output)

When limited to length, the VHDL language description of the module is omitted.

The module consists of a 6 bit binary counter and a 16 bit accumulator. When working, the 16 bit accumulator adds the sum of 32 times of instantaneous speed output values, counted by the 6 bit counter. When the counter counts to 32, the result of the accumulator is taken in the first 11 bits, and the average of the 32 speeds can be obtained. The action of the counter and accumulator is triggered by the calculation process cal_signal of the instantaneous speed calculation module. When the falling edge of the signal appears, the accumulator adds the value of n_mea to the value in its registers, and finally puts the result into the register of the accumulator. After compiling, the average speed calculation module takes up 40 logic units, occupying 1.4% of the total resources of the chip. The module work sequence is shown in Figure 5.48, where

n_ave = average rotational speed output of the average speed calculation module
cal_signal = symbol of the calculation process signal, output by speed calculation module (input)
n_mea = instantaneous speed value, output by the speed operation module (input)
adder_a = accumulator register (internal signal)
count = accumulative counter register (internal signal)

The speed difference module consists of two 11-bit registers and a subtractor, triggered by the output signal cal_signal of instantaneous speed calculation module. When the cal_signal rising edge appears, the cal_signal triggers the two registers. One of the triggers gets another register's contents and another register obtains the speed deviation of the new value. Then the subtractor subtracts the contents of the two registers

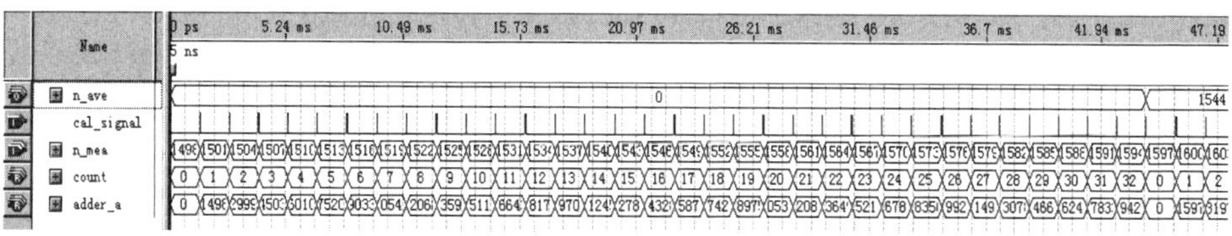

Figure 5.48 Working sequence of the average speed calculation module.

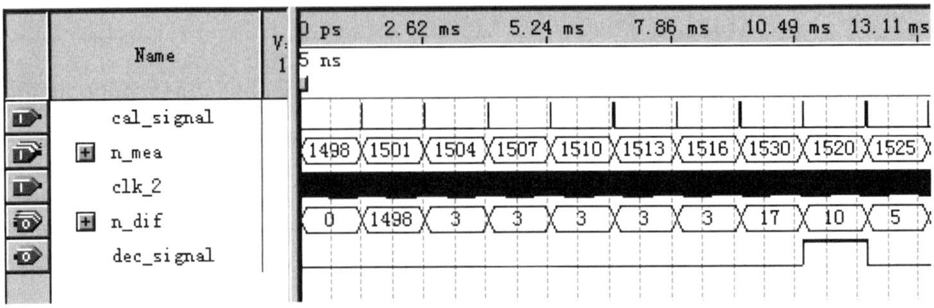

Figure 5.49 Sequence of the speed difference module.

and obtains the data of the speed difference. Because the common subtraction is the implementation of the unsigned operation, in the process of subtraction, the two values need to be compared first and the subtrahend and minuend need to be determined to ensure the process will not overflow.

The speed difference module occupies 58 logic units. After compilation, it occupies 2% of the total resources of the chip. The sequence of the speed difference module is shown in Figure 5.49, where

cal_signal = symbol signal of the calculation process output by the speed calculation module (input)
n_mea = instantaneous speed value output by the speed operation module (input)
clk_2 = 5 MHz clock signal (input)
n_dif = absolute value of the rotational speed difference (output)
dec_signal = deceleration marker (output)

When the rotational speed of the crankshaft is reduced or unchanged, the dec_signal output is "1" and when the crankshaft speed rises, the dec_signal output is "0."

The first two speed difference calculation processes in Figure 5.48 show that the value in the register starts as 0, so the data in the first two differential processes are invalid. In an actual working process, the speed will not change abruptly because of the starting process, so there is no such phenomenon.

5.5.4 Design of the Control Pulse Generation Module for the Injector

5.5.4.1 The Function, Input, and Output of the Injector Control Pulse Generation Module

The function of the injector to control the pulse generation module is to generate an injector control pulse based on the calculated or inquired control regulation. The input includes the injection law data, the cycle measurement data of the crankshaft position signal, the pulse of the crankshaft position sensor, and the top dead center pulse. The fuel injection law is generated by the upper control module and is stored in the double port RAM. These data include:

(1) The setting value *A* of the timing fuel supply.
The value is 10 bit, ranging from 0 to 1023, and the unit is a 6/64° crank angle. The value indicates the crankshaft angle difference between the set timing fuel supply

α and 66° before the top dead center of the cylinder. The value is

$$A = (66 - \alpha)/(6/64) \tag{5.11}$$

The reason why the unit of A is a 6/64° crankshaft angle is that it is convenient for the subsequent calculation of the timing value.

(2) When the value of the crankshaft position signal equals any of 1, 21, 41, 61, 81, 101, the fuel supply parameters of the corresponding cylinder in the double port RAM are read.

(3) According to the obtained fuel supply parameters, the calculation and timing values of the specific fuel supply time are calculated by the following formula, and the data are input to the corresponding registers.

(a) Count of fuel supply time, pos_count:

$$\text{pos\_count} = A\ (9 \text{ downto } 6) + (n - 1) * 20 + 1 \tag{5.12}$$

where

A (9 downto 6) = binary high 4 bit value of the set value A of the fuel supply timing

n = order of fuel supplies, where $n = 1$ for the first cylinder fuel supply, $n = 2$ for the fifth cylinder fuel supply, $n = 3$ for the third, $n = 4$ for the sixth, $n = 5$ for the second, $n = 6$ for the fourth cylinder fuel supply

(b) Timing value of the fuel supply time, pre_inj_start_timer:

$$\text{pre\_inj\_start\_timer} = ((A\ (5 \text{ downto } 0) \times \text{mea\_value}) \gg 6) \tag{5.13}$$

where

A (5 downto 0) = binary low 6 bit value of the set value A of the fuel supply timing

mea_value = cycle number of the clock signals in a crankshaft position cycle signal

$\gg 6$ = represents a 6 bit operation to the right, equivalent to being divided by 64

(c) Timing value of the pre-injection at the end, pre_inj_end_timer:

$$\text{pre\_inj\_end\_timer} = \text{pre\_inj\_start\_timer} + \text{w1} \tag{5.14}$$

where

w1 = timing value corresponding to the pulse width of the pre-injection

(d) Timing value of the main injection at the start time, main_inj_start_timer:

$$\text{main\_inj\_start\_timer} = \text{pre\_inj\_end\_timer} + \text{w2} \tag{5.15}$$

where

w2 = time difference between the pre-injection and main jet timing value

(e) Timing value of the main injection high-voltage drive at the end time, main_inj_hi_end_timer:

$$\text{main\_inj\_hi\_end\_timer} = \text{main\_inj\_start\_timer} + \text{w4} \tag{5.16}$$

where

w4 = timing value of the driving time of the main jet

(f) Timing value of the main injection at the end time, main_inj_end_timer:

$$\text{main\_inj\_end\_timer} = \text{main\_inj\_start\_timer} + \text{w3} \quad (5.17)$$

where

w3 = timing value of the main injection time

(4) When the count value of the crankshaft position signal is equal to that of the set value of the fuel supply time, the trigger timer starts timing.
(5) When the timing value of the timer is equal to the timing value of the fuel supply time, the fuel supply control signal is set to "1."
(6) When the value of the timer is equal to the timing value at the end time of the pre-injection, the fuel supply control signal is set to "0."
(7) When the value of the timer is equal to the timing value at the start time of the main injection, the fuel supply control signal is set to "1."
(8) When the value of the timer is equal to the timing value at the end time of the main injection, the step-down pulse generation timer is started.
(9) When the value of the timer is equal to the timing value at the end time of the main injection, the fuel supply control signal is set to "0" and the work of the step-down pulse generation timer is finished.
(10) The fuel supply control signal and the output pulse signal of the step-down pulse generation timer are carried out and operated, and then the output is the control signal of the injector.

The above is the injection control signal generation process but neglects the compensation for the opening delay time of the injector nozzle. In order to ensure the accurate control of the actual injection rule, when the fuel injection control signal occurs, the delay of the injection should be compensated and the fuel supply control signal should be issued in advance. There are two ways to calculate time compensation.

5.5.4.1.1 Shortening Timing Compensation Method

The principle of this method is to shorten the timing phase accurately before the start of the injection, in order to achieve the purpose of sending the fuel supply control pulse in advance. Since the injection delay data are described in time units during the timing in fuel injector, the shortening timing compensation method is the easiest. However, the length of time before the injection process starts is uncertain, and the timing process is less than the time of turning the crankshaft over 6°, so when the diesel engine is at the rated speed, this process is very short. To reduce the timing value to compensate is very difficult, so the time of the timing process has to be shortened.

Because the injection delay time of the injector is usually less than 1.8 ms, the timing procedure can exceed 1.8 ms when the length of the counting process of the crankshaft position signal is reduced, and then compensation by shortening the timing can proceed. The timing of the compensation is illustrated as shown in Figure 5.50.

By calculation, when the diesel engine is in the controller design of a maximum speed of 2000 r/min, the corresponding crank angle of 1.8 ms is 21.6°. Therefore, four counting periods of the crankshaft position signal can be reduced to increase the timing time of the crankshaft so that it can turn over the 24° crank angle. Under the maximum speed of 2000 r/min, this time is 2 ms and is enough for the compensation process.

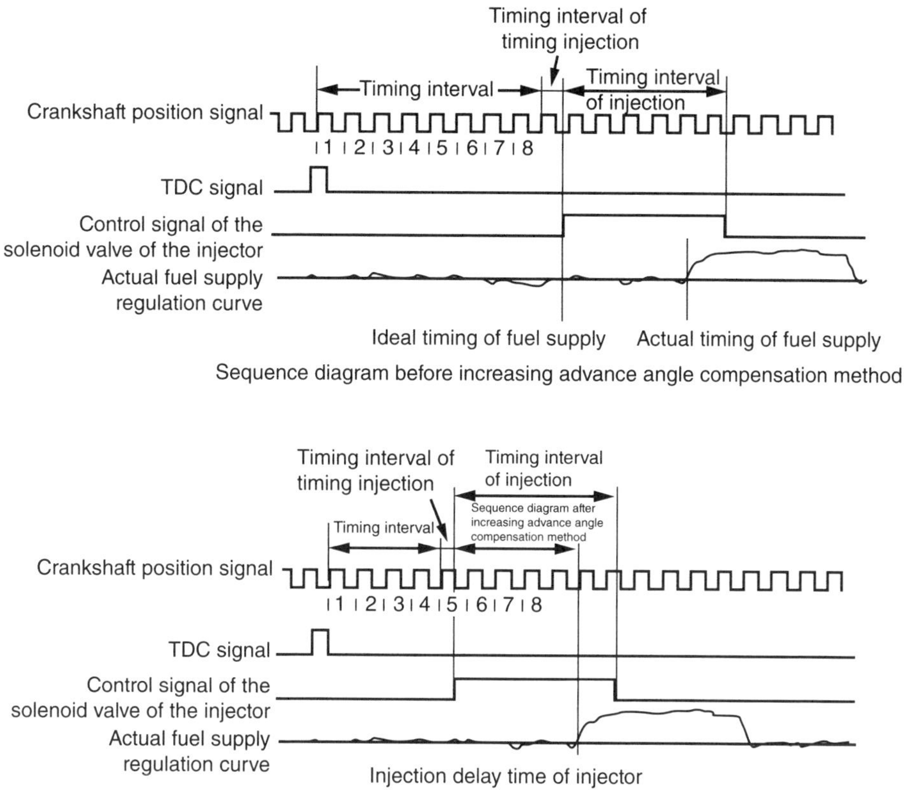

Figure 5.50 Sequence diagram of the shortening time compensation method before and after compensation.

When the delay compensation is used in this way, the counting and timing formula are as follows:

$$\text{pos\_count} = A(9 \text{ down to } 6) + (n - 1) * 20 - 3 \tag{5.18}$$

where

$$\text{pre\_inj\_start\_timer} = (((A(5 \text{ downto } 0) + 4) \times \text{mea\_value}) \gg 6) - \text{wdelay}$$

$$\text{wdelay} = \text{to the injection delay of the injector}$$

The formulas for other parameters are not changed.

5.5.4.1.2 Increasing the Advance Angle Compensation Method

Another method is to change the advance angle compensation. By subtracting a certain value of B of the fuel injection advance angle, a certain advanced time t1 will exist between the actual corresponding time and the target injection advance angle. If the advanced time value is greater than the injection starting delay, t1 > tdelay, and by the time delay t2 = (t1 – tdelay), the injection advance angle compensation can be achieved. The process is shown in Figure 5.51.

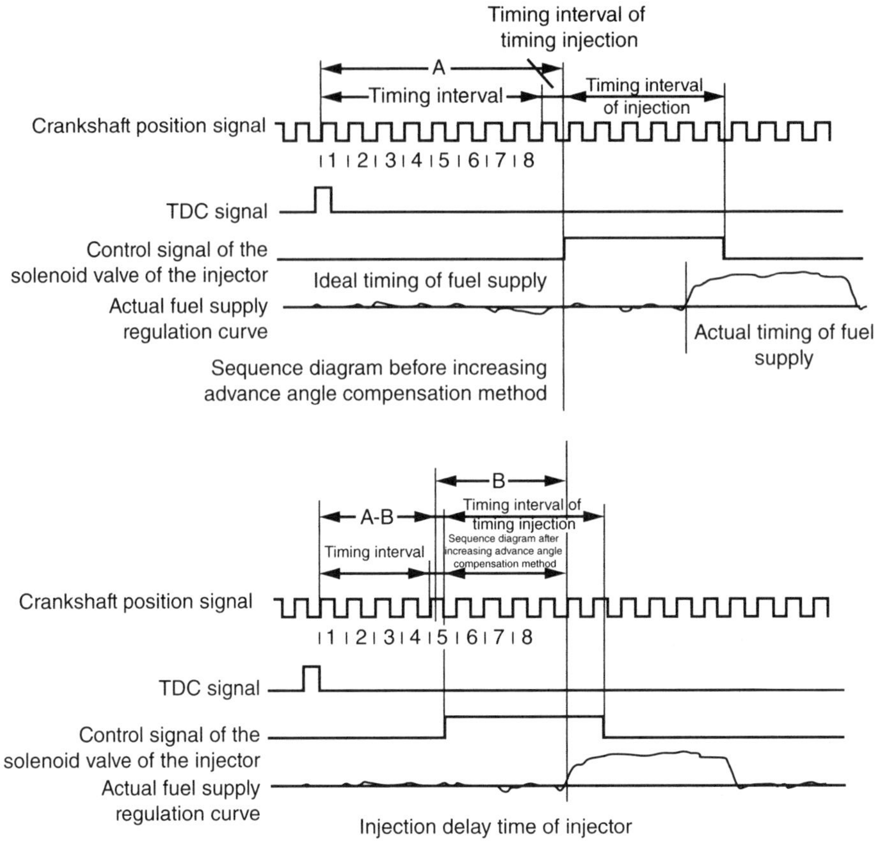

Figure 5.51 Sequence diagram before and after increasing the advance angle compensation method.

When the injection advance angle is advanced by B, the amount of time ahead of the injection time t1 is shown by

$$t1 = (B \times 6/64)/(360 \times n\_mea/60) = B/(n\_mea \times 64)\ (s) \tag{5.19}$$

where

n_mea = instantaneous speed of the crankshaft

If $B = n\_mea/8$,

$$t1 = (n\_mea/8)/(n\_mea^{*}\ 64) = 1/512\ (s) = 1.953\ (ms)$$

According to the existing calibrating test of the injector, 1.953 ms is greater than all the injectors' starting delay time. Therefore, it can be used to compensate the injection delay of the injector. The timing value of the corresponding 5 MHz clock at this time is 1.953 * 5000 = 9766.

This method is used to compensate the injection starting delay, and the calculation formulas for counting and timing are as follows:

$$A' = A - (\text{n\_mea}/8)$$

$$\text{pos\_count} = A'\,(9\ \text{downto}\ 6) + (n-1)^{*}\,20 + 1 \tag{5.20}$$

$$\text{pre\_inj\_start\_timer} = (((A'(5\ \text{down to}\ 0)) \times \text{mea\_value}) \gg 6) + 9766 - \text{wdelay} \tag{5.21}$$

If w5 = 9766−wdelay,

$$\text{pre\_inj\_start\_timer} = (((A'(5\ \text{down to}\ 0)) \times \text{mea\_value}) \gg 6) + \text{w5} \tag{5.22}$$

The formulas for other parameters are not changed.

5.5.4.2 The Realization of the Control Pulse Generation Module of the Injector

According to the working principle mentioned above, because the generation of the injector control pulse is a sequential process, the pulse generation module of the fuel injector can be completed through the sequential state machine. First, according to the sequence process, the whole work process is divided into several states, and the parameters of the output signal in each state are determined, and then the next state encoding is set.

With this method, the continuous process of the injection control pulse can be realized according to the principle of the injection control pulse generation given in Section 5.5.2.

6

Research on Matching Technology

The matching of the common rail system and the corresponding diesel engine is a complex and important work. The matching of the common rail system and diesel engine is the optimization design and matching of the common rail system component structure and the common rail system control parameters based on the performance of the diesel engine, to achieve an excellent diesel engine performance. The matching technology of common rail system and diesel engine mainly includes the matching technology of the common rail system components and the optimization and injection control MAP optimization calibration technology.

6.1 Component Matching Technology of the Common Rail System

The matching of the common rail system components and diesel engine mainly includes: matching design of the high-pressure fuel pump and diesel engine, matching design of the common rail and diesel engine, and matching design of the fuel injector and diesel engine.

6.1.1 Matching Design of the High-Pressure Fuel Pump

According to the fuel consumption rate of the diesel engine at the rated operating condition, the amount of fuel per cycle required for the diesel engine can be calculated:

$$V_c = \frac{10^6 b_e P_e \tau}{120 n \rho} \tag{6.1}$$

In the formula, V_c is the amount of circular fuel injection, mm^3/str; b_e is the fuel consumption rate of the calibration power point, g/(kW h); P_e is the calibration power of diesel engine, kW; n is the rotational speed of the diesel engine, r/min; ρ is fuel density, kg/m^3; and τ is the diesel engine stroke number.

The total fuel supply amount of the high-pressure pump includes fuel injection Q_{inj}, high-pressure recovery fuel Q_r, and fuel leakage Q_l. The maximum circulation fuel supply of the high-pressure pump can be obtained by

$$Q_{pump} = Q_{inj} + Q_r + Q_l \doteq iQ_{inj} \tag{6.2}$$

Common Rail Fuel Injection Technology in Diesel Engines, First Edition.
Guangyao Ouyang, Shijie An, Zhenming Liu and Yuxue Li.

In the formula, i takes the value of $3 \sim 5$. It can be seen from Eq. (6.2) that the high-pressure fuel pump must provide several times more fuel supply than the fuel injection volume to meet the needs of the system. In addition, the high-pressure pump must match the maximum speed of the diesel engine.

6.1.2 Matching Design of the Rail Chamber

The rail chamber is used to store high-pressure fuel. The design of the rail chamber mainly takes account of the suppression of pressure fluctuation caused by injection of the high-pressure fuel pump and fuel injector, and the reduction of the pressure of the rail chamber and the following time of the target rail pressure. When the volume of the rail chamber increases, the pressure fluctuation tends to be mild, but the pressure setting time in the rail chamber increases. Therefore, the choice of the volume of the rail chamber should take into account the total volume of the diesel engine and the circulating fuel supply of the high-pressure pump, so that in the process of fuel supply and injection, the pressure fluctuation in the chamber is small and the response of the rail pressure changes quickly.

6.1.3 Matching Design of the Injector

The electronically controlled injector is a key component in the high-pressure common rail system. Its structural parameters directly affect diesel engine combustion, and then cause a significant impact on the performance of the diesel engine. Therefore, matching the design of the fuel injector is most critical. In matching the design of the injector and diesel engine, the electronically controlled injector should not only satisfy the requirement of the circulating fuel injection, but also more importantly it must meet the structural parameters of the injector nozzle, such as the number of nozzle orifices, nozzle diameter, nozzle length and nozzle angle, and optimize matching of the combustion chamber. The electronically controlled injector has many structural parameters and complex interaction and restriction relationships with the injection parameters. It is time-consuming and costly to carry out a large number of program trials and trial work. The computational fluid dynamic (CFD) working process software can accurately simulate the influence of injector structure parameters and injection parameters on the combustion and emission performance of high-pressure common rail diesel engine, but to get the best solution, inevitably a higher computation cost will be consumed.

DoE technology can not only show the relationships among parameters, optimize the design effectively, but can also save laboratory cost, shorten the product development cycle, and improve product quality. At present, advanced DoE technology is dominated by DoE and technological support enterprises in Europe and in the United States, such as IAV in Germany, AVL in Austria, and Ford in the United States. The application and development of DoE technology in China are slow and basically stop at the conventional test method. Modern DoE technology is a frontier subject based on mathematical modeling, statistical theory, and computer aided modeling, based on model optimization. In this chapter, DoE technology is introduced into matching the design of the electronically controlled injector and the combustion chamber of a high-pressure common rail diesel engine. Based on the experimental calculation matrix for the DOE design injector structure parameters and injection parameters, the numerical calculation is made by using FIRE software, the sensitivity and the interaction effect of different parameters is studied, and then the response surface approximation model is applied to construct the optimization target function of combustion and emission. Finally, the genetic algorithm is applied to optimize the approximate model, and the optimal combination of structural parameters and injection parameters of the injector is determined.

The rated speed of the diesel engine is 1500 r/min, the rated power is 186 kW, and the technical specifications are shown in Table 6.1. The combustion chamber of the diesel engine is a straight port deep ω shape with a moderate swirl strength, and has an offset of 2.8 mm from the center line of the cylinder. The structure size of the injector and the installation location of the combustion chamber are shown in Figure 6.1.

Taking this diesel engine as the target, and the economy, emission, and combustion noise as the indexes, the matching optimization is made on six injection system parameters: the injection starting point, the number of the injection orifices, the diameter of the orifice, the injection angle, the ratio of the length to the diameter of the nozzle, and the height of the nozzle. The optimization design process is shown in Figure 6.2.

Table 6.1 Technical specification parameters of a certain type of diesel engine.

Engine type	6-cylinder, V shape, water cooling, 4-stroke, direct injection
Cylinder diameter × stroke (mm)	128 × 140
Crank/connecting rod	0.255
Rated speed (r/min)	1500
Rated power (kW)	186
Compression ratio	15

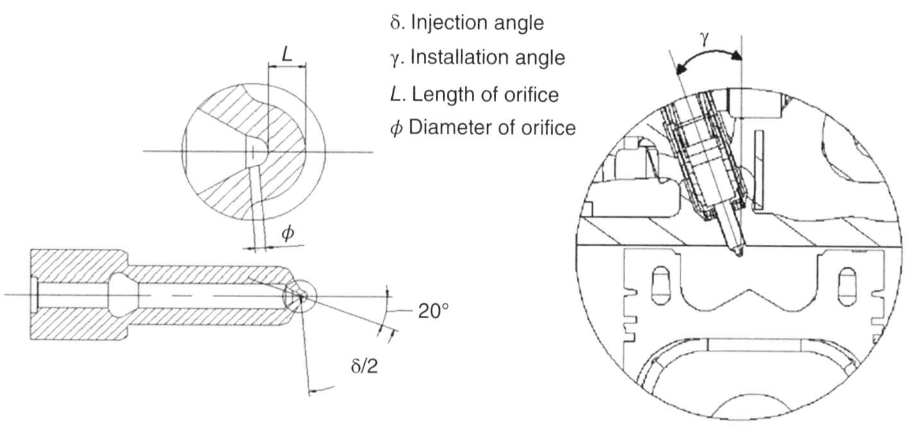

Figure 6.1 Structure sketch of the nozzle and its installation position.

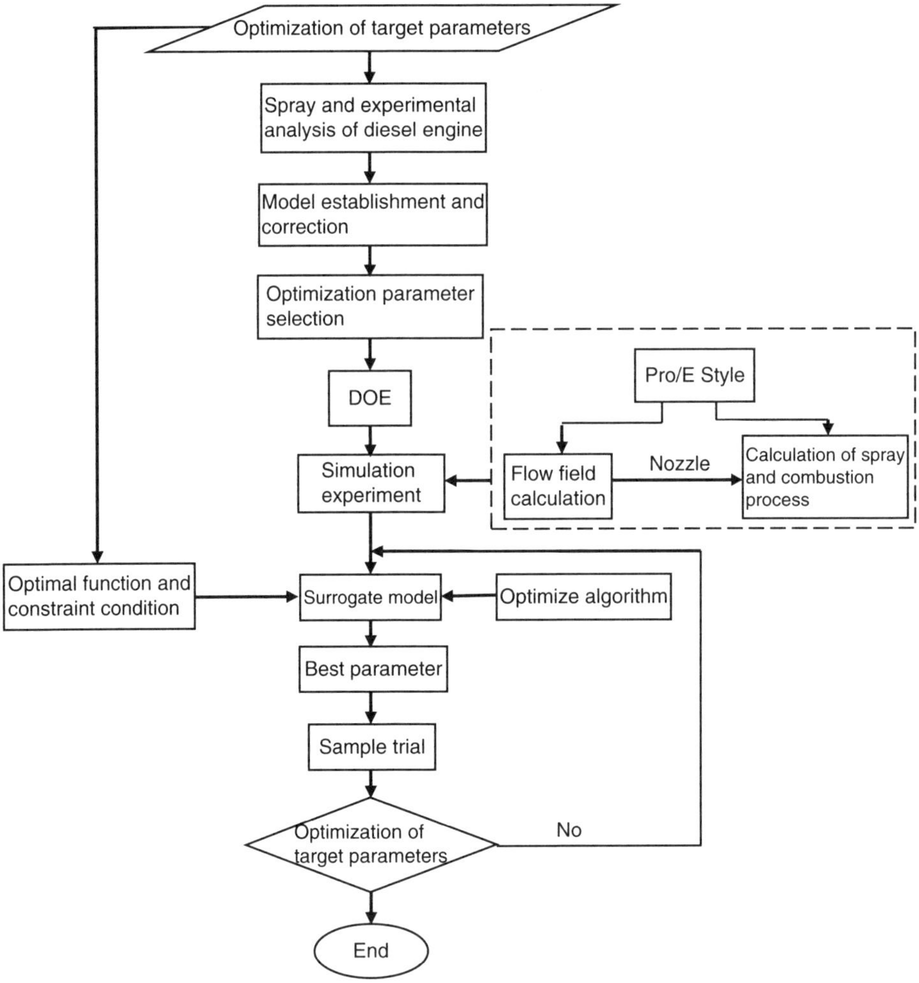

Figure 6.2 Parameter matching design process of the high-pressure common rail diesel engine injection system.

6.1.3.1 Modeling and Verification of Diesel Engine Spray and the Combustion Simulation Model

The spray mixing process in the diesel engine is a complex gas–liquid two-phase flowing process with heat and mass multidimensional transient transfer. Combustion is a compressible and complex multiphase flow process occurring under high temperature and high pressure. Therefore, only the multidimensional model can accurately reflect the essence and the change regulation of spray mixing and the combustion process. The CFD software FIRE is used to simulate the spray and combustion process of the high-pressure common rail diesel engine. The main physical and chemical submodels are shown in Table 6.2.

In the KH-RT model, the first oil droplet breaking time is determined by the empirical constant C2 in the model. C2 is related to the turbulence level of the initial jet, depending

Table 6.2 Main physical and chemical submodels.

Turbulence model	$k - \xi - f$ turbulence model
Wall model	Composite wall function method
Fuel atomization model	KH-RT spray breaking model
Fuel evaporation model	Dukowicz evaporation model
Fuel impingement model	Walljet0 impingement model
Combustion model	ECFM-3Z combustion model
NO_x emission model	Zeldovich module
Carbon smoke emission model	Kennedy–Hiroyasu–Magnussen module

Table 6.3 Nozzle design schemes.

Scheme	Number of orifices	Orifice diameter (mm)	Jet angle (°CA)	*l*/*d*
1	6	0.2	140	3.5
2	6	0.15	140	4.6
3	8	0.2	140	3.5
4	8	0.13	140	5.4
5	6	0.2	110	3.5
6	6	0.2	110	4.5
7	8	0.13	110	5.4
8	8	0.13	110	6.9

on the nozzle structure and injection pressure. The larger the C2, the longer the spray breaking time is, the larger the average diameter of the droplet is, and the larger the penetration distance of the oil beam is, which affect the formation and combustion process of the mixed gas in the cylinder.

This project has designed eight sets of different nozzle structures of non-axisymmetric nozzles, as shown in Table 6.3, an established spray flash photography bench, and has obtained the shape of the nozzle and the spray development process. Therefore, taking the test data of the spray penetration distance as the verification standard, after calculating and comparing the different C2 values, the value of C2 for each nozzle plan is determined, and the relationship between the C2 and nozzle structure parameters is deduced. Then the spray development process of different structural nozzles can be accurately simulated. The function of the structure of C2 with the number of orifices, the orifice aperture, the ratio of the length to the diameter, and the angle of the jet is as follows:

$$C_2 = a_1 \times d^{a_2} \times (l/d)^{a_3} \times (1 + \cos\alpha)^{a_4} \tag{6.3}$$

In the formula, a_i is the formula coefficient, d is the diameter of the jet orifice, l/d is the ratio of the length to the diameter of the orifice, α is the angle between the nozzle and the axis of the injector, and the calculation formula is as follows:

$$\cos\alpha = \cos\gamma \times \cos(\delta/2) - \sin(\delta/2) \times \sin\gamma \times \cos\theta \tag{6.4}$$

In the formula, γ is the installation angle of the injector, δ is the jet angle, and θ is the angle between the calculation orifice and the plane abscissa of the outlet plane, which is related to the number of jet orifices.

The formula (6.3) is a non-linear mathematical model. By giving the initial value of the determined parameters in the model, the least squares iteration method is used to regress the equation, and the specific mathematical model between the variable parameter and the target parameters is established. In Matlab, the lsqcurvefit function is used to draw up the non-linear curve of the least squares method, and the optimal sum coefficient of the obtained C2 equation is as follows: $a_1 = 31.2419$, $a_2 = -0.1806$, $a_3 = 0.1796$, and $a_4 = 0.3761$.

The working processes of the cylinder in two operating conditions are simulated and the results are compared with the test results. The two conditions are: scheme A, the number of nozzle orifices is 8, the orifice diameter is 0.13 mm, ratio of the length to the diameter is 5.4, the jet angle is 140° under the 80% rated condition, and the starting injection point is 10 degBTDC; and scheme B, the number of nozzle orifices is 6, the orifice diameter is 0.2 mm, the ratio of the length to the diameter is 4.5, the jet angle is 110°, and the starting injection point is 14 degBTDC. Figure 6.3 is a comparison of the simulation and test results of the pressure curves and cumulative heat release rates under two working conditions, which shows that the modified spray model has a higher simulation accuracy for the combustion process of different nozzles.

6.1.3.2 Optimal Parameters and Objective Functions

In this case, by optimizing the parameters of the common rail injection system, the NO_x emissions, soot emissions, fuel consumption, and combustion noise of the diesel engine can reach a better level at a rated load. The parameters of the injection system mainly include the fuel injection parameters (injection pressure, injection starting point, etc.), the nozzle structure parameters, and the installation position of the injector. The optimum parameters of the injection system are as follows: the injection starting point, the number of orifices, the diameter of the orifice, the angle of the jet, the ratio of the length to the diameter of the orifice, and the protruding height of the nozzle. The main objective of optimizing the parameters of the injection system is to reduce the fuel consumption, to improve the efficiency, to keep low NO_x and soot emissions, and low combustion noise. According to the target attainment method in the multivariable processing method, the target function f is

$$f = \frac{1000}{\left(\frac{NO_x}{(NO_x)_0}\right) + \left(\frac{\text{soot}}{(\text{soot})_0}\right)^2 + \left(\frac{dp/d\varphi}{(dp/d\varphi)_0}\right) + \left(\frac{\text{BSFC}}{(\text{BSFC})_0}\right)^{20}} \tag{6.5}$$

In the formula, $(NO_x)_0$, $(\text{soot})_0$, $(dp/d\varphi)_0$, and $(\text{BSFC})_0$ are respectively the target values of NO_x, soot, the pressure increase rate, and the fuel consumption rate.

6.1.3.3 Simulation Experiment Design (DOE)

The central composite design method (CCD) is used for experimental design. The central composite design method is extended by the 2^n full factor design. Because there is

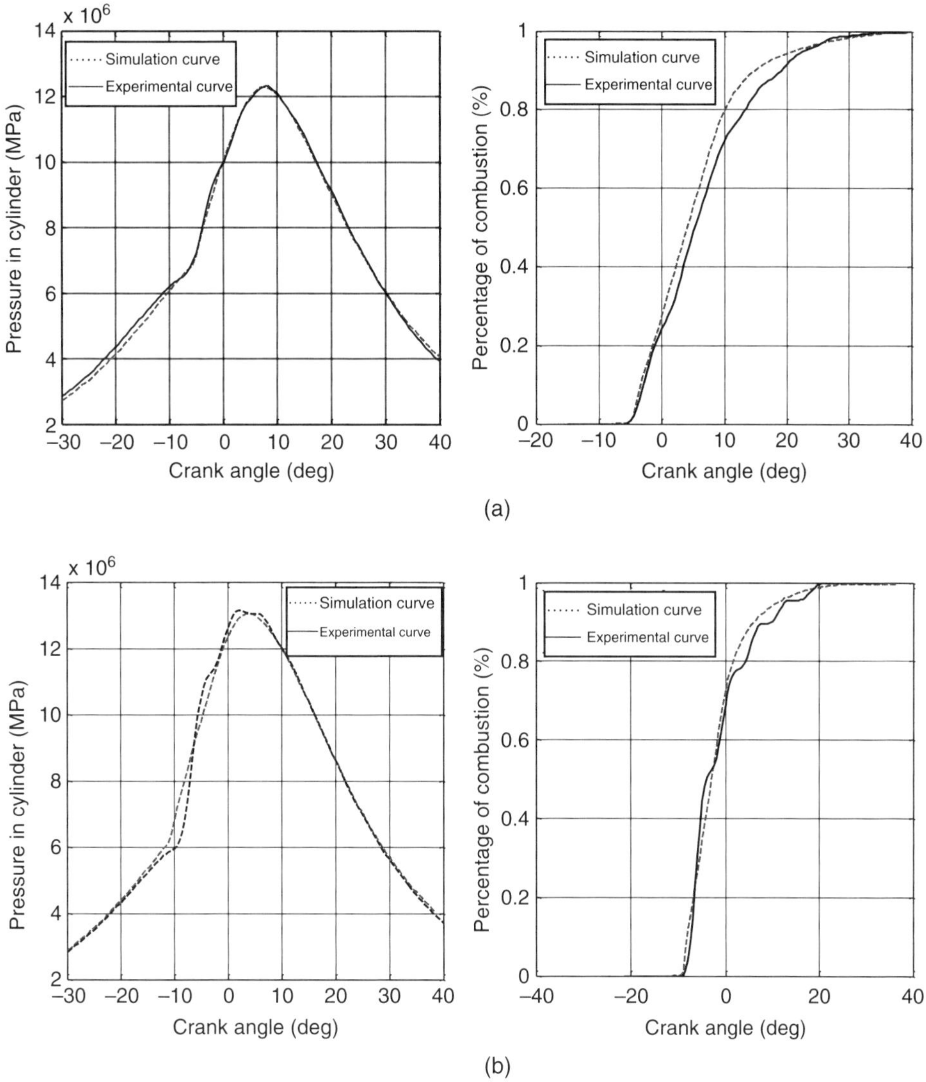

Figure 6.3 Comparison of the simulation results and the test results of the pressure and combustion percentage in the cylinder.

not enough information to reflect the two pure coefficients of the two-order model in the least squares method in the 2^n test design, in order to get enough information in this area, the design point is usually reduced by adding additional center and axis points (add a design variable $\pm\alpha$ at the center point); α usually takes the value 1 or $\sqrt[4]{n}$. The number of test times is $2^n + 2n + 1$, where n is the factor. The use of the CCD can ensure that more accurate solutions can be obtained in all aspects when the response surface model is optimized. There are six factors in this design, 77 test schemes, and the level of each factor is as shown in Table 6.4.

Table 6.4 Level setting of the parameter factors of the common rail system design.

Factors	Maximal value	Minimal value	Central point	α
Injection starting point φ (degBTDC)	20	4	12	1
Number of orifices n	8	6	7	1
Diameter of orifice d (mm)	0.13	0.20	0.165	1
Jet angle δ (°)	150	110	130	1
Ratio of length to diameter of the orifice l/d	6	3	4.5	1
Protruding height of the nozzle h (mm)	6	3	4.5	1

According to the experimental design, three-dimensional geometric modeling of the internal circulation area of the injector with different nozzle structures is built by using Pro/Engineer; then the calculation grid is generated and a two-phase flow calculation is carried out by generating the .Stl file into FIRE software. The flow coefficient of each nozzle, the law of injection, and the nozzle files are obtained, and the coupling calculation with the spray and combustion model is carried out. Finally, in the Fire software 2D result, the curves of NO_x, soot, and cylinder pressure with crank angle can be obtained, and the pressure rise rate and fuel consumption rate can be obtained by processing the p–φ graph.

6.1.3.4 Establishment of an Approximate Model for the Response Surface

The approximate model is a kind of model that predicts the unknown point response value by using the response information of known points. Its essence is a mathematical model of fitting the discrete data using an approximation method with the constraints of fitting precision and prediction accuracy. The polynomial response surface model is based on statistical methods and mathematical methods and expresses implicit functions by approximating a polynomial with an explicit expression form (not limited to polynomials). Fundamentally, the response surface method is a set of statistical methods used to find the best response values that consider the variation or uncertainty of the value of the input variable. In this section, the quadratic polynomial response surface model is used to construct an approximate model. The mathematical expression is as follows:

$$f(x) = a_0 + \sum_{i=1}^{n} a_i x_i + \sum_{i=1}^{n} a_{ii} x_i^2 + \sum_{j<1}^{n} a_{ji} x_j x_i \tag{6.6}$$

In the formula, n is the number of design variables, $x_{i,j}(i,j = 1, 2, \ldots, n)$ is the design variable, and a_0, a_i, and a_{ji} are the polynomial coefficients.

Based on the results of the simulation test, the formula of the quadratic polynomial response surface model between the objective function f and the injection system parameters constructed by the approximation module of iSIGHT software is as follows:

$$
\begin{aligned}
f(x) = & -473 + 7.89 \times \delta + 3638 \times d - 137 \times h - 22 \times (l/d) \\
& + 5.37 \times n + 2.64 \times \varphi - 0.006 \times \delta^2 - 8252 \times d^2 \\
& + 11.2 \times h^2 + 3.28 \times (l/d)^2 - 17.58 \times \delta \times d + 0.056 \times \delta \times h \\
& - 0.0362 \times \delta \times (l/d) - 0.33 \times \delta \times n + 0.024 \times \delta \times \varphi \\
& + 61.2 \times d \times h - 28.64 \times d \times (l/d) + 102.735 \times d \times n \\
& - 21.55 \times d \times \varphi + 1.0089 \times h \times (l/d) + 3.368 \times h \times n \\
& + 0.157 \times h \times \varphi - 0.508 \times (l/d) \times n - 0.0074 \times (l/d) \times \varphi \\
& - 0.4 \times n \times \varphi
\end{aligned} \tag{6.7}
$$

In iSIGHT, the validity of its response is judged by the determination coefficient R^2_{adj} of the model and the value of the adjusted decision coefficient B. When the R^2 and R^2_{adj} are closer to 1, the fitting of the approximate model is better. The determination coefficient R^2 and the adjustment coefficient R^2_{adj} of the model are 0.974 39 and 0.912 81, respectively, indicating that the fitting effect of the approximate model is preferable.

6.2 Parameter Optimization and Result Analysis of the Injection System

6.2.1 DoE Optimization

Through the ANOVA (analysis of variance) of the simulation results, the results of the Pareto diagram and main effect diagram can be obtained and the optimization plan of the injection system parameters can be obtained from the main effect diagram.

Figure 6.4 is the Pareto plot and main effect of the target function f of the interaction between the parameters of the injection system and each parameter. It is shown from the figure that the two main factors that have the greatest influence on the target function f are the diameter of the jet orifice and the jet angle. The optimal combination of the parameters of the ejector system, as shown in Table 6.5, can be obtained from the main effect graph when the target function f is used as the evaluation index.

The result of DoE optimization is the approximate solution in the test level rather than the global optimal solution. The optimization parameters are discrete points instead of continuous space, so it is necessary to carry out global optimization.

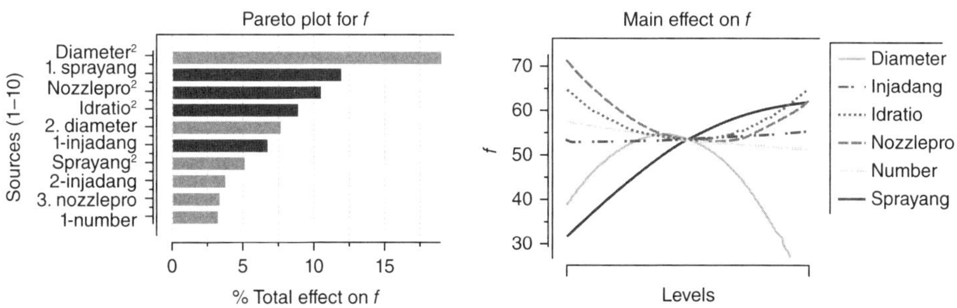

Figure 6.4 Pareto plot and main effect diagram of the injection system parameters to f.

Table 6.5 Optimal scheme of the parameter of the injection system with the target function f.

Orifice diameter (mm)	Number of orifices	Ratio of length to diameter	Jet angle (°)	Protruding height of nozzle (mm)	Injection starting point (degBTDC)
0.165	6	3	150	3	4

6.2.2 Global Optimization Based on the Approximate Model

According to the structural characteristics of the combustion system of the TBD234 diesel engine, the constraint conditions for the starting point, the number of injection orifices, the diameter of the orifice, the angle of injection, the ratio of the length to the diameter of the orifice, and the protruding height of the nozzle have been determined. The mathematical expression of the optimization problem of the injection parameters of the common rail system is defined as follows:

$$\max f = f(x)$$

$$s.t. \begin{array}{c} 110 \le \delta \le 150 \\ 0.13 \le d \le 0.20 \\ 3 \le h \le 6 \\ 3 \le l/d \le 6 \\ 6 \le n \le 8 \\ -20 \le \varphi \le -4 \end{array} \tag{6.8}$$

Based on the approximate model, the optimization of the injection system parameters is carried out by using the multi-island genetic algorithm and the non-linear programming by quadratic Lagrangian (NLPQL). There are 6080 steps calculated in this optimization and the final optimization results are shown in Table 6.6.

For the objective function f based on the economy, the final optimization result is to choose a smaller orifice diameter, more orifices, and a smaller orifice length–diameter ratio, to improve the quality of the atomization spray. The larger spray angle and smaller nozzle protruding height can ensure that the fuel beam in the combustion chamber has a reasonable placement, improves the utilization rate of the combustion chamber, and ensures an air mixture formation with good quality. With the starting point of 8 degBTDC, it ensures the full combustion near the TDC and improves the combustion efficiency, which can effectively reduce the fuel consumption rate and control the emission and noise level reasonably.

Table 6.6 Global optimization results of injection system parameters based on the target function f.

Orifice diameter (mm)	Number of orifices	Ratio of length to diameter	Jet angle (°)	Protruding height of nozzle (mm)	Injection starting point (degBTDC)
0.145	7	3	150	3	8

Table 6.7 Simulation and optimization comparison of the two schemes.

Scheme	NO_x (10^{-6})	Soot (10^{-6})	*gf*	d*p*/d*φ*	*f*		
					Simulation value	Optimization value	Error (%)
A	115.2	0.1516	216	0.78	87.6	88.1	0.59
B	174.7	0.1447	207	0.96	84.8	81.7	3.78

6.2.3 Optimization Results Analysis

The results obtained by using a genetic algorithm combined with a sequential quadratic programming algorithm and the results obtained by the DoE optimization method are different from each other. Sequential number A is used to express the optimization plan of the DoE method and sequence number B is used to express the optimization scheme of the genetic algorithm. The simulation calculations of the injection process and the working process in the cylinder of the two schemes are carried out, as shown in Table 6.7. The two optimization schemes were compared and analyzed. Finally, the nozzle of the B scheme was selected and processed, and the matching test was carried out on the test bench to verify the optimization effect and the correctness of the whole optimization process.

It can be seen from the calculation results of the two schemes and the calculated value of objective function f that the carbon smoke emissions of the two schemes are similar, and the fuel consumption rate gf of B was low, but its comprehensive performance index (objective function f) is smaller than that of A, which is mainly due to the emission of the NO_x scheme of A is relatively low. However, from the original intention of building the objective function f, scheme A is not desirable.

In Figure 6.5, the simulation results of the schemes A and B are compared with the results of the matching test. Their pressures, temperatures, and the percentage curves of the combustion in the cylinder are in good agreement. In Table 6.8, the test data of B and the original test are compared. The fuel consumption rate of the diesel engine is 192 g/kW h, which is 5% lower than that of original machine, NO_x and carbon smoke emissions are, respectively, 10% and 18% lower than those of the original machine, and the air noise is slightly higher, but is also in the allowed range. Therefore, in this section, using the genetic algorithm and optimization method of the quadratic programming algorithm to optimize the design parameters of the injection system is acceptable. It can quickly search the global optimization of the space and get a credible optimization result, which is more desirable than a single application of the DoE design method for single point analysis. The optimization algorithm based on the genetic algorithm combined with the sequential quadratic programming algorithm can search the global optimization of the space quickly and can provide a reference for the actual combustion process optimization control, which greatly reduces the workload of the matching test of the electronically controlled injector and the diesel engine combustor.

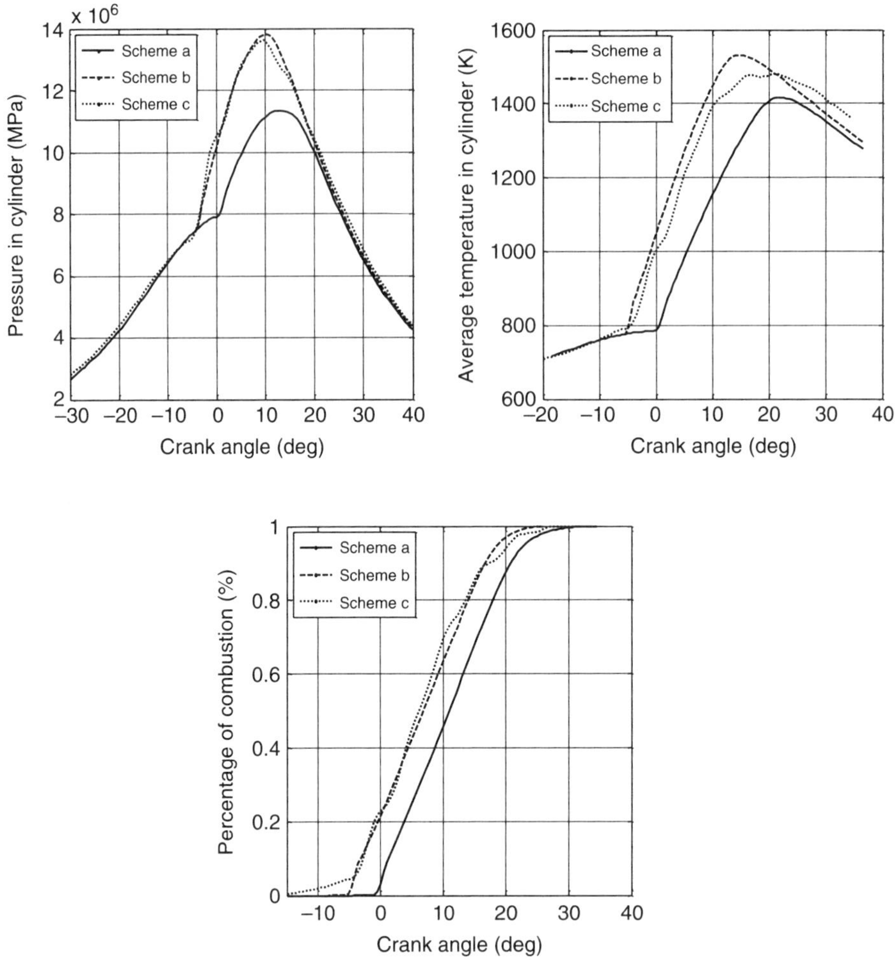

Figure 6.5 Simulation results of the pressure, average temperature, and combustion percentage in A and B cylinders and the test results of scheme B.

Table 6.8 Comparison of test data between scheme B and the original machine.

Test parameters	Exhaust temperature (°C)	Noise (dB)	NO_x (ppm)	Smoke (impermeable %)	Fuel consumption rate (g/kW h)
Original	553	106	1163	10.9	203
Scheme B	542	109	1057	9.0	192

6.3 Optimization Calibration Technology of the Jet Control MAP

6.3.1 Summary

The injection parameters of the common rail diesel engine mainly include pre-injection timing, pre-injection volume, main pre-injection interval, main injection timing, rail pressure, post-injection timing, and post-injection volume. The parameters of high-pressure common rail diesel injection control are stored in the form of data files (MAPs, Curves, Constants) in the electrical control unit (ECU). When running, the corresponding data are read according to each working condition to control the execution unit in real time. These data MAPs are made by a large number of calibration experiments, modification, optimization of data, and then by the corresponding calibration software. Different diesel engines have different injection control MAPs. Therefore, for different diesel engines, the matching of MAP data must be calibrated. The calibration of the common rail system and diesel engine has become an important part of the development of the high-pressure common rail diesel engine system. The cycle and accuracy of calibration work directly affect the development cycle and quality of the common rail diesel engine.

6.3.2 Optimal Calibration Method

There are three commonly used optimization methods: manual optimization, off-line steady-state optimization, and on-line automatic optimization.

Manual calibration is a traditional calibration method. The calibration engineers adjust their control parameters according to their experience, and then test and record the comparison test results. This method is time-consuming and laborious, the experience of the demarcation engineer has a great influence on the result, and in addition the result of the test is not optimal.

Off-line steady-state optimization calibration is the calibration test and optimization process separately. First, the calibration test is carried out. According to certain test rules, after obtaining experimental data, the engine performance model is established and then optimized, and finally the optimal control parameters are obtained. Prahabkar first came up with the off-line steady-state calibration method. Its basic idea is to optimize the analysis of fuel consumption and emission value as a function of speed, torque, and engine control parameters under the condition of warming up, using curve-fitting technology. Based on the mathematical model of the engine and the emission curve of fuel consumption, the optimization objective function and constraint function are determined. The conjugate gradient method is applied to determine the optimal parameters of the engine. The advantage of this method is that it has a better optimization result than the traditional manual method and the calibration is more accurate. However, it does have disadvantages because it has many test points and a long test period.

The on-line automatic optimal calibration method was proposed by Cassidy. It is the application of an automatic calibration system to realize the optimization and calibration of engine control parameters. On the basis of known conditions such as the engine performance optimization target, constraint condition, control variable range, and boundary condition, based on the set test cycle, optimization software is used to optimize the control parameters automatically. The automatic optimization calibration combines the test process with the optimal calculation, optimizes experiments based on optimization problems, automatically tests, measures, and collects data, optimizes the test results on-line, and controls the experimental process according to the optimization results until the optimal solution is obtained. Once the test process begins, the process of automatic optimization will be realized without the need of human intervention. Its advantage is that the test time is short and the precision is high.

6.3.3 Optimization of Target Analysis

Optimizing the calibration target is the basis and task of matching, optimizing, and calibrating the parameters of the injection control. The analysis of the optimizing calibration target is to determine the optimization target, control parameters, and constraints. Because of the need to save energy, it is important that the diesel engine must have good fuel economy, as fuel economy is the primary goal. Therefore, fuel economy is generally taken as the goal of optimization.

At present, the requirements for reducing emissions and pollution of diesel engines are stricter global. Therefore, the constraint conditions for the optimization and calibration of a diesel electronic control system can be determined as the quantity of various emissions lower than the limits of emission regulations. The electronically controlled system of diesel engine optimization constraints can be divided into mandatory constraints and structural constraints. The former is the diesel emission under the legal emissions law (mainly for NO_x, CO, HC, PM); the latter is mainly the cylinder peak pressure, exhaust temperature, cooling water temperature, and the strength limit of the maximum torque. Generally, the limited diesel emission is chosen as the optimal constraint condition. Other constraints can be selectively added to the optimization problem according to the actual requirements.

6.4 Off-line Steady-State Optimization Calibration of the Common Rail Diesel Engine

Due to the complexity of the electronically controlled common rail system of the diesel engine, the coupling effect between the electronically controlled injection parameters and the inherent contradiction of engine performance parameters, the optimized calibration of the electronically controlled common rail system of the diesel engine usually takes a longer period at a higher cost. The manual online calibration method has a high calibration cost and low calibration quality, which cannot meet the requirements of engineering practice. In addition, the advanced automatic calibration systems in foreign countries are too expensive. The model based on the off-line steady-state optimization calibration method can solve the above problem well. The method includes three steps. First, select a limited number of reasonable test points through experimental design. Then use test data to establish a mathematical model between engine performance parameters and control parameters. Finally, use the model to optimize the control parameters of the electrically controlled system. The model-based off-line steady-state optimization calibration method can not only make the test point relatively small but can also be used for optimization, which will reduce the consumption of engine test resources. In this section, the model based on the off-line steady-state optimization calibration method is used to calibrate and optimize the economy of a common rail diesel engine. The test results show that the calibration quality and efficiency can be improved by using this method.

6.4.1 Mathematical Model for Optimization of the Electric Control Parameters

At present, the optimization of electronically controlled common rail system parameters generally take fuel economy as the optimization objective, take the discharge quantity lower than the limit value of the emission regulation as the constraint, and optimize the calibration of the parameters, such as the advanced injection angle and the rail pressure of the main injection. The mathematical model of the fuel economy optimization problem is as follows:

$$\min\ b_e(n, Q, \theta, P_{cr})$$

$$s.t. \begin{array}{r} T_p \le T_{p\,\mathrm{lim}} \\ P_b \le P_{b\,\mathrm{lim}} \\ \mathrm{NO}_x \le (\mathrm{NO}_x)_{\mathrm{lim}} \\ \lambda \le \lambda_{\mathrm{lim}} \end{array} \tag{6.9}$$

In the formula, b_e is the fuel consumption rate; n is the diesel engine speed; Q is the circulating fuel injection volume; P_{cr} is the rail pressure; θ is the main injection timing; T_p is the exhaust temperature; $T_{p\mathrm{lim}}$ is the maximum limit value of exhaust temperature; P_b is the combustion pressure; $P_{b\mathrm{lim}}$ is the combustion pressure limit; $(\mathrm{NO}_x)_{\mathrm{lim}}$ is the emission limit value of NO_x; λ is the smoke degree; andλ_{lim} is the smoke limit value. The optimization target is b_e and the optimization parameters are P_{cr} and θ. Namely, by calibrating the optimization parameters P_{cr} and θ, the diesel engine can achieve the lowest fuel consumption rate at every working point (n, Q) under the premise of meeting the constraint conditions.

6.4.2 Experimental Design

The time and cost of optimizing the test for each node of each MAP of ECU is very high. However, through the design of the test, the most effective points are chosen to be tested and higher profits can be obtained at a lower cost in a short time. The experimental design is to optimize the possible test schemes and arrange the test factors scientifically and rationally. The purpose of the experimental design is to determine how to collect data for statistical inference. In order to predict the response relationship between the electric control parameters and the diesel engine performance accurately, the test points are supposed to be representative and able to cover all the working conditions of the diesel engine if possible. There are four kinds of methods: the orthogonal experiment method, the space-filling method, the classical method, and the optimum seeking method.

The common rail diesel engine studied in this paper is a diesel generator set. Ten working points are selected as the design test points, corresponding to 10–100% load conditions of the common rail diesel engine, respectively, and the operation point selection

Table 6.9 Test design working point.

Working point	1	2	3	4	5	6	7	8	9	10
Circulating fuel injection (mm^3/str)	10	20	35	55	75	100	125	150	175	200
Rotational speed (r/min)						1500				

Table 6.10 Level and factor setting of each test point.

	Factor	Level				
A	Rail pressure (bar)	900	1000	1100	1200	1300
B	Injection advance angle (°CA)	6	8	10	12	14

is shown in Table 6.9. At each working point, two factors, the main injection advance angle and the rail pressure, are changed, with each factor being set in up to five levels, The factors and levels are set as shown in Table 6.10.

6.4.3 Establishment of the Performance Prediction Response Model

The goal of establishing the performance prediction response model is to establish the mapping relationships between the engine performance parameters (fuel consumption rate, emissions, exhaust temperature, etc.) and the parameters of the electric control system, such as rail pressure, the main injection advance angle, and so on, based on the experimental data. In order to achieve an accurate and optimal calibration, the response model must be accurate and have good prediction generalization ability. The commonly used modeling methods include polynomial (POLY2), multistrip function (PolySpline), growth model (MMF), free node spline function (FK2), and so on.

Ten sets of test data were obtained for a common rail diesel generator set test in 10 working conditions. The independent variables of the data in the group are the control parameters such as the rail pressure P_{cr} and the main injection advance angle θ. The dependent variables are the performance parameters of the diesel engine, such as the fuel consumption rate b_e, NO_x, and others. In the group, the relationship between the change of b_e with rail pressure P_{cr} and the main injection advance angle θ is established using the quadratic polynomial, and the expression is as follows:

$$b_e = aP_{cr}^2 + b\theta^2 + mP_{cr} + n\theta + kP_{cr}\theta + c \tag{6.10}$$

In the formula, a, b, m, n, k, and c are regression coefficients.

The above regression polynomial fitting is carried out on the 10 sets of data at 10 working points, respectively, and the change function of the fuel consumption rate b_e with rail pressure and the main injection advance angle can be obtained at different operating points; the change function of NO_x rail pressure and the main injection advance angle can also be obtained. Taking the sixth working condition as an example, the fuel consumption rate and NO_x emissions are as follows:

$$b_e = 0.00009P_{cr}^2 - 0.0054\theta^2 - 0.28P_{cr}^2 - 4.73\theta + 0.0045P_{cr}\theta + 389.28 \tag{6.11}$$

$$NO_x = -0.001P_{cr}^2 - 0.89\theta^2 + 3.16P_{cr} + 143.31\theta - 0.035P_{cr}\theta - 1637.82 \tag{6.12}$$

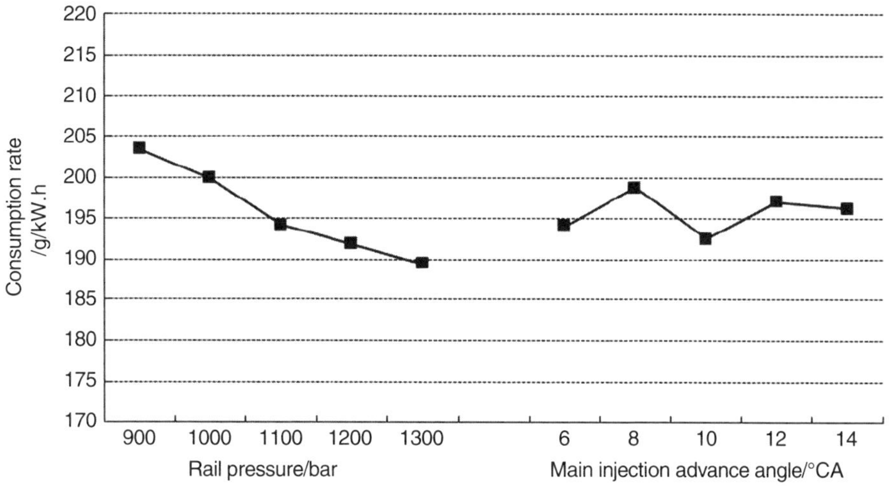

Figure 6.6 Influence trend of the rail pressure and main injection advance angle on fuel consumption.

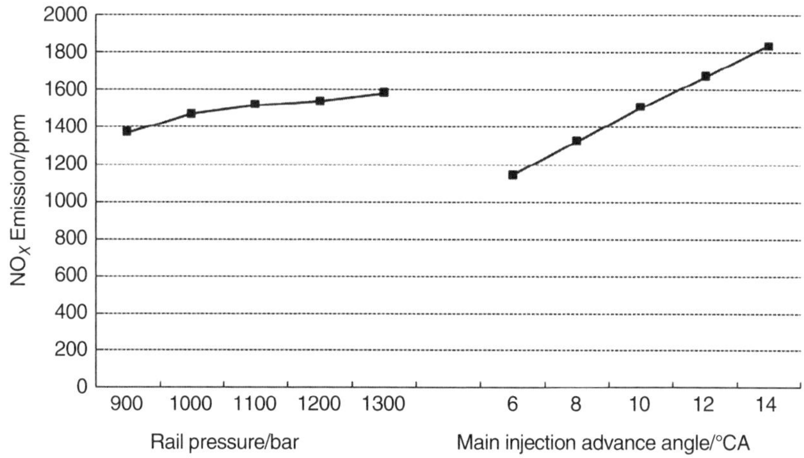

Figure 6.7 Influence trend of the rail pressure and main injection advance angle on NO_x emission.

Through the range analysis of the orthogonal test results of the sixth working point, the influence trend of rail pressure and the main injection advance angle on fuel consumption and NO_x emissions can be obtained, as shown in Figures 6.6 and 6.7.

6.4.4 Optimal Calibration

After establishing a good response model according to the experimental data, a suitable optimization algorithm can be selected for global optimization of the control parameters. At present, the most commonly used optimization algorithms are the Lagrange multiplier method, the genetic algorithm, and so on. In this paper, the genetic algorithm and sequential quadratic programming algorithm are selected to optimize the rail pressure and main injection advance angle under 10 selected working conditions.

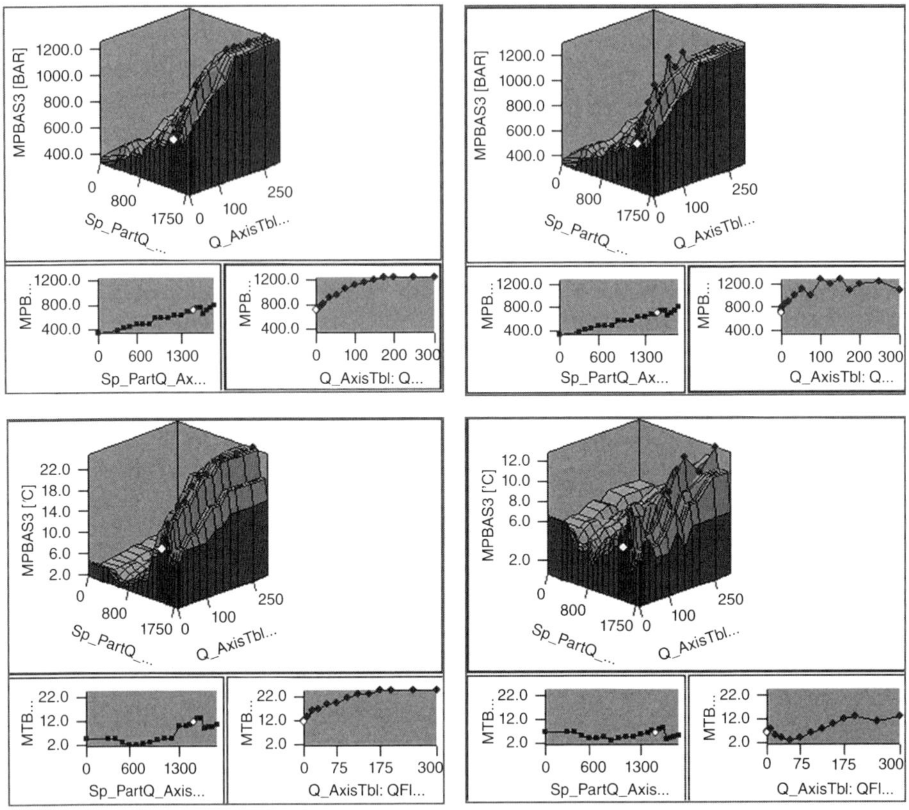

Figure 6.8 Comparison of the rail pressure and main injection advance angle MAP before and after economy optimization.

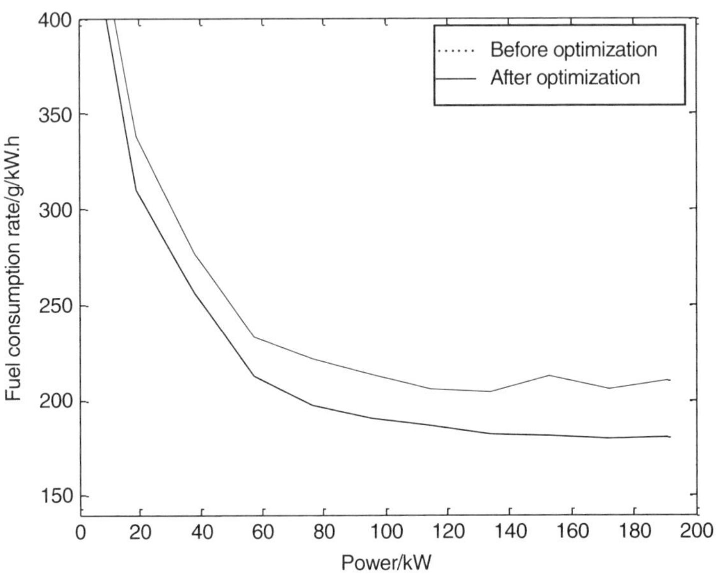

Figure 6.9 Comparison of fuel consumption curves of load characteristics before and after calibration and optimization.

For example, under the sixth working condition, the optimal rail pressure and main injection advance angle are 1300 bar and 6 °CA respectively. Then the optimal rail pressure MAP and the main injection advance angle MAP under all working conditions can be obtained by interpolating the rail pressure and the main injection advance angle under each working condition. The comparison of P_{crmap} and θ_{MAP} after the calibrated optimization is shown in Figure 6.8.

6.4.5 Test Result

The load characteristics of a common rail diesel engine before and after calibration and optimization are compared in Figure 6.9. It can be seen that, after the optimization of A and B, the fuel economy of the engine is obviously improved. The practice results show that the model-based off-line steady-state optimization calibration method can greatly reduce the number and time of the test and is very effective in improving the calibration quality and efficiency.

7

Development of the Dual Pressure Common Rail System

With the further stricter emission control and the need for full operating optimization, some shortcomings of the existing common rail system are gradually emerging. One is that the present common rail injection pressure, at under 160 MPa, is difficult to solve the contradiction between NO_x and PM emissions. If the injection pressure is above 180 MPa, the above contradiction can be solved. On the other hand, the uncontrollable fuel injection law of the second-generation common rail system is difficult to coordinate the emission control and combustion optimization, and can only be achieved through multiple injection methods such as pre-injection and post-injection. Based on the development of modern common rail technology, the ultra-high-pressure injection common rail system with controllable fuel injection is the inevitable way to solve the problems of marine diesel engine emissions and the optimization of all working conditions.

Combined with the development needs of the common rail electric jet technology, based on the current development of the common rail electronically controlled fuel injection technology in China and based on the full absorption of the advantages of the high-pressure common rail system and the pressure storage pressurized common rail system, it is very necessary to carry out research work on the new dual pressure common rail system (DPCRS).

The design idea of the new DPCRS is derived from the rail pressure enhanced common rail system of BOSCH. Although BOSCH has introduced the fourth generation N4 type fuel injector based on the idea of rail pressure enhancement, the system structure is not suitable for design and development in China. Its pressurized piston is integrated in the injector body, resulting in densely covered fuel channels and chambers in the fuel injector. Meanwhile, in order to control the action of the pressurized piston and injector, two solenoid valves need to be installed in the injector, resulting the difficulty in solenoid wiring. The machining and manufacturing precision of such a complex structure is difficult to be realized in China. It is necessary to study and develop an ultra-high DPCRS adapted to domestic conditions.

The new DPCRS is based on the current high-pressure common rail system, and a fuel pressure supercharger is installed between the common rail chamber and the injector. According to the change of engine condition, the fuel supply to the injector can be carried out under two pressures (base pressure of 100 MPa and high pressure of 200 ~ 250 MPa), respectively, so that the engine can run stably under the condition of optimized performance.

Common Rail Fuel Injection Technology in Diesel Engines, First Edition.
Guangyao Ouyang, Shijie An, Zhenming Liu and Yuxue Li.

The main features of the dual pressure system are:

(1) The DPCRS can realize three working modes in principle:
   a. *Partial load operation mode.* This is the common operating condition of most engines. The load change range is wide in partial load, so the corresponding circulating fuel injection varies greatly. The engine speed is low and the fuel supply per cycle is small. The base pressure injection part of the fuel injection pressure can be appropriately reduced, and then the rest of the fuel is pressurized to jet. With the increase of load, the required injection quantity increases. The base pressure injection can be used directly or the base pressure injection can be used first; then the high-pressure injection is carried out. During a base pressure fuel supply, the low fuel pressure is beneficial to improve the working reliability of the system and reduce the consumption power of the fuel injection system at the same time, which is conducive to a further economy improvement. This is very important for some engines that operate for a long time, such as the high-power diesel engines for ships and locomotives.
   b. *Rated load operation mode.* A dual pressure system uses the sequential control method in the injection process of a work cycle. First, the base pressure jet is followed by high-pressure injection, and the injection rate formed is optimized by controlling the start interval angle of the two injections, so that economy and emission are reasonably balanced and the engine has a good working performance. In this case, the ratio of the base pressure and high pressure can be adjusted to meet the demand of the fuel injection.
   c. *High load (110% rated load) operation mode.* This working condition is less used, but its condition must be met, especially in emergency situations. In this operation mode, the dual pressure system can use high-pressure injection or base pressure and a high-pressure mixed injection. In a relatively short period of time, a large quantity of fuel is needed to meet the combustion requirement with good injection quality. This will ensure that the engine has excellent power and economy.

   In conclusion, the DPCRS further develops the potential of the common rail system pressure–time regulation theory, which enables the engine to achieve excellent performance under all operating conditions. It has universal applicability.
(2) The dual pressure system retains the main components of the existing high-pressure common rail system, such as the electronically controlled injector and the adjustable rail pressure rail system, so that in the DPCRS all adjustable links in the existing high-pressure common rail system are still retained, such as the adjustment of the fuel rail pressure, injector multiple injection, etc., which makes it have a stronger, more flexible adjustment ability. It provides a very favorable condition for further improving the performance of the engine.
(3) The dual pressure system in the fuel supercharger uses a series parallel circuit, which can achieve the base pressure fuel supply, the medium- to high-pressure sequential fuel supply, the high-pressure fuel supply, and other fuel supply methods. In operation, the system is based on the base pressure fuel, which improves the reliability of the system compared to the existing high-pressure common rail system. At the same time, the design of the supercharger and fuel injector with a separate structure

have been invented. Not only can its performance (an achievable pressure forming technology and injection rate) be comparable with the most advanced BOSCH fourth generation N4 products, but also its mounting arrangement is more flexible and convenient, and the manufacturing difficulty and cost are reduced, which makes it easy to promote its use. Compared with the pressure storage pressurization type common rail system, the advantages of the dual pressure system are that under the same maximum injection pressure, the dual pressure system improves the pressure of the low-pressure rail and thus reduces the pressure ratio, which reduces the flow rate of the control fuel channel, the load of the supercharger piston, and the leakage of fuel, while maintaining the manufacturing accuracy requirement. At the same time, the low-pressure rail can directly load part of the task of the fuel supply, simplifying the structure of the system.

7.1 Structure Design and Simulation Modeling of the Dual Pressure Common Rail System

The new type of DPCRS designed in this paper is based on the current high-pressure common rail system. Only the electronically controlled supercharger is added as the other components remain the same, so only the design of the electronically controlled supercharger is needed according to the needs of the system. First, the structure of the new type of electronically controlled supercharger is designed, its structure features are analyzed, and the working principle of the system is introduced. On this basis, a DPCRS simulation model is set up.

7.1.1 Design of the Dual Pressure Common Rail System Supercharger

The supercharger is the key part of the DPCRS. The performance and control flexibility of the supercharger play a decisive role in the DPCRS to achieve dual pressure injection and injection regulation. This section describes the design of fuel pressure superchargers that are separated, single-in or single-out from the injector, or have the function of a series and parallel connection with a base-pressure and a high-pressure fuel. The structure of the supercharger is shown in Figure 7.1,and the DPCRS assembly is shown in Figure 7.2.

The main concepts commonly used in the DPCRS are given below:

Base-pressure fuel. The fuel in the common rail chamber.

High-pressure fuel. The high-pressure fuel transferred from the pressurized base-pressure fuel by the supercharger to the injector.

Base-pressure fuel channel. The route of the base-pressure fuel flows through. The flow direction is to the rail chamber, the inlet of the supercharging device, the one-way valve, the central channel of the pressurized piston, the fuel outlet of the supercharger, and the injector.

High-pressure fuel channel. The route of the high-pressure fuel flows through. After the one-way valve is closed, the high-pressure fuel pressurized by the supercharger goes through the center channel of the piston and the fuel outlet of the supercharger, and finally supplies the injector.

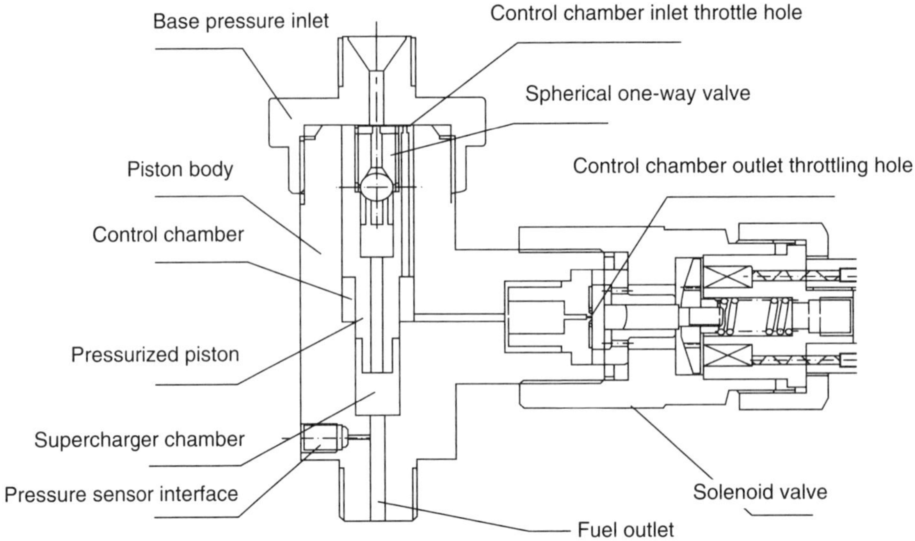

Figure 7.1 Supercharger structure of the dual pressure common rail system.

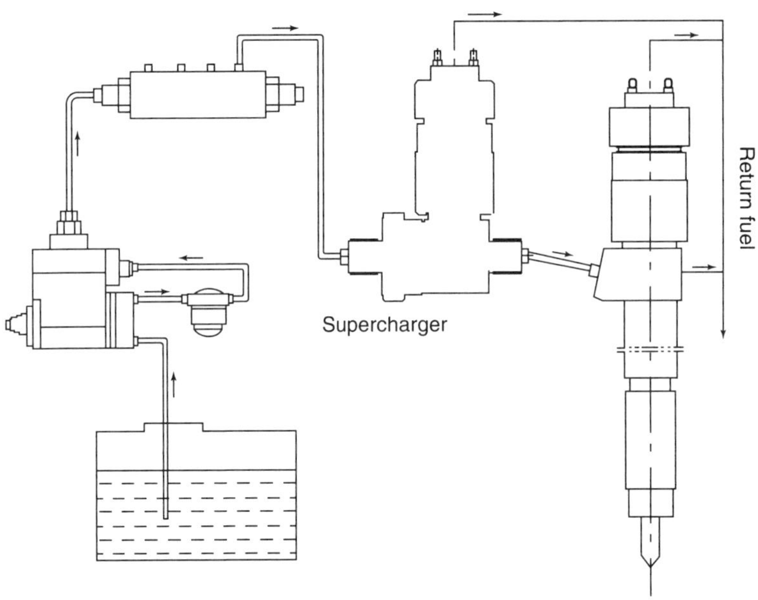

Figure 7.2 System assembly of the dual pressure common rail system.

The single inlet and the single outlet fuel channel. The base-pressure fuel and the high-pressure fuel all come from the common rail chamber and enter the supercharger by the inlet of the supercharger. No matter whether the fuel is pressurized or not, it flows out through the fuel outlet of the supercharger and supplies the injector, so the channels are called single inlet and single outlet.

Series and parallel circuit. When the system injects the same base pressure or high pressure, although the fuel flows through the same channel, the base-pressure fuel and the high-pressure fuel are independent of each other. At this time, the fuel roads are parallel. When the system is mixed with base-pressure and high-pressure fuel, the base-pressure fuel affects the high-pressure fuel. At this time, the two fuel roads are connected in series. Therefore, the base-pressure fuel road and the high-pressure oil road are series and parallel fuel roads.

Base-pressure and high-pressure jetting interval time (T). The algebraic value of the supercharger solenoid valve opening time is reduced by the opening time of the injector solenoid valve. When T is positive, pressurization goes first and then injection; when it is negative, injection is carried out first; when it is zero, pressurization and injection start at the same time.

The principle of the supercharger:

Base-pressure fuel supply. Fuel is supplied to the injector by the base pressure fuel channel; at this time the supercharger is only equivalent to a one-way valve.

High-pressure fuel supply. The initial state of the supercharger, the top ends of the control room, plenum, and the pressurized piston are filled with the same base-pressure fuel. The pressure of the pressurized piston is counteracted by the liquid pressure at both ends and the pressurized piston is in a state of rest. The supercharger solenoid valve is switched on after receiving the pressurization signal. The fuel in the control chamber flows back into the fuel tank, resulting in the pressure drop in the volume. The pressure of the pressurized piston is unbalanced and moves to the side of the small end, and the check valve is closed. The volume of fuel in the chamber is rapidly compressed and the pressure rises. The high-pressure fuel is supplied to the injector and the fuel injector is timely injected. After the supercharger solenoid valve is closed, the fuel in the control room, through the inlet throttle of the control room, obtains the fuel supply in the rail chamber and recovers the pressure. At the same time, the fuel injected into the supercharger chamber is supplemented by a one-way valve and resets the pressurized piston using a reset string with a reset spring, until the pressure of the pressurized piston is balanced.

The structure characteristics of the DPCRS:

Integration of the one-way valve and supercharged piston. This structure is the key to achieve the functions of the single inlet, single output, and base pressure and high voltage series and parallel connection and makes the system fuel channels simplified and compact.

The movement guide channel of the unidirectional valve ball is designed to form a blossom shape and improves the circulation performance of the base pressure fuel roads.

The control room fuel outlet throttle valve is designed as an independent part. The supercharger body and the valve depend on the plane compaction to seal the base pressure fuel in the control chamber, and thus it is easy to replace. This valve is the key to control the pressure of the control room. It determines the supercharging performance and response speed of the supercharger. It is designed as an independent component and is easy to make series so as to facilitate the experimental research.

The central channel of the pressurized piston has the effect of deformation compensation to improve its sealing property.

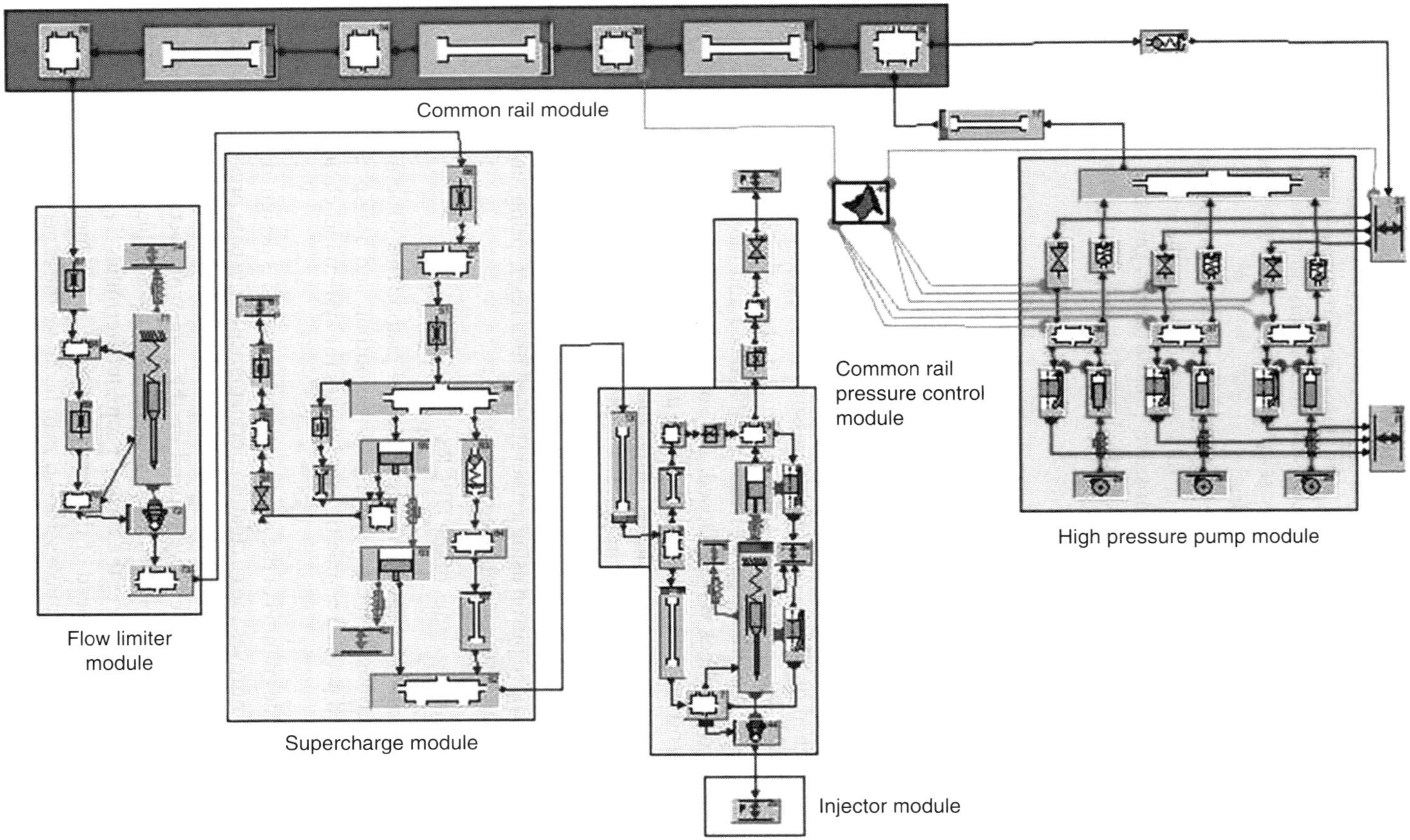

Figure 7.3 Simulation model of the dual pressure common rail system.

The supercharger is separated from the injector, and the installation arrangement is flexible, which reduces the difficulty and cost of manufacturing.

Lower pressurization reduces the flow of the fuel control channel, reducing the load of the pressurized piston and leakage of fuel, thus reducing the requirements of manufacturing precision.

7.1.2 Modeling of the Dual Pressure Common Rail System

The working process of the DPCRS involves many aspects, such as electromagnetic, hydraulic, and mechanical movement. Various parameters interact with and influence each other; the relationships are very complex, but there is no explicit analytical relationship with a single parameter. Therefore, it is very difficult to handle and analyze the specific effects of the specific structural factors of the above relationships by experiment. The workload is huge and the cost is high. In this section, based on the study of the DPCRS structure, the simulation model of the DPCRS is established by HYDSIM. The model is shown in Figure 7.3.

7.2 Simulation Study of the Dual Pressure Common Rail System

The main technical specifications of the DPCRS designed in this section are as follows:

System pressure requirements. Maximum rail pressure 100 MPa, maximum pressurization pressure 200 MPa.

System fuel requirement. Circulating fuel injection amount $q_{inj} = 160\,mm^3$. When the rail pressure is 130 MPa, an electronically controlled injector controls the return fuel and leakage return fuel to occupy 56% of the total injection volume. Under this working condition, the fuel supply volume of the primary injection chamber is $250\,mm^3$.

System response speed. Supercharger solenoid valve needs to cooperate with the injector control solenoid valve, so the response speed of the supercharger solenoid valve should not be lower than the response speed of the solenoid valve.

Based on the system simulation model established in the last section, the simulation program is used to simulate the dynamic working process of the system. The flexibility of the injection control and the dynamic response characteristics of the system are studied. Then the simulation model is used to analyze the effect of the injection law on the injection characteristics and to compare it with the related test results, thus verifying the reliability of the model. A sensitivity analysis of the system structure parameters is carried out and the main structural parameters of the system optimization are determined. In the process of simulation, it is found that there is a pressure oscillation due to the supercharger in the system. The method of restraining the pressure oscillation is discussed and the system is trial manufactured.

7.2.1 Study of the Dynamic Characteristics of the System

The dynamic characteristics of the DPCRS are mainly studied in the process of fuel flow and pressurization. During these processes, the fluid pressure changes dramatically.

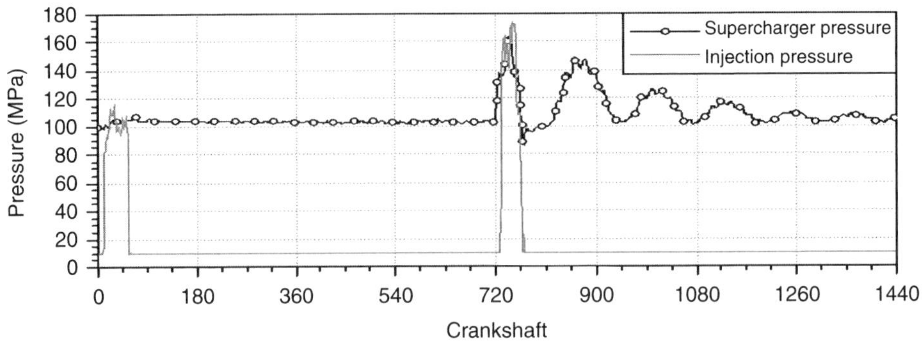

Figure 7.4 Pressure in the supercharger volume and injection rate.

Therefore, the influence of pressure on the fuel characteristics needs to be considered. Because the dynamic process of the system in only several milliseconds of pressure is simulated, the effect of temperature change is not considered.

7.2.1.1 Simulation of the Dynamic Characteristics of the System

The calculation results of the pressure of the supercharger chamber and the corresponding injection law are shown in Figure 7.4. When the pressurization ratio is 2, the pressure of the supercharger nozzle after pressurization is close to 180 MPa and the effect of pressurization is obvious. After the second high-pressure injection, there is still an obvious pressure fluctuation in the supercharger chamber, which may be caused by oscillation of the pressurization piston. The pressure fluctuation generated by the supercharger basically ends when the second working cycles start. The simulation results show that the pressure oscillation does not cause the injector to generate the second injection. The main reason is that the pressure wave has been oscillating above the base pressure value, while the dual pressure system adopts an electronically controlled injector. The working principle of the electronically controlled injector can ensure a good closing performance of the needle valve.

In order to analyze the influence of the pressurization ratio on pressurization under a different base pressure, the settings of simulation are as follows: the pressurization ratio of the supercharger is 1.4, 2, 3, and 4 (other structural parameters of the supercharger remain unchanged) and the base pressure (MPa) is 40, 60, 80, and 100.

From Figure 7.5, it can be seen that when the theoretical pressurization ratio is determined, the base pressure increases linearly with the increase of the base pressure. Under the same base pressure, the high pressure increases with the increase of the pressurization ratio, but with the increase of pressure ratio, the distance between the two adjacent pressure lines becomes closer. In order to analyze the relationship between the theoretical pressurization ratio and the actual ratio, each actual ratio under the theoretical pressure ratio is calculated. The actual pressurization ratio is, under the condition of setting the base pressure and the theoretical pressure ratio, the calculated ratio of the peak value of the high pressure to the set base pressure. From Figure 7.6, it can be seen that with the increase of the theoretical pressurization ratio, the degree of the actual pressurization ratio deviates from it, and does not increase linearly with the theoretical pressure ratio. After the theoretical pressurization ratio reaches 2, the effect of pressurization does not become obvious by increasing the ratio. For this reason, the simulation

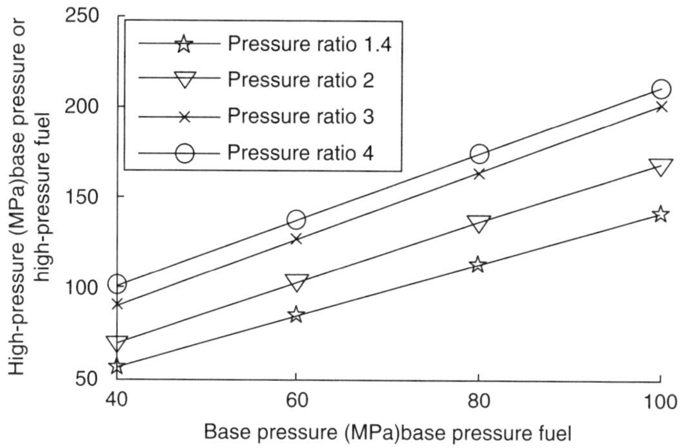

Figure 7.5 Influence of the theoretical pressurization ratio on pressurization.

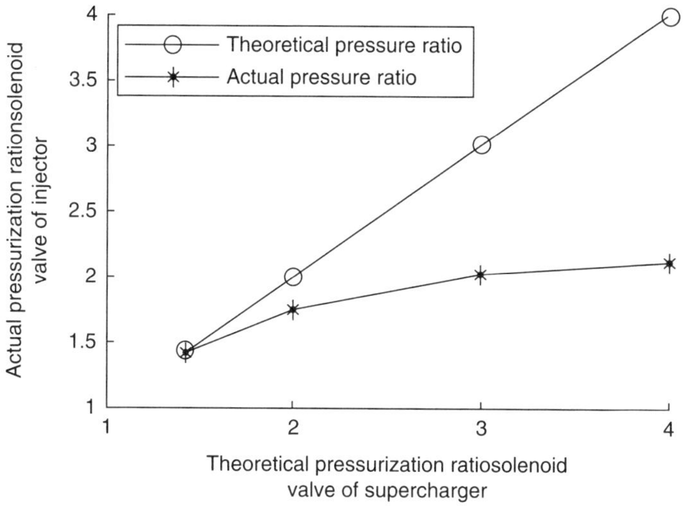

Figure 7.6 Theoretical pressurization ratio and actual ratio.

of the dual pressure system and the optimization design of the supercharger are all based on a theoretical pressurization ratio of 2.

In theory, by adjusting the timing sequence of the supercharger and injector (Figure 7.7a), both the base pressure and the high pressure can be achieved in one injection, and different shapes of injection rules can be obtained through the combination of the base-pressure and high-pressure injection timing. In order to verify the effect of the control time sequence on the law of fuel injection, the injection law when ΔT_{int} is equal to 0.2, 0, and −0.2 ms was calculated, respectively. The results in Figure 7.7b shows that when $\Delta T_{int} = 0.2$ ms, the solenoid valve of the supercharger works first, then the high-pressure fuel is injected into the system, and the fuel injection law is shown to be rectangular; when $\Delta T_{int} = 0$ ms, the solenoid valves of the supercharger and injector

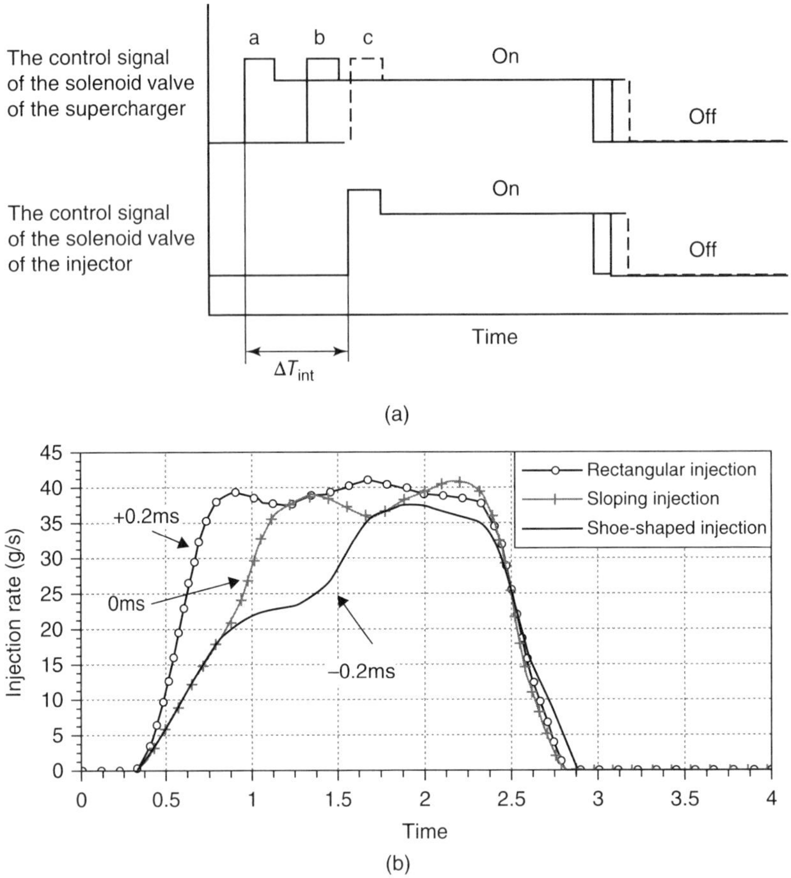

Figure 7.7 Control timing of the supercharger and injector and the corresponding injection rate.

open at the same time, the system starts to jet the base pressure fuel, the supercharger starts pressurization, the displacement of the supercharger piston is larger than the fuel injection quantity, and the injection pressure increases gradually, and, at this time, the injection law is shown to be sloping; when $\Delta T_{int} = -0.2$ ms, the injector solenoid valve opens first and the system starts to jet the base-pressure fuel. After 0.2 ms, the supercharger starts working, and the injection pressure increases gradually because the volume of the supercharger piston is larger than that of the fuel injection. At this time, the injection rule is shown to shoe shaped. The calculation results show that the system can change the shape of the fuel injection rate.

When the DPCRS works at the base pressure state, the supercharger can be simplified as a one-way valve. The injection rules under the conditions of the same high-pressure common rail system and the same DPCRS, namely the same rail pressure and the same fuel injector control timing, are compared (Figures 7.7 and 7.8). The results show that the DPCRS can respond to the requirement of multiple injection as easily as the common rail system, and fully inherits the flexibility of the current high-pressure common rail injection system.

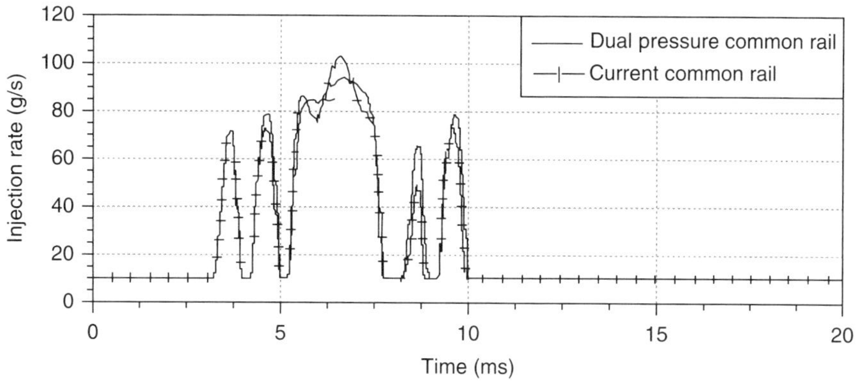

Figure 7.8 Comparison of the injection law of the dual pressure common rail and current common rail systems.

7.2.1.2 Sensitivity Analysis of the Structural Parameters of the Supercharger

The dynamic simulation of the working process of the system is carried out discussed above. The flexibility of injection control is studied. The influence of the fuel injection rule on the injection characteristics is analyzed and compared with the related test results, which proves the feasibility and reliability of the designed system principle and structure. In order to master the influence regulation of the system structure parameters on the output characteristics of the system and determine the parameter space of the structural optimization, a sensitivity analysis of the structural parameters is needed.

(1) Influence of the pressurization ratio on the performance of the system. The pressure variation of the supercharger chamber is calculated at the pressurization ratio of 2, 3, and 4, respectively. The results are shown in Figure 7.9a. The pressure peak value increases with the increase of the pressurization ratio, but the peak value does not increase proportionally. At the same time, the greater the ratio, the lower is the pressure fluctuation frequency and the slower is the attenuation. Pressurization is a transient process, which is accompanied by the injection of fuel injectors and all kinds of wave mixing in the flow field. These reasons may cause the high pressure not to increase proportionally with the theoretical pressurization ratio.
(2) Influence of the sectional area of the pressurized piston on the performance of the system under the same pressurization ratio. The pressure of the supercharger chamber is calculated when the small end area of the pressurized piston is 50, 60, and 70 mm^2, respectively. The results are shown in Figure 7.9b.
(3) Influence of the area of the inlet fuel throttle in the control room of the supercharger device on the system performance. The pressure changes of the supercharger chamber are calculated when the diameter of the inlet throttle is 0.3, 0.4, and 0.5 mm, respectively. The results are shown in Figure 7.10a. With the increase of the diameter of the throttle orifice, the effect of the supercharger is obviously reduced. When the diameter increases, the pressure release process of the control chamber is slowed down, so the diameter of the throttle hole should be minimized. Changing the diameter of the throttle hole in the control chamber will change the frequency of the pressure fluctuation in the supercharger chamber.

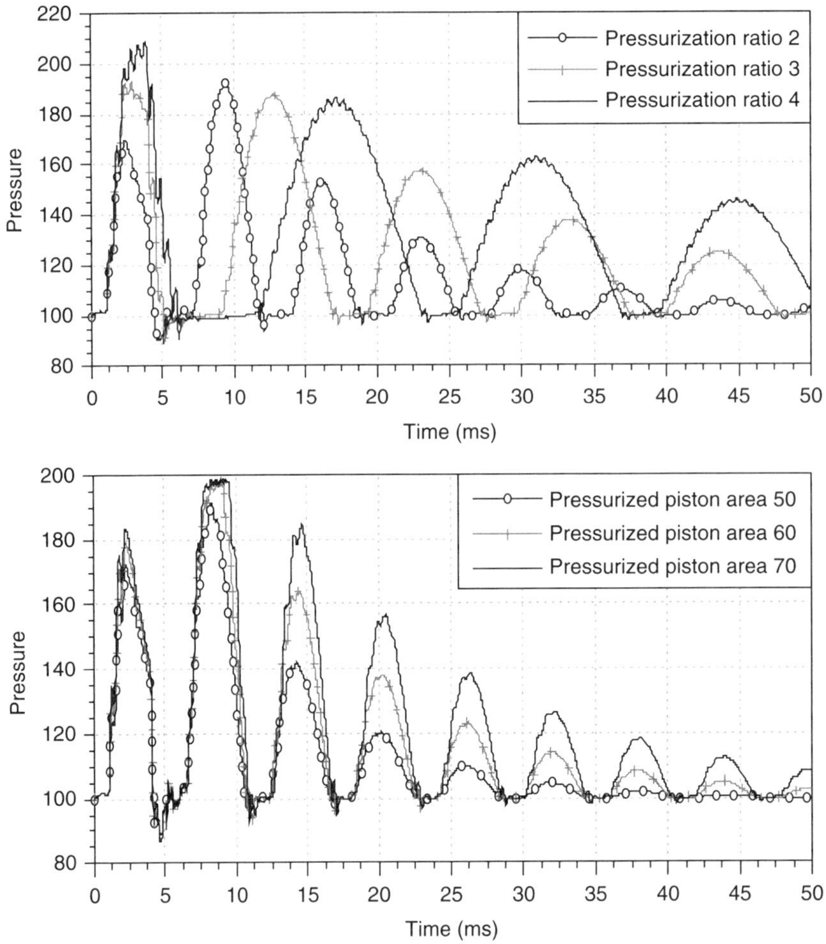

Figure 7.9 Supercharger chamber pressure of different pressurization ratios and different pressurized piston areas.

(4) Influence of the fuel outlet throttle in the supercharger device control room on the system performance. The pressure changes of the supercharger chamber are calculated when the diameter is 0.4, 0.6, and 0.8 mm, respectively. The results are shown in Figure 7.10b. With the increase of the diameter of the fuel outlet orifice, the pressurization effect is stronger. The main reason is that the speed of the control pressure release accelerates and the pressurization piston obtains greater acceleration. The pressure fluctuation amplitude of the supercharger chamber increases and the frequency is constant. However, the larger the diameter of the outlet orifice, the higher are the performance requirements of the solenoid valve. Therefore, the diameter of the throttle hole should be determined according to the performance of the selected solenoid valve.

(5) Influence of the fuel passage length in the control chamber of the supercharger device on the system performance. The pressure changes of the supercharger chamber in the control chamber are calculated when the length is 16, 32, and

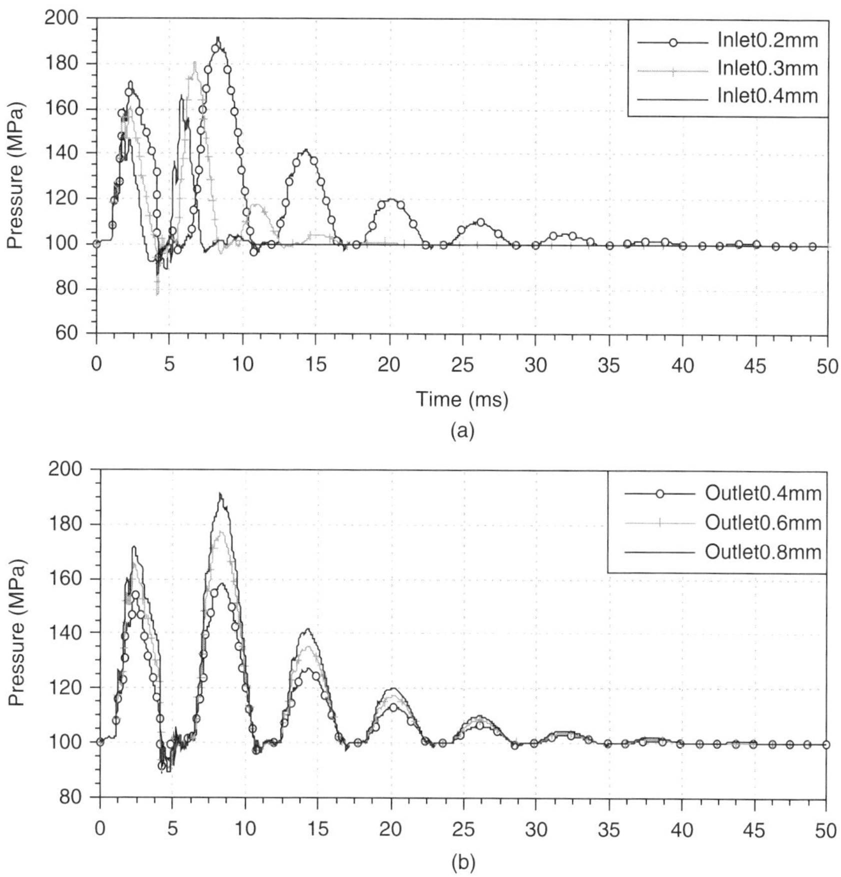

Figure 7.10 Comparison of pressure in the supercharger chamber in control chambers with different fuel throttling holes and fuel throttle diameters.

48 mm, respectively. The results are shown in Figure 7.11. The length of the fuel outlet has no obvious effect on the pressurization effect.

(6) Influence of the length of the fuel inlet channel in the control chamber on the pressurization. The pressures of the supercharger chamber are calculated when the length of the inlet passage of the control chamber is 20, 40, and 60 mm respectively. The results are shown in Figure 7.12. With the increase of the length of the inlet channel, the maximum pressurization of the supercharger chamber increases and the duration of the pressure oscillation is prolonged.

(7) Influence of the change of the ratio of the pressurization piston area to the sectional area of the nozzle on the performance of the system. The constant condition is that the pressurization ratio is 2, the small end area of the pressurization piston is 64 mm^2, the diameter of the nozzle is 0.2 mm, the pressurization and the injection start at the same time, and the injection duration is 2 ms. By changing the number of orifices, the pressure changes of the supercharger chamber are calculated, respectively, when the number of orifice numbers is 4, 6, and 8 (corresponding to the ratio of the piston small end area to the nozzle area being 509, 340, and 255).

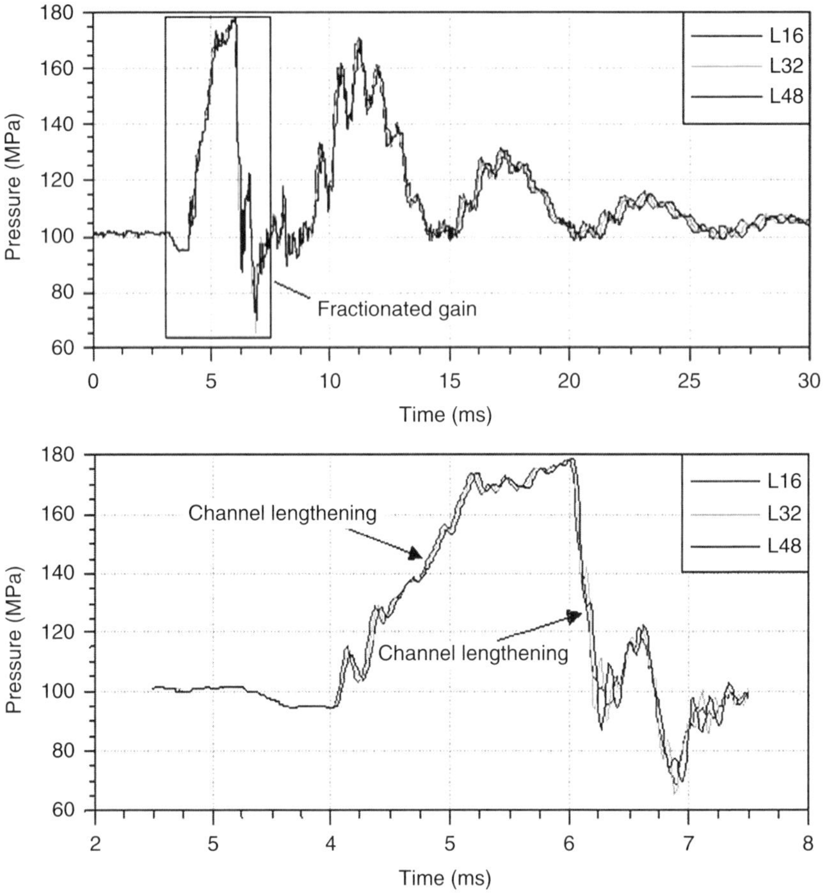

Figure 7.11 The influence of the length of the fuel passage in the control chamber on the pressurization.

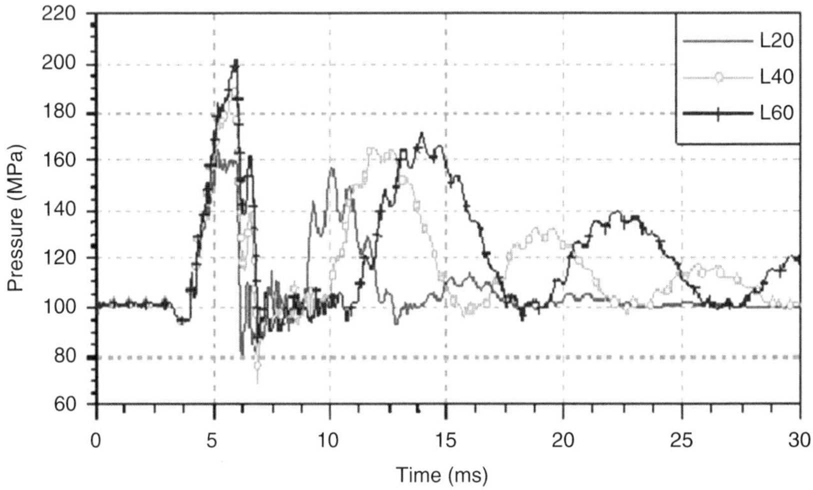

Figure 7.12 Influence of the inlet channel length of the control chamber on pressurization.

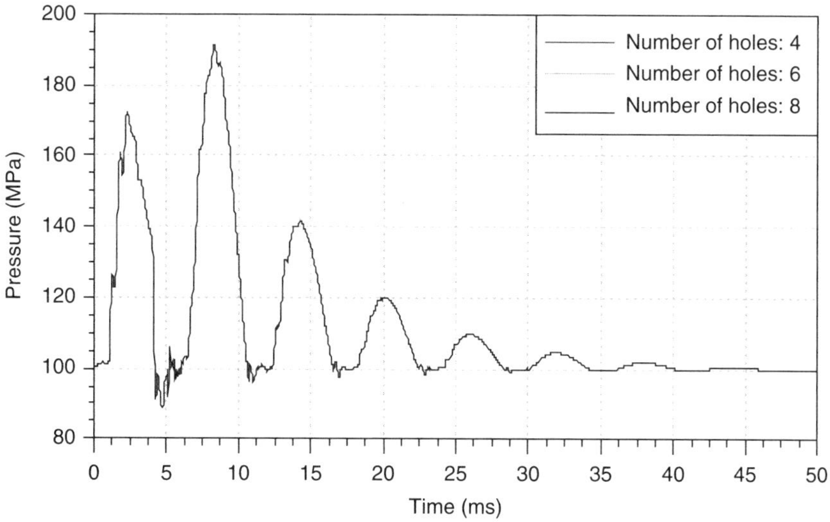

Figure 7.13 Comparison of pressure of the supercharger chamber with the area ratio of different pressurization piston and nozzle section areas.

The calculation results are shown in Figure 7.13. The area ratio in a certain range has no effect on the pressure of the supercharger chamber.

(8) Influence of the volume of the control room on the performance of the system. The pressure changes of the supercharger chamber are calculated when the control chamber volume is 500, 1000, and 1500 mm^3, respectively, and the results are shown in Figure 7.14 (the partial enlargement map). With the increase of the control chamber volume, the effective pressurization pressure of the supercharger chamber is increased, and the trend of reduction slows down, but has little effect. However, it has some effect on stabilizing the pressure fluctuation in the supercharger chamber.

(9) Influence of the volume of the supercharger chamber on the performance of the system. The pressure changes of the supercharger chamber are calculated when the volume of the supercharger chamber is 320, 640, and 960 mm^3, respectively, and the results are shown in Figure 7.15. With the increase of the volume of the supercharger chamber, the maximum pressure of the pressurized chamber is reduced, the amplitude of pressure fluctuation decreases, and the frequency is constant.

(10) Influence of the stiffness of the reset spring on the performance of the system. The pre-tightening force of the pressurization piston reset spring is zero before pressurization. The pressure changes of the supercharger chamber are calculated when the spring stiffness is 50, 100, and 150 N/mm, respectively. The result is shown in Figure 7.16. The increase of the spring stiffness consumes some kinetic energy of the pressurization piston and reduces the effective pressurization. After completion of the pressurization, the time needed for the pressure recovery of the supercharger chamber is shortened. The stiffness of the spring should take into consideration the effect of the supercharging and the response of the system.

Through the above analysis, the main structural parameters affecting pressurization characteristics are determined as follows: the ratio of pressurization, the

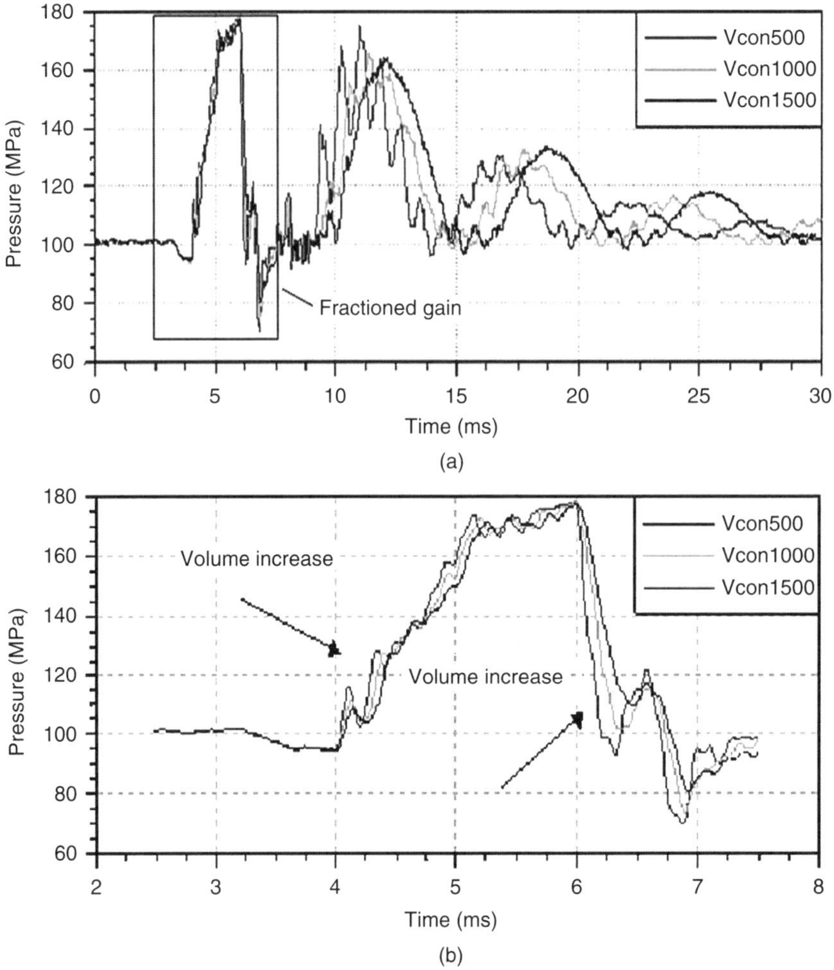

Figure 7.14 Comparison of supercharger chamber pressures in different control room volumes.

section area of the pressurization piston, the control chamber inlet and outlet orifice diameters, the volume of the supercharger chamber, the volume of the control chamber, and the stiffness of the pressurization piston reset spring.

7.2.1.3 Study on Pressure Oscillation Elimination of the Supercharger Chamber in the Dual Pressure Common Rail System

Through the simulation and calculation of the system, it was found that there is pressure oscillation in the supercharger chamber after pressurization, and the amplitude of the second pressure peak is equivalent to that of the first pressure peak. The oscillation continues until the end of a working cycle. As this pressure fluctuation may affect the stability and reliability of the injector, measures should be taken to eliminate or weaken the fluctuation. Using the WDD (wave dynamics and dampening) method of wave dynamics and damping, the following two measures are adopted: (i) a one-way valve is set

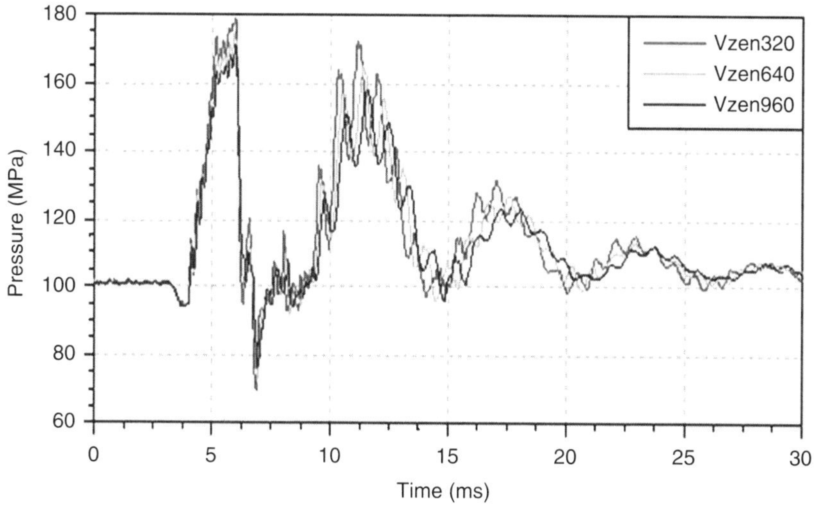

Figure 7.15 Comparison of supercharger chamber pressures in different pressurized chamber volumes.

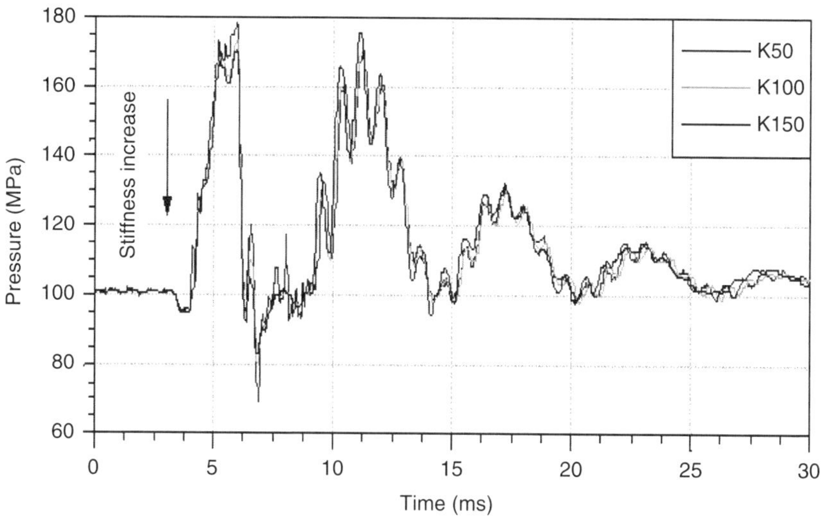

Figure 7.16 Pressurization of different pressurization piston spring stiffnesses.

between the supercharger and the injector and (ii) a damping chamber is set between the supercharger and the injector. The system structure is shown in Figure 7.17.

7.2.1.3.1 Scheme I

Before and after setting of the one-way valve, the pressure of the supercharger chamber and the pressure of the injector accumulator are compared through the simulation calculation. The situation is shown in Figure 7.18. After a one-way valve is added, the pressure oscillation duration of the supercharger chamber is shortened and the pressure oscillation of the accumulator chamber is basically eliminated. However, the pressure of

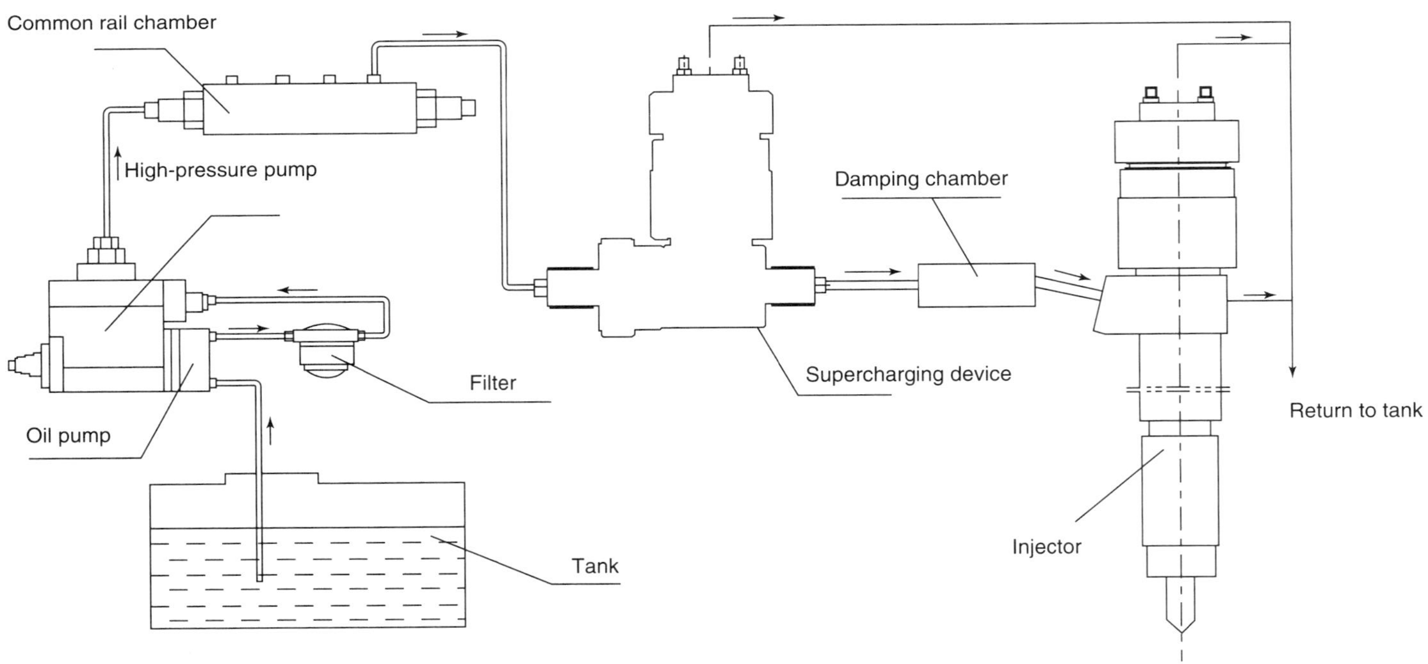

Figure 7.17 Dual pressure common rail system with damping cavity.

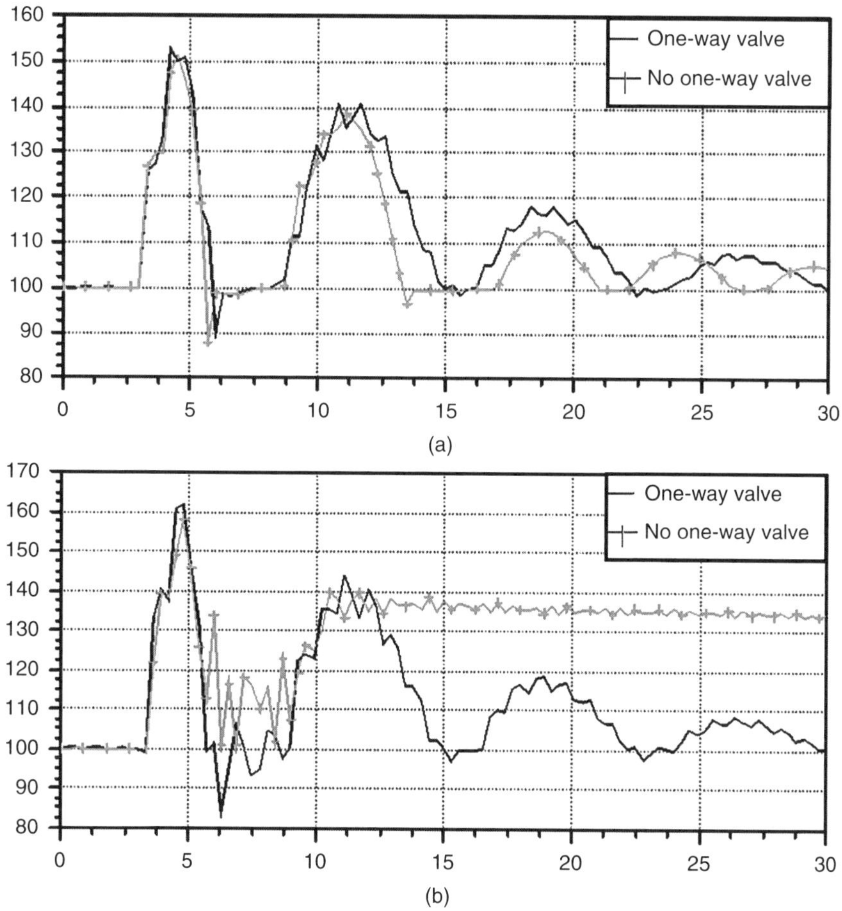

Figure 7.18 Pressure changes in the supercharger chamber of the front and back of the one-way valve and the accumulator chamber of the injector.

the accumulator chamber of the injector cannot restore the pressure of the rail chamber, which will affect the next injection.

7.2.1.3.2 Scheme II

After setting the damping chamber, the pressure of the supercharger chamber and the injection chamber pressure are calculated when the volume of the damping cavity is 1500 and 500 mm^3, and then without the damping cavity, respectively. The selection of the volume of the damping chamber mainly considers the volume of the supercharger chamber and that of the connection line. The situation is shown in Figure 7.19. When the damping chamber is added, the pressure of the supercharger chamber is basically the same as that of the injector, and the peak value and the duration of the oscillation are significantly reduced. The larger the volume of the chamber, the more obvious the effect is. After the damping cavity is added, the effective pressure is weakened.

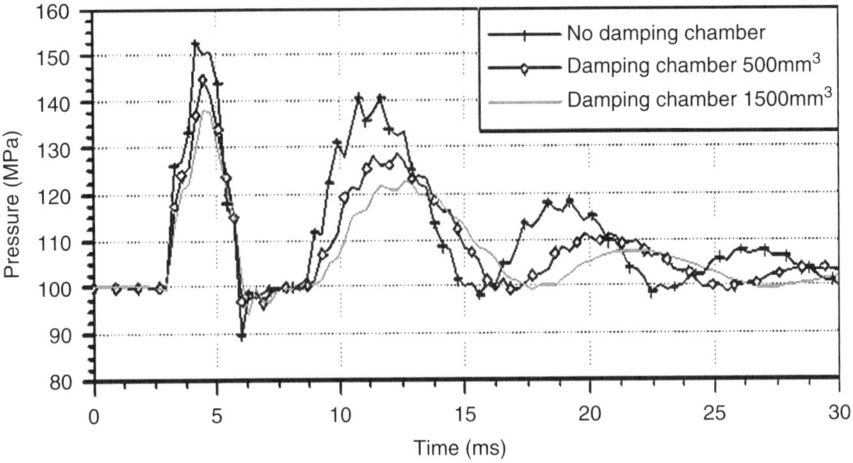

(a)

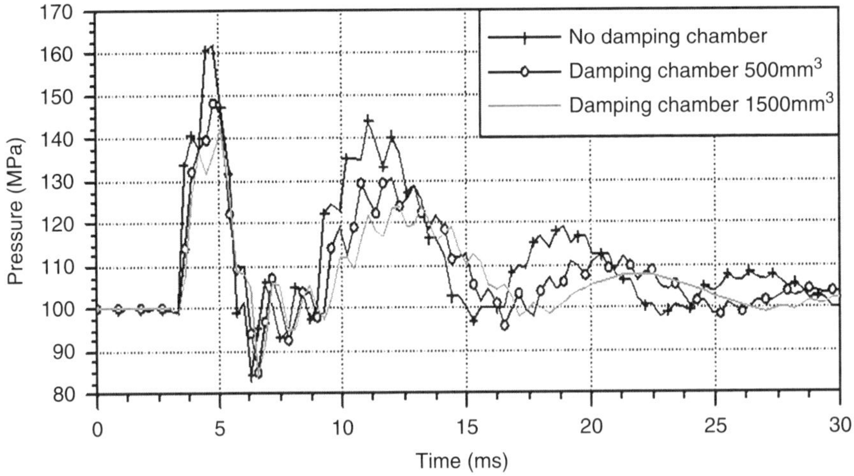

(b)

Figure 7.19 Pressure change of the supercharger chamber and injector accumulator chamber with different damping chamber volumes.

The above two methods of restraining pressure oscillation are weighed and considered, and the damping cavity is finally decided. The main reason is that the one-way valve blocked the passage of pressure back to the pressure of the chamber before high-pressure injection, making the next injection pressure out of control. The damping chamber can suppress the pressure oscillation. Although it has a certain effect on the pressurization effect, it can be compromised.

7.2.2 Prototype Trial Production

According to the results of the calculation and analysis, a prototype of a supercharger is made. The prototype is shown in Figure 7.20. In order to prevent the influence of

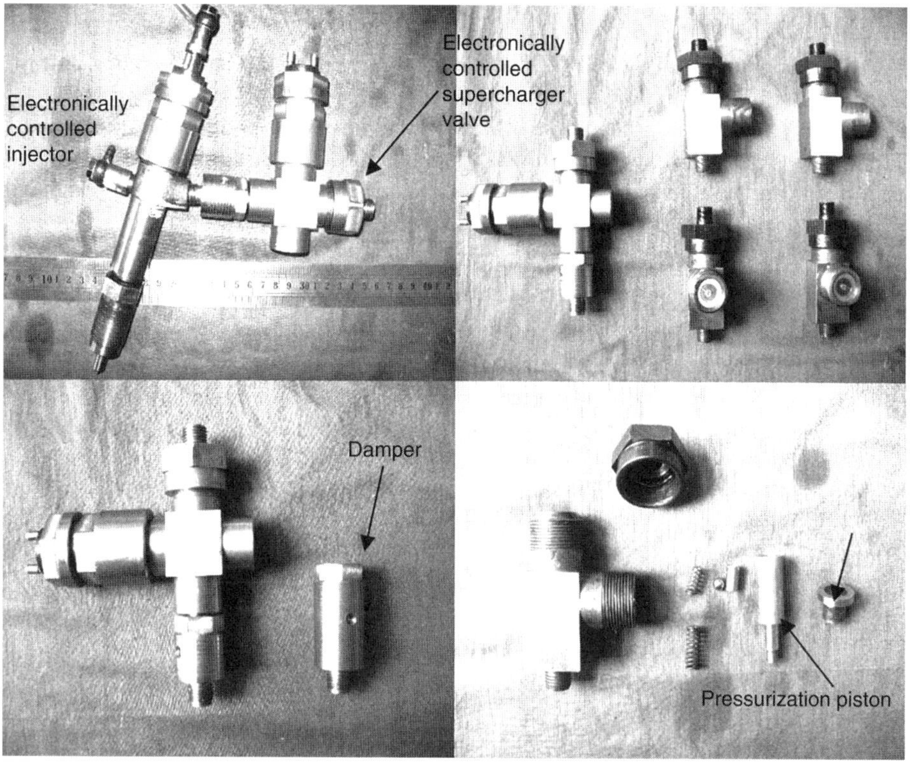

Figure 7.20 Dual pressure common rail system supercharger prototype.

machining accuracy and other factors on pressurization pressure, and to ensure that the pressurization pressure is above 180 MPa, three kinds of supercharging ratios (2, 3, and 4) are processed under the condition that the other structural parameters of the supercharger remain unchanged.

In order to eliminate the reduction of the supercharging effect in the process of pressurization due to the deformation of the high-pressure fuel channel, the fuel outlet orifice of the electronically controlled supercharger is directly butted with the high-pressure fuel inlet orifice of the electronically controlled injector, shortening the high-pressure fuel path.

7.3 Control Strategy and Implementation of the Dual Pressure Common Rail System

The working mode of the DPCRS is different from that of the current high-pressure common rail system. The electronically controlled supercharger and the injector need both independent control and timing matching. Therefore, the injection control module of the DPCRS needs to design a control and drive circuit that can drive the dual solenoid valve and with a timing sequence relationship that is adjustable.

7.3.1 Control Strategy of the Dual Pressure Common Rail System

Through the above analysis, the DPCRS can realize three kinds of injection law, rectangular injection, slope injection, and shoe-shaped injection, and can achieve three kinds of working modes in principle, namely the part load operation mode, operation mode, and high load (110% load) operation mode. Each operation mode needs a corresponding injection rule and control strategy to achieve it. Therefore, the DPCRS needs a control mode corresponding to its operation mode.

(1) Partial load control model. Taking the base pressure injection as an example, the control unit only needs to send the fuel injection control signal of the injector according to the trigger signal of the top dead center, and the fuel injection control signal is amplified by the driving circuit to control the action of the solenoid valve of the fuel injector. At this time, the control mode of the current high-pressure common rail system is exactly the same. The time series of the base pressure jet control is shown in Figure 7.21.
(2) Rated load control mode. In this operating mode, according to the requirements of the engine, the control unit calculates the required fuel injection quantity and injection law and determines the duration of base-pressure injection and high-pressure injection according to the injection law. Accordingly, the control unit is triggered by the TDC trigger signal, the control signal is sent to the supercharger solenoid valve and the fuel injector solenoid valve, and the control signal is amplified by the driving circuit to control the solenoid valve action. The time series of the rated load control is shown in Figure 7.22.
(3) High load control mode. Under this operating mode, the dual pressure system can use high-pressure injection or a base-pressure and high-pressure mixed injection. In this operating mode, the control unit according to the requirements of the engine calculates the required injection quantity and according to the injection quantity the high-pressure injection duration is determined. Accordingly, the control unit is triggered by the TDC trigger signal and the control signal is sent to the supercharger solenoid valve and the fuel injector solenoid valve; the control signal is amplified by the driving circuit to control the solenoid valve action. The control sequence of the high load control mode is shown in Figure 7.23.

The operation mode of the DPCRS is not entirely restricted by any kind of fuel injection rule. According to the need of engine combustion, a shoe-shaped injection can be carried out under various load modes, and the fuel quantity and fuel injection rule can

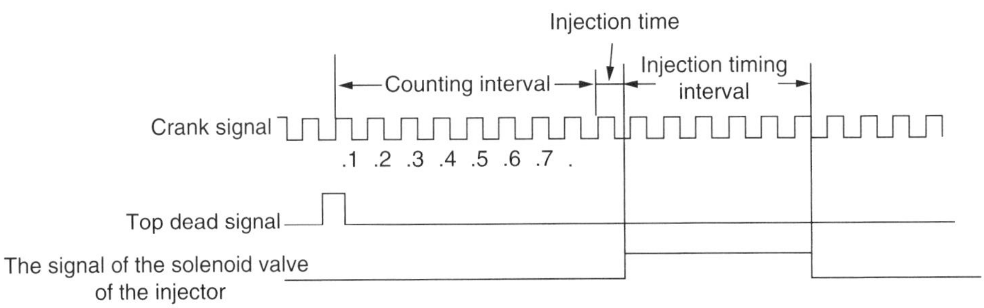

Figure 7.21 Time sequence diagram of a low load control mode.

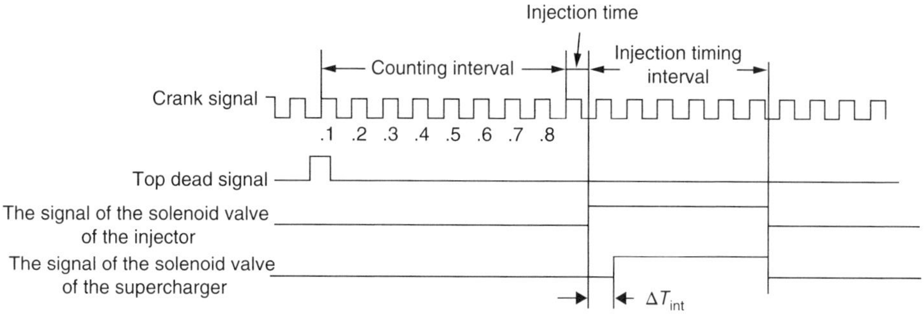

Figure 7.22 Time sequence diagram of a rated load control mode.

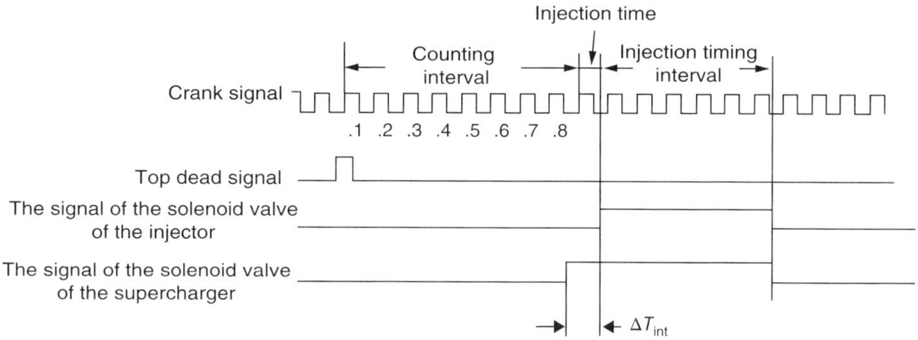

Figure 7.23 Time sequence diagram of a high load control mode.

be adjusted according to requirements. The core of the above control mode is the control of the supercharger solenoid valve and the fuel injector solenoid valve, thus designing a control and power amplifying circuit that can drive double electromagnetic valves and with timing sequence relations adjusted as needed.

7.3.2 Hardware and Software Design of the Controller Based on the Single Chip Microcomputer

7.3.2.1 The Basic Composition of the Control System

In this chapter, the control circuit of the dual pressure common rail injection based on a PIC microcomputer is designed. The basic hardware composition of the control circuit is shown in Figure 7.24. The main components of the control system are introduced as follows:

(1) The core of the controller is composed of a PIC16F877A single chip microcomputer (SCM) and peripheral circuit. In an MPLAB integrated development environment (IDE), the control program is programmed in C language.
(2) The serial communication circuit is used to communicate with the SCM and the PC machine.
(3) The magnetoelectric sensor is used to collect the top dead center signal on the camshaft and adjust the sine wave signal to a square pulse signal through the pulse

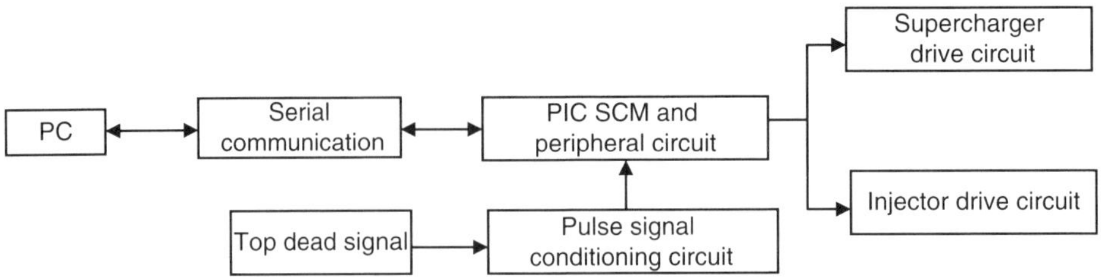

Figure 7.24 The basic hardware composing the control system.

signal conditioning circuit. Then this pulse signal triggers the output injection control signal of the SCM.

(4) The +5 V DC power supply is used for the pulse signal conditioning circuit and the optoelectronic isolator in the drive circuit. The +15 V DC power supply is used in the optoelectronic isolator, which is used to drive the field effect tube. The +50 ~ +120 V adjustable DC power supply and +24 V DC power supply are used to provide a high/low voltage power supply for the drive circuit.

7.3.2.2 Performance of Control Chip and Its Circuit Design

7.3.2.2.1 The Circuit Design of the Minimum System of the Single Chip Microcomputer

The SCM is the core hardware of the whole control system. This article selects the PIC16F877A SCM.

According to the functional characteristics and test needs of each pin, the smallest system of the SCM is designed, including the power supply part, crystal oscillator part, and reset switch part. The circuit principle is shown in Figure 7.25. The capacitance of the power supply is mainly used for filtering and voltage regulation, the resistance is used for current limiting, and the diode is used for power indication.

When the wire jumper is short connected to the lower ends of J3 and J4, the crystal oscillator is selected as the system clock and the external RC oscillation is selected as the system clock when the upper two ends are jumped.

During the manual reset, a low-level signal can be added to the MCLR reset end in the execution of the program to reset. The manual reset is set up in order to force the SCM to reset and make the program start running again.

7.3.2.2.2 Design of the Serial Communication Circuit

The input and output of the SCM are TTL levels, where high +5 V, low 0 V, and the RS-232 levels of the computer serial port are high −3 ~ −15 V, low +3 ~ +12 V. The electrical characteristics of the two are inconsistent, so the communication between the computer and the SCM needs to use the hardware circuit to realize the level conversion. At the transmitter, the TTL level is converted to the RS-232 level by the driver and the RS-232 level is converted to the TTL level at the receiver.

In this paper, the Max232 of MAXIM Company is used as the level conversion chip. Max232 is a dual transmission/reception converter with two transmitters and two receivers on one piece and uses a single power supply (+5 V). The principle of the serial communication circuit is shown in Figure 7.26. The RC6 pin of the microcontroller is the T2in pin for the sending terminal to connect to Max232. The chip RC7 pin is the

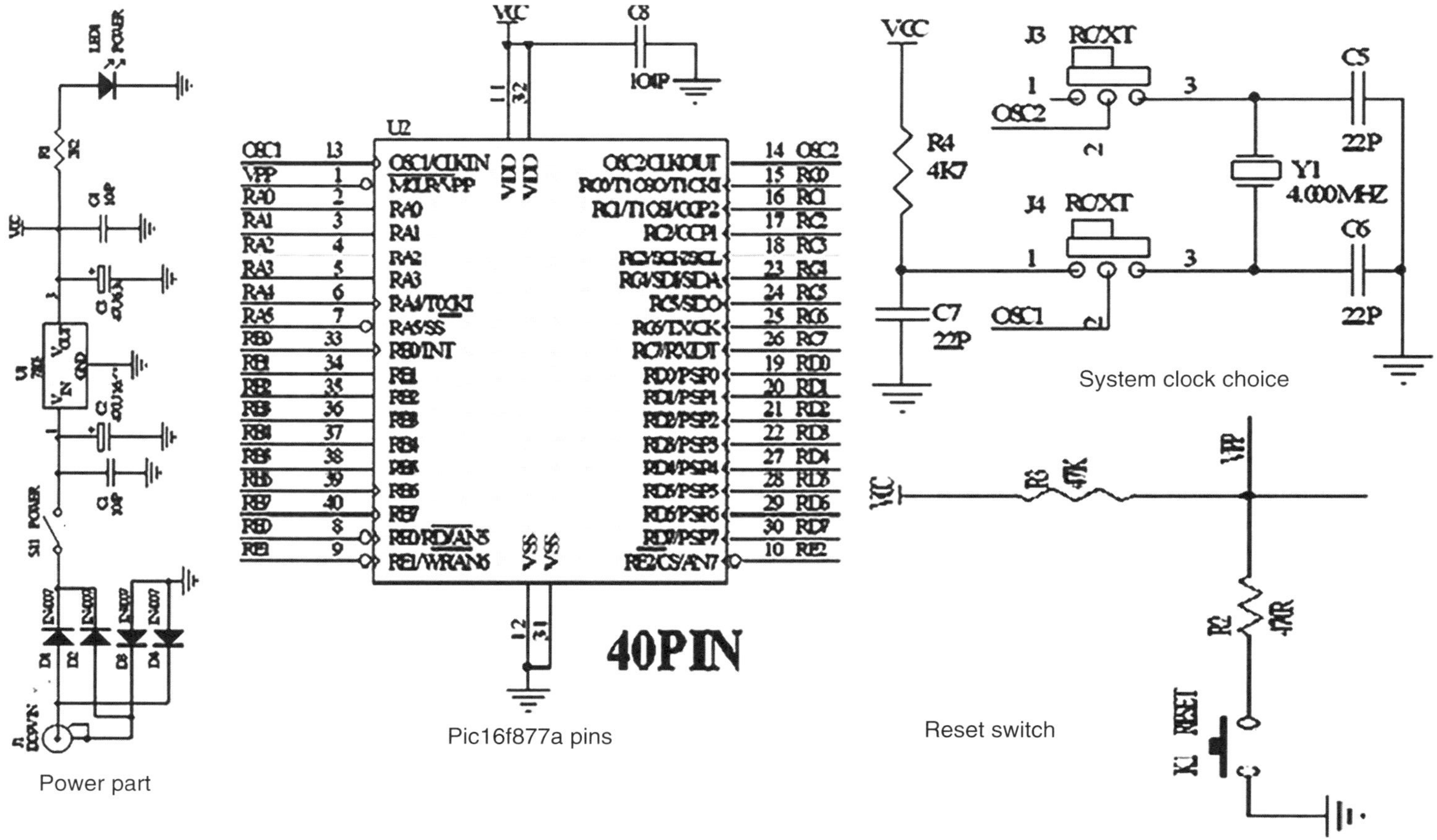

Figure 7.25 The smallest system of the SCM.

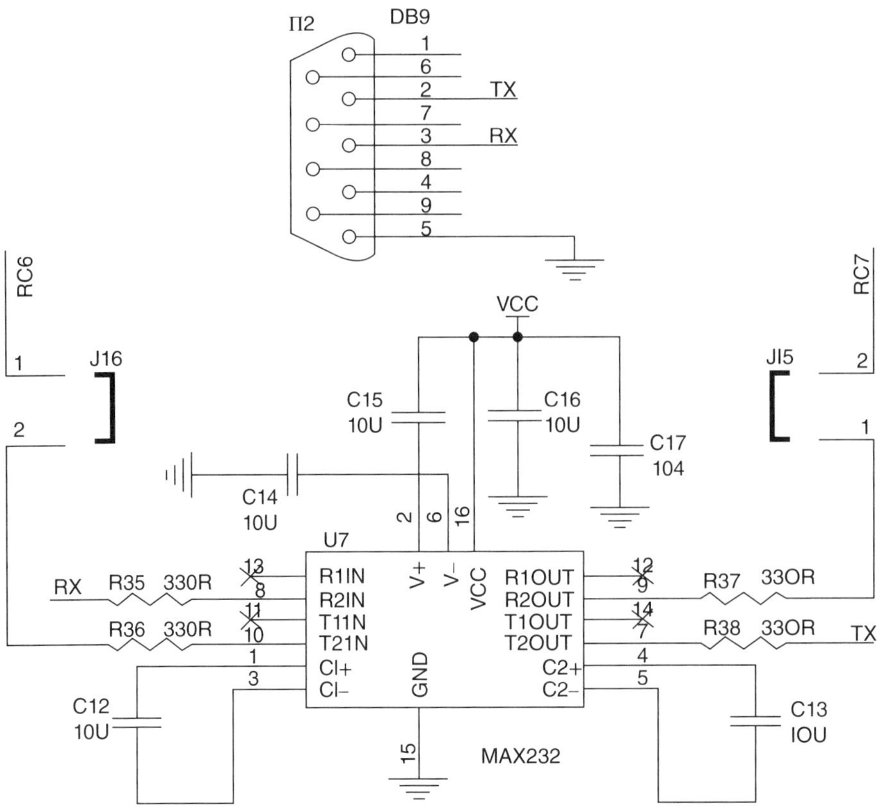

Figure 7.26 SCM serial communication.

R2out pin of the receiving terminal and the T2out and R2in of Max232 are, respectively, RS-232 and RX pins. The capacitance is mainly used for power conversion.

7.3.2.2.3 Pulse Signal Conditioning Circuit

This design uses the magnetoelectric sensor to collect the top dead center signal of the camshaft and converts the sine wave signal to the pulse signal through the signal conditioning circuit, triggering the output signal of the microcontroller. The following are requirements for the conditioning circuit:

(1) Demand for voltage amplitude. Due to the use of the magnetoelectric sensor, when the speed is not the same, the output signal amplitude will show a large difference, but the voltage controller needs a voltage range with a logic level standard. Therefore, the signal conditioning circuit should be able to limit the voltage amplitude in the standard range in order to prevent the emergence of the phenomenon of a high voltage pulse or leakage.
(2) Rising time requirement. The signals of the position of the crankshaft and the upper stop point are all pulse signals. In the process of working, the rising edge is the working point. Because the voltage in the actual circuit cannot be abruptly mutated, there is a rising time of the signal. In order to avoid the timing error caused by this unstable state, the rise of the pulse signal is required to be as short as possible.

(3) Signal isolation requirements. The crankshaft position signal and the upper stop signal occur through the magnetoelectric sensor. Because the electromagnetic interference of the environment of the diesel engine is very serious, in order to prevent the external disturbance affecting the controller, it is necessary to adopt the high-speed photoelectric isolation device to isolate the two signals, in order to reduce the disturbance to the controller.
(4) Phase requirement. Since the original signal is processed to the control core after multistage processing, the phase difference between the input and output of the signal conditioning circuit is always there. However, in order to ensure the stability of the oil supply control pulse, the phase difference of the signal conditioning circuit must be stable and compensatable.

Based on the requirements of the above four aspects, the pulse signal conditioning circuit used in this paper is shown in Figure 7.27.

The circuit is divided into four parts. In the first part, a single power amplifier LM358 and capacitors C1, C2, R2, and R3 are used to form a tw0-order Bessel filter. Among them, C1 is 0.0047 μF, C2 is 0.0068 μF, R2 is 8.25 kΩ, and R3 is 14.7 kΩ.

The second part is a voltage comparator composed of a single voltage comparator LM393. LM393 can work normally under the condition of a $2 \sim 18$ V single power supply and transforms the sine wave into a square wave.

The third part is an isolating circuit composed of the optocoupler 6N139. The two power supplies VCC1-GND1 and VCC2-GND2 at both ends of the optocoupler are electrically isolated. The optical signal is used to isolate the voltage signals at both ends. The optical isolation device 6N139 is a high-frequency Optocoupler. The high-frequency response characteristic is better than the ordinary optocoupler. The output signal through the optocoupler is passed through the fourth part of the circuit, the voltage comparator, in order to reduce the rising time of the output signal.

7.3.2.3 Programming of Control System

The whole software system includes: a system initialization program, a high and low voltage control pulse generator, a serial communication program, and an external interrupt program. In this chapter, the PICC program is written in the MPLAB-IDE environment and the PICC18 of HI-TECH Company is chosen as the compiler.

7.3.3 Drive Circuit Design

The control signal output of the SCM is the weak electric signal of +5 V, which cannot control the solenoid valve directly. Therefore, it is necessary to add the driving circuit between the control signal and the solenoid valve.

7.3.3.1 Design Requirements of the Driving Circuit

This section outlines the requirement to design two kinds of drive circuits, one that drives the solenoid valves and the other that drives the GMM valves. Although the solenoid valve is used in the dual pressure system test, the test of the GMM actuator also needs to be observed. This is a brief introduction to the designed drive but we can find any differences through it. The two types of solenoid valves require different driving current waveforms (Figure 7.28), so it is necessary to design a drive circuit with different structures.

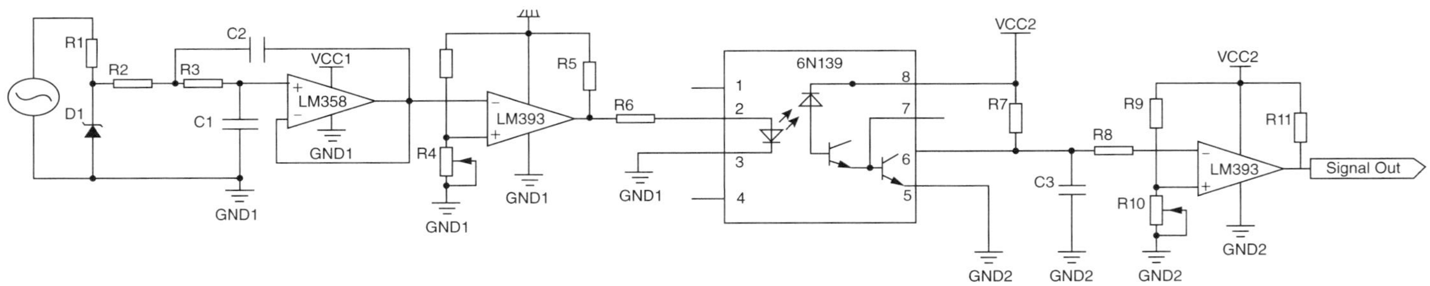

Figure 7.27 Pulse signal conditioning circuit.

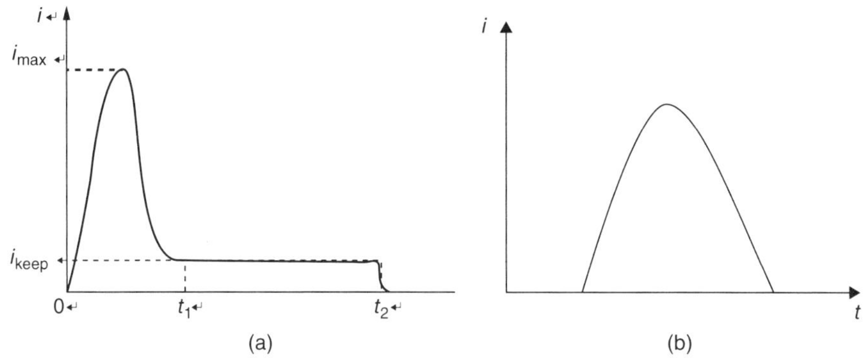

Figure 7.28 Driving current of solenoid (a) and GMM (b).

The electromagnetic force produced by the solenoid valve is inversely proportional to the air gap between the armature and the electromagnet. Due to the largest air gap in the initial state, the same electromagnetic force is required for a larger electromagnetic current compared with the suction state, while in the suction state a smaller current is needed. Therefore, in order to improve the opening speed of the solenoid valve and reduce its power consumption in the state of sucking, there are certain requirements for the driving current waveform of the solenoid valve. It is required that the starting current can be quickly loaded and the solenoid valve can be quickly actuated. The maximum starting current should meet the requirements of starting the solenoid valve. It is necessary to reduce the current and reduce the impact between the armature and the limit plate before solenoid valve actuation. When the solenoid valve is completely actuated, the current should be kept stable and the current should be as small as possible under the condition that the reliable actuation is guaranteed. This can improve the closing response speed of the solenoid valve, reduce the calorific value of the solenoid valve, and increase the life of the solenoid valve. The GMM actuator requires a stable current to maintain the stability of the excitation magnetic field, thus ensuring the stability of the output displacement.

In order to ensure that the solenoid valve can be quickly sucked up, the armature must get enough energy in a short time. To release this energy in a very short time (about 0.3 ms), the instantaneous power is very large. Considering the energy loss caused by the heating of the solenoid valve coil, the actual driving energy is larger than W, so the actual instantaneous power is much larger. Because of the intermittent work, the average power required for the solenoid valve is actually quite small. Obviously, the average power and its matching ordinary power cannot meet the requirement of this special large current pulse load, so the drive circuit needs to be designed separately, so that it can keep the current running smoothly.

7.3.3.2 Design of the Power Drive Circuit

7.3.3.2.1 Power Drive Circuit of the GMM Actuator

The GMM actuator is driven by a single voltage mode. The circuit structure is shown in Figure 7.29a. Figure 7.29b is a simulation output current. In this circuit,

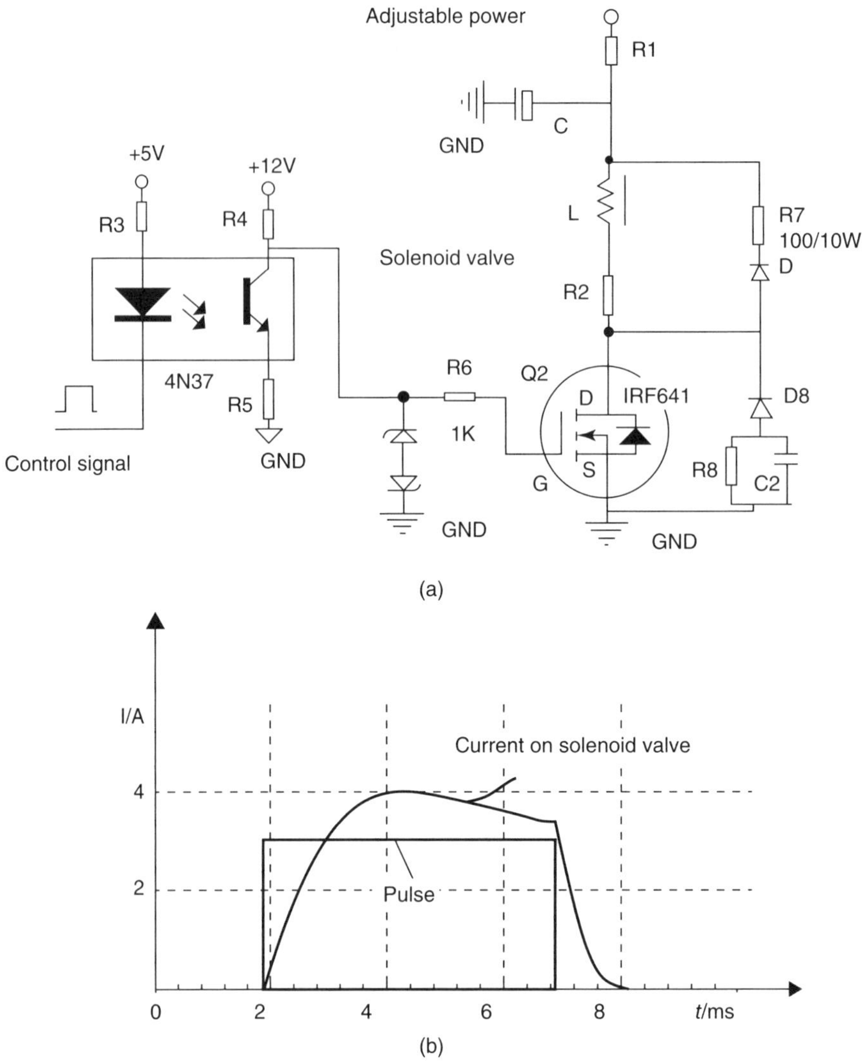

Figure 7.29 Drive mode (a) of the GMM actuator and its output current (b).

a $+50 \sim +150$ V adjustable DC power supply is used to drive the voltage. The switch state of the metal oxide semiconductor field-effect transistor (MOSFET) tube IRF641 is controlled by the pulse signal, in order to control the switch of the solenoid valve. The pulse signal is output by the PIC16F877A SCM. The pulse width can be controlled by serial communication according to the needs of the test. In order to prevent interference, the photoelectric coupler 4N37 is used to separate the control signal from the driving electrical signal and to separate the output port of the photoelectric isolator from the other grounding, in orderto enhance the anti-interference ability of the system.

The MOSFET is called the MOS tube, which is a kind of unipolar voltage controlled conductive device. It has self-closing ability and the switching speed is fast. It has a high input impedance, a low driving current, a wide safety working area, no second-induced breakdown, high-frequency performance, high transconductance, and the leakage current is a negative temperature coefficient, has good thermal stability, and has the ability of an automatic and uniform temperature distribution. The IRF641 is used in this drive circuit. The maximum leakage voltage of the MOS tube can reach 400 V, which is greater than 120 V. The gate source voltage VGS threshold of 2 ~ −4 V and a turn off and pass time of less than 18 ns meet the design requirements. The MOS tube is affected by the voltage or current during the switch process, so the RCD protection circuit is configured. Because the solenoid valve is the inductive load, when the power pipe is turned off, the mutation of the leakage current will produce the overshoot of the drain voltage, which is higher than the external power supply, causing the pipe to break through. Therefore, it is necessary to use the current circuit to protect the pipe. In this paper, a diode–resistance current circuit is used to protect the power pipe.

The operation process is as follows: when the trigger signal is captured by the MCU CCP1 module, the control pulse is sent. When the pulse is high, the input diode of the optocoupler 4N37 leads to light. The triode CE cut-off, the gate of the IRF641 (G pole), is high and the power pipe is completely open. The energy stored in the capacitor C is released so that the current on the coil passing through the solenoid valve rises rapidly. Meanwhile, R1 limits the mutation of the power supply current and limits the current fluctuation of the power supply to a smaller range. When the pulse signal is low, the optoelectronic isolator 4N37 is connected and the power pipe is cut off, so that the current is cut off in order to make the solenoid valve close quickly. At this time the power is charged to C by R1, but as the voltage of C increases, the current of the power supply is gradually reduced, preparing for the next drive. The capacitance C is usually based on the value of the formula $C = L^{*}$ (1.4/R1) 2, which can make the current on the solenoid valve reach a larger peak in a shorter time. Under the premise that the capacitance discharge energy is enough to drive the solenoid valve, the size of the discharge capacitance should be reduced as much as possible in order to shorten the charge and discharge time of the capacitor.

7.3.3.2.2 Power Drive Circuit of the Solenoid Valve

The high-voltage input voltage of the power drive part is 120 V, the internal resistance of the solenoid valve is 1 Ω, the inductance is 0.9 mH, the minimum open current is 15 A, the minimum holding current is 4 A, and the driving current waveform of the solenoid valve needs to meet the requirements of the previous section. According to the above requirements, this paper uses a dual-power dual-switch drive circuit; the circuit diagram of the drive circuit is shown in Figure 7.30.

The advantages of the dual-power dual-switch drive circuit are that the driving current is stable, the high and low driving voltage is converted smoothly, the switching loss of the switch device is small, and the current and voltage impact is small. The disadvantage is that two power sources are used and the number of switch tubes is increased by one, which increases the complexity of the system. All power MOSFETs

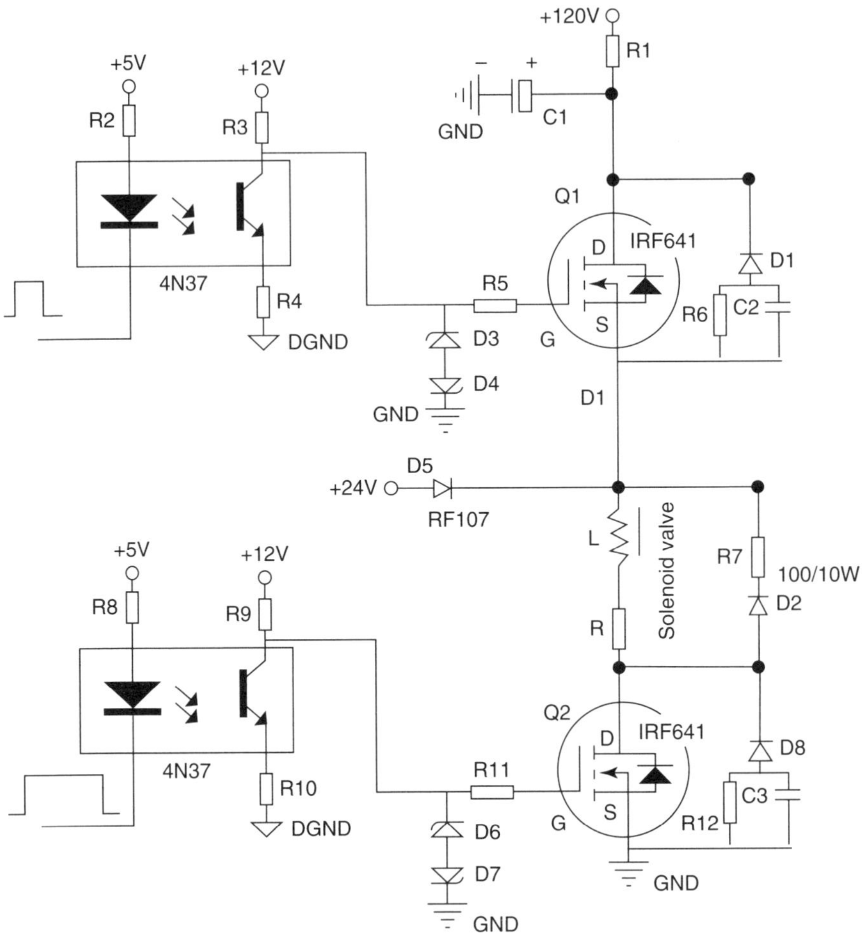

Figure 7.30 A schematic diagram of the structure of a dual voltage drive circuit.

will be impacted by the switch voltage and current, so we need to configure the RCD absorption circuit. The driver circuit of the high switch tube needs two isolated power supplies.

The work process is as follows: the SCM simultaneously sends out two control signals, where one way controls the high voltage and the other way controls the low voltage. When both of the two control signals are high logic, the MOS tube Q1 and Q2 are all connected. At this time, the 24 V low voltage is cut off through the diode D5 and the 120 V high voltage acts on the solenoid valve, which makes the driving current rise and speed up the opening speed of the solenoid valve. The starting process of the solenoid valve is the same as the single voltage driving mode. After the solenoid valve is turned on, the circuit turns to the low-pressure operation. When the high-voltage control pulse is low and the low-voltage control pulse is high, the 24 V low voltage provides the power to the solenoid valve until the low-voltage control pulse also becomes low The dual-voltage dual-switch drive circuit realizes high and low voltage switching through two control signals in order to provide a small maintenance current and to reduce energy

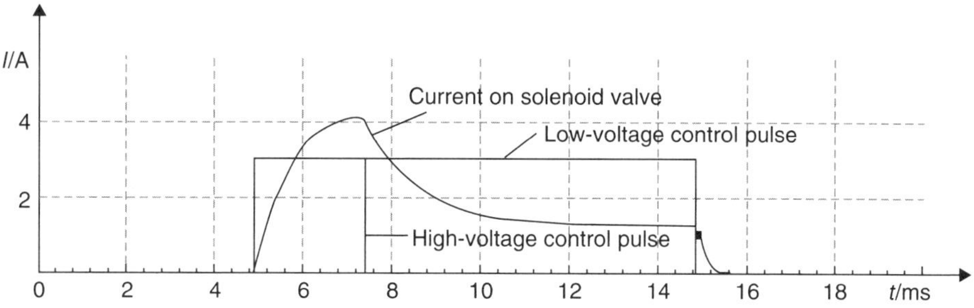

Figure 7.31 Drive current waveform.

consumption. Simulation of the output current of the drive circuit and the control signal is shown in Figure 7.31.

7.4 Experimental Study on the Dual Pressure Common Rail System

Through the research in the previous chapters, the simulation model of the DPCRS has been set up and the running law of the dual pressure system has been mastered by simulation calculations. On the basis of the optimization design, the trial production of the main parts has been completed. A complete set of the dual pressure common rail electronic control fuel injection system has been constructed. In order to check the accuracy of the simulation model and test the performance of the dual pressure system, it is necessary to carry out the necessary experimental research. Through experiments, we should investigate the performance indexes of the DPCRS, including: the pressurization pressure of the electronic supercharger, the adjustability of the fuel injection law, the control oil consumption, and the spray characteristics of the supercharger.

7.4.1 Test of Pressurization Pressure and Injection Law

7.4.1.1 Test Platform for Pressurization Pressure and Fuel Injection

Figure 7.32 is a composition diagram of the test system for the working characteristics of a DPCRS. The system can be used to measure the pressurization pressure, the fuel injection rate, and the solenoid valve control signal. In the process of experiment, serial port communication between the computer and the control circuit can be used to change the control time sequence of the solenoid valve and the supercharger solenoid valve, and then we can study the influence of the control time sequence on the fuel injection rate. The main test equipment for the system is shown in Table 7.1. The DPCRS test platform is shown in Figure 7.33.

The EMI fuel injection law tester is the key instrument for the performance test of the DPCRS. The basic characteristics are as follows:

(1) Trigger signal: 1/rev and 3600/rev (provided by the test-bed).
(2) Injection speed: 30–3000 times/min.
(3) Measurement accuracy: 0.6 mm^3 = 0.0006 ml.
(4) Measurement range of fuel injection: 0–600 mm^3/cycle.

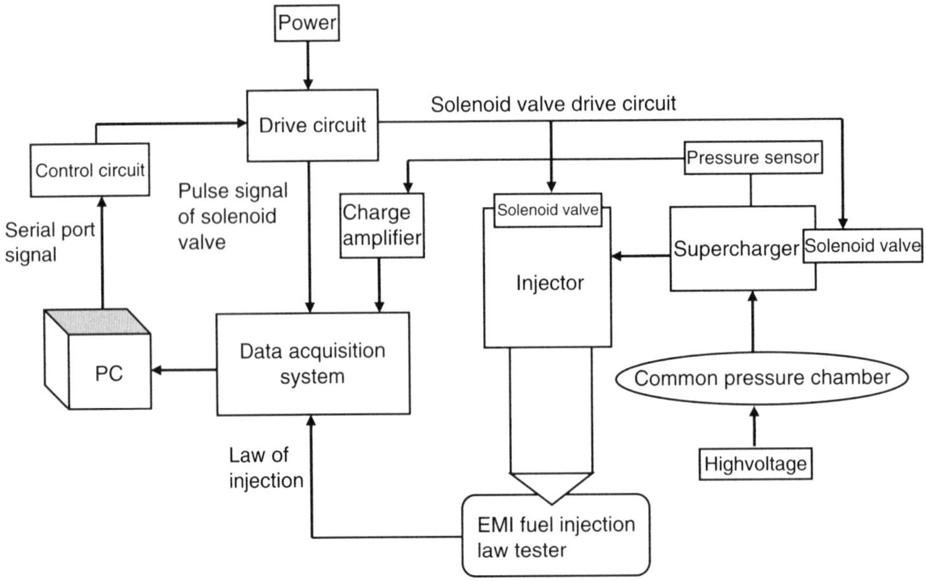

Figure 7.32 Schematic diagram of a dual pressure common rail system test device.

Table 7.1 Test equipment list.

Name	Model	Performance index	Manufacturer
Power supply	DNC-F-150/03	Output voltage 60–150 V	Brilliance Power Equipment Co., Ltd.
drive circuit			Developed by self
Piezo crystal pressure sensor	JC-1	Charge sensitivity 36 pC/g	Fuzhou Radio Instrument Factory
Charge amplifier data acquisition system	FDH-4	Output 10 mV/unit	Yangzhou Radio No. 2 Factory
			Nanjing Steam Turbine Co., Ltd.
EMI fuel injection law tester			

(5) The maximum, minimum, mean, and mean variance of the results are given by multiple injections, such as 100 or 1000 times, which can be set by the user and can be used to evaluate the consistency or stability of a single cylinder (or a single injector) for continuous operation.
(6) The measurement results can be displayed on its LCD screen and on the terminal computer screen.

The piezo pressure sensor is used to measure the pressure in the test. In order to get an accurate pressure value, the sensor needs to be calibrated. Fitting the calibrated data of the sensor is shown in Figure 7.34.

Figure 7.33 Test bench of the DPCRS.

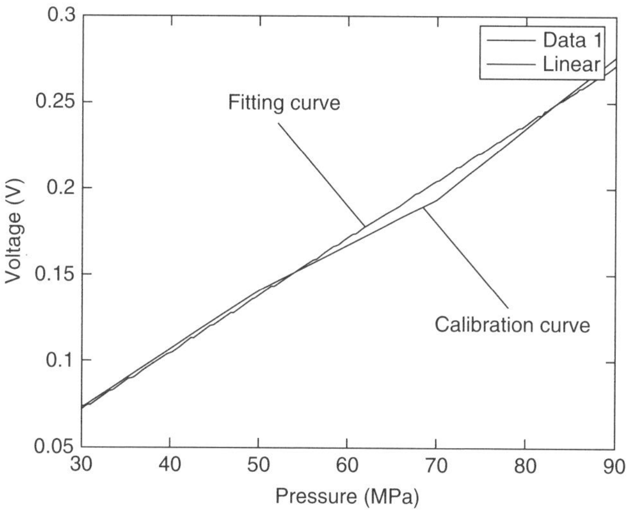

Figure 7.34 Sensor calibration curve.

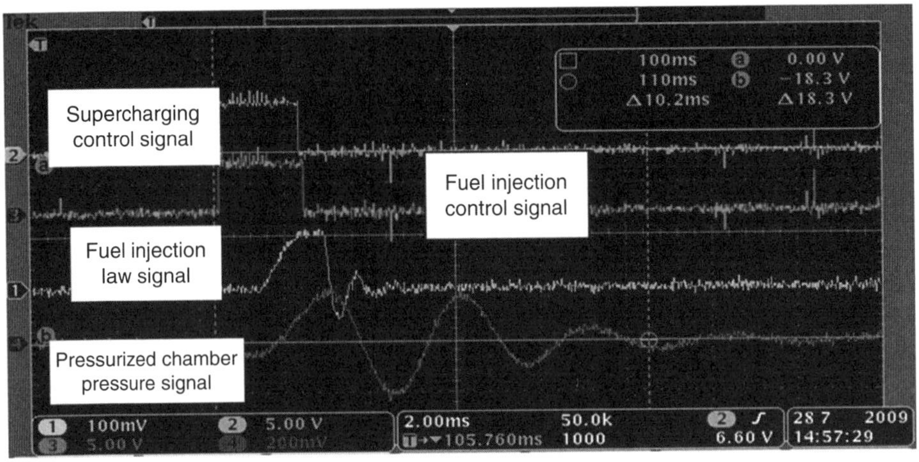

Figure 7.35 Test results.

The DPCRS, which is assembled by three kinds of structural turbocharger, is named system A, system B, and system C. The turbocharging ratio of the system is the only difference, where the ratio for system A is 2, for system B is 3, and for system C is 4.

The test platform results are shown in Figure 7.35.

7.4.1.2 Simulation and Test

The performance of the DPCRS is analyzed by the simulation program and the output characteristics of the system are calculated, that is, the pressure and the law of fuel injection. The output characteristics of the prototype are obtained from the test. Comparing the simulation results with the experimental results, we can verify the accuracy of the simulation program and also find the actual performance of the system, which helps to further optimize the system design. We only use the comparison between the simulation and the experimental results of scheme C.

(1) Comparison of pressurized pressure
Operating conditions. The pressure of the rail cavity is 100 MPa, the duration of the injection control pulse is 2 ms, and $\Delta T_{\text{int}} = +1$ ms. The results of the comparison are shown in Figure 7.36. The peak value of the test booster is up to 215 MPa, the peak value of the simulation supercharging is 230 MPa, and the test value is lower than the simulation value, which is mainly due to the insufficient machining precision of the couple parts. The continuous running of the system leads to a high oil temperature, which affects the effect of supercharging. The pressure rise rate of the simulation is higher than the test value, mainly because the theoretical response speed of the system is higher than the actual value.

(2) Comparison of the law of fuel injection
Comparison of working conditions. The rail pressure is 100 MPa, the injection control pulse width is 2 ms, the control time series $\Delta T_{\text{int}} = +1, 0, -0.5,$ and -1 ms. The results of the comparison are shown in Figure 7.37.

The simulation results show that changing the timing of turbocharging and injection control will change the shape of fuel injection. From the measured results, the shape of

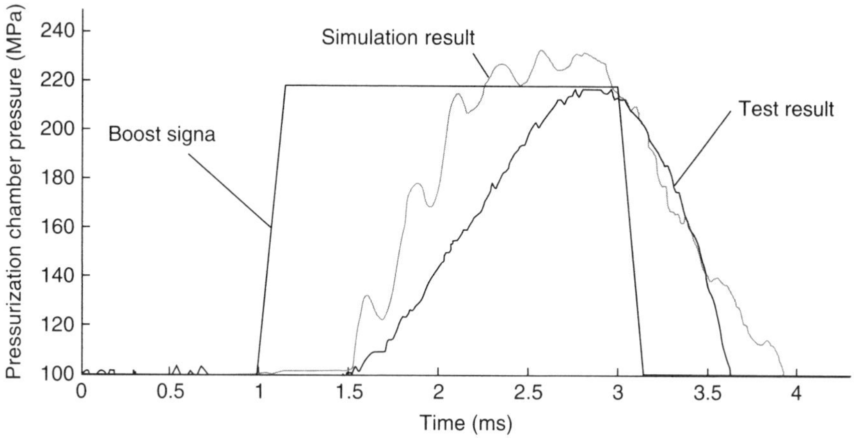

Figure 7.36 Comparison of the amplified pressure of the test results and simulation results.

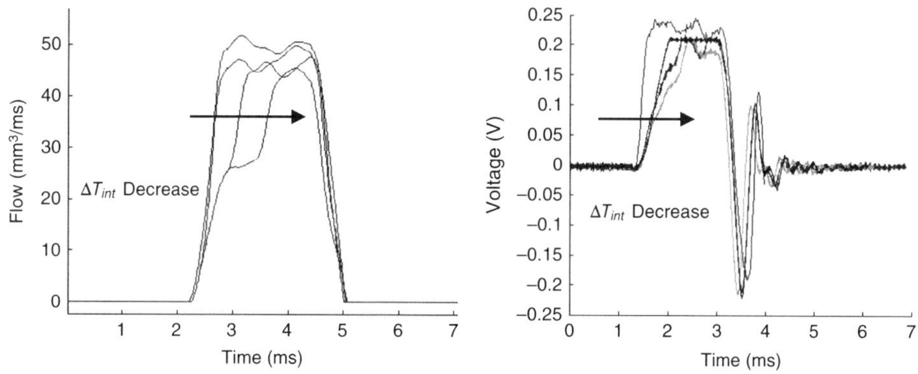

Figure 7.37 Comparison of the injection rate of the test results and simulation results.

the fuel injection rate is successfully adjusted by adjusting the control time series, and the credibility of the simulation model is further confirmed.

7.4.1.3 Effect of the Turbocharging Ratio on Pressure and Fuel Injection Law

This experiment is mainly to study the effect of the turbocharging ratio on the pressure rise and the fuel injection rate under different control time when determining the rail pressure and fuel injection duration.

Comparison working conditions are rail pressure of 100 MPa and injection pulse duration of 2 ms.

(1) $\Delta T_{\text{int}} = +1.0\,\text{ms}$.

When the turbocharging time is 1 ms ahead of the injection time, the comparison test results of turbocharging and the fuel injection rate are as shown in Figure 7.38. The arrows point, respectively, to system A, system B, and system C. With the increase in the turbocharging ratio, the peak of pressurization pressure is increased, where the peak of pressurization pressure for system A is 170 MPa, for system B is 195 MPa, and for system C is 217 MPa. As a result of the advance booster,

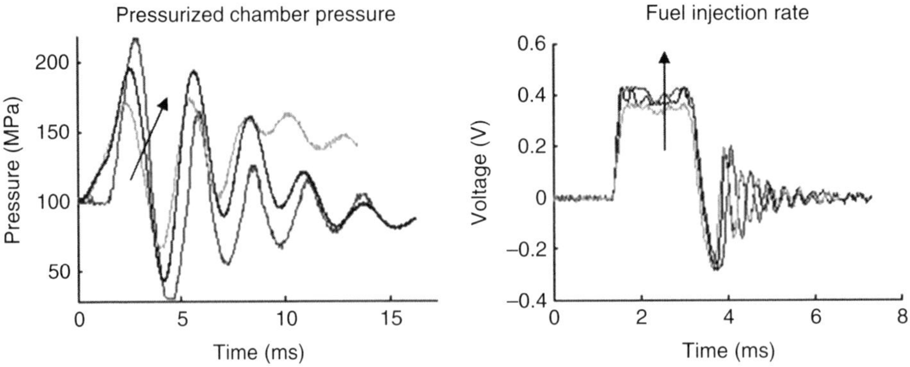

Figure 7.38 $\Delta T_{int} = +1.0$ ms, comparison of the pressurized pressure and injection rate.

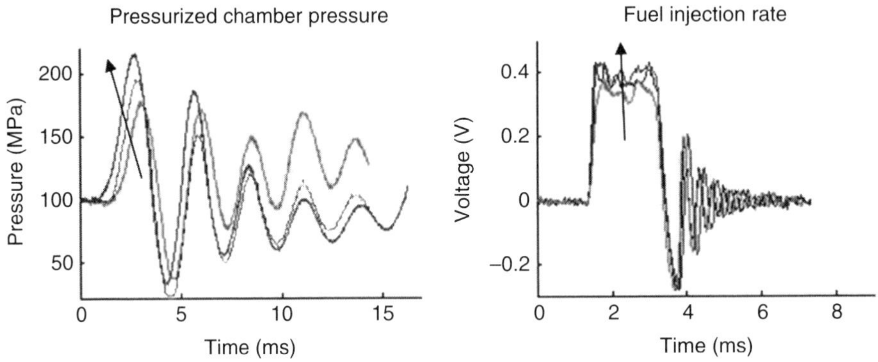

Figure 7.39 $\Delta T_{int} = +0.5$ ms, comparison of the pressurized pressure and injection rate.

the fuel injection rate of the three systems is rectangular, which is in agreement with the simulation results. Oscillation in the injection rate curve is the inherent phenomenon of the EMI fuel injection law tester. The pressurization pressure oscillates, but it does not affect the injector's normal operation. Because the end of the shock is fast, it will basically end in 20 ms, and it will not affect the next injection.

(2) $\Delta T_{int} = +0.5$ ms.
When the turbocharging time is 0.5 ms ahead of the injection time, the comparison test results of the turbocharging and the fuel injection rate are as shown in Figure 7.39. With the increase in the turbocharging ratio, the peak of pressurization pressure is increased, where the peak of pressurization pressure for system A is 176 MPa, for system B is 195 MPa, and for system C is 215 MPa. The injection rate of the three systems is rectangular. There is no obvious change in the shape of the pressure and fuel injection rate compared with $\Delta T_{int} = +1.0$ ms.

(3) $\Delta T_{int} = +0.0$ ms.
When the turbocharger solenoid valve and the injector solenoid valve are turned on at the same time, the test results of the turbocharging and fuel injection rate of each system are compared to those in Figure 7.40. The arrows are, respectively, pointing to system A, system B, and system C. With the increase in the turbocharging

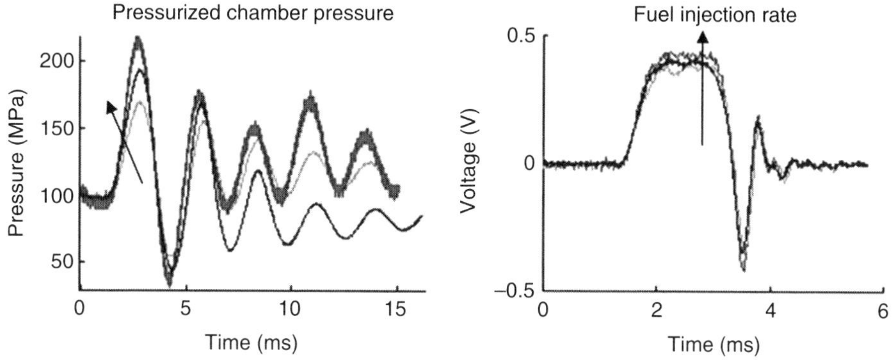

Figure 7.40 $\Delta T_{int} = +0.0$ ms, comparison of the pressurized pressure and injection rate.

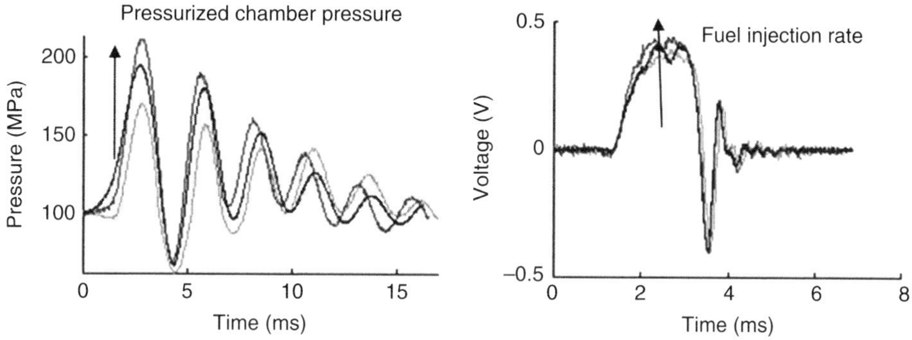

Figure 7.41 $\Delta T_{int} = -0.1$ ms, comparison of the pressurized pressure and injection rate.

ratio, the peak of pressurization pressure is increased, where the peak of pressurization pressure for system A is 169 MPa, for system B is 193 MPa, and for system C is 215 MPa. At this time the shape of the fuel injection rate changes, the slope of the initial phase of the injection is reduced, and it is approximately rectangular.

(4) $\Delta T_{int} = -0.1$ ms.
The solenoid valve of the injector is turned on first, the turbocharger solenoid valve is turned on after a delay of 0.1 ms, and the test results of the system turbocharging and fuel injection rate are compared to those i7.41. With the increase in the turbocharging ratio, the peak of pressurization pressure is increased, where the peak of pressurization pressure for system A is 170 MPa, for system B is 194 MPa, and for system C is 210 MPa. At this point the shape of the fuel injection rate is sloping.

(5) $\Delta T_{int} = -0.5$ ms.
The solenoid valve of the injector is turned on first, the turbocharger solenoid valve is turned on after a delay of 0.5 ms, and the test results of the system turbocharging and fuel injection rate are compared to those in Figure 7.42. With the increase in the turbocharging ratio, the peak of pressurization pressure is increased, where the peak of pressurization pressure for system A is 169 MPa, for system B is 192 MPa, and for system C is 215 MPa. At this point the shape of the fuel injection rate is sloping.

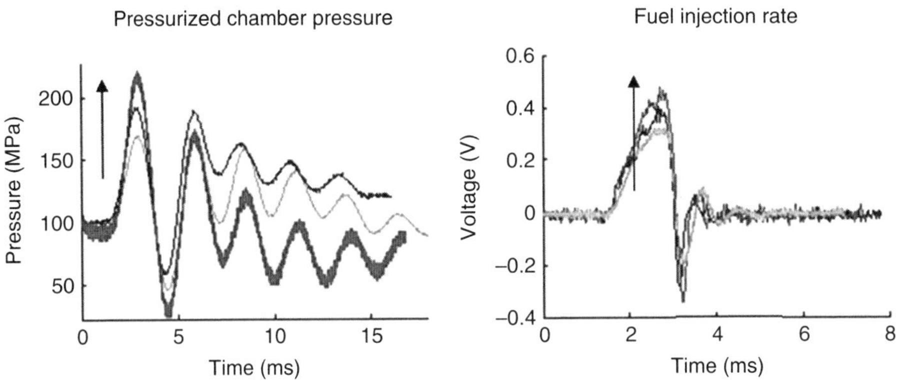

Figure 7.42 $\Delta T_{int} = -0.5$ ms, comparison of the pressurized pressure and injection rate.

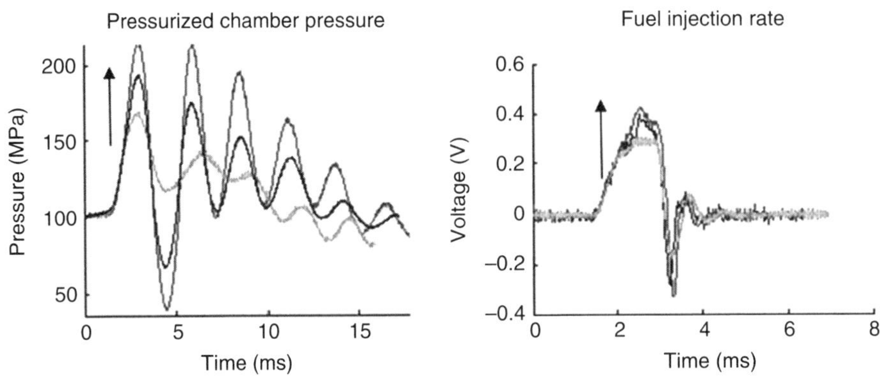

Figure 7.43 $\Delta T_{int} = -1.0$ ms, comparison of the pressurized pressure and injection rate.

(6) $\Delta T_{int} = -1.0$ ms.
The solenoid valve of the injector is turned on first, the turbocharger solenoid valve is turned on after a delay of 1.0 ms, and the test results of the system turbocharging and fuel injection rate are compared to those in Figure 7.43. With the increase in the turbocharging ratio, the peak of pressurization pressure is increased, where the peak of pressurization pressure for System A is 167 MPa, for System B is 192 MPa, and for System C is 215 MPa. At this point the shape of the fuel injection rate is boot-shaped.

The results of the above six cases are summed up in Table 7.2. The data from the table show that all the systems have an obvious booster effect, but the turbocharging ratio measured by each system is lower than the original booster ratio. The change in the control timing of the solenoid valve and the supercharger solenoid valve in the system will have an obvious effect on the shape of the fuel injection rate. When ΔT_{int} moves from positive to negative, the shape of the injection rate is gradually transferred from the rectangle to the boot shape.

Table 7.2 Comparison of test results.

	System name	ΔT_{int}						Designed booster ratio	Actual booster ratio
		+1.0	+0.5	+0.0	−0.1	−0.5	−1.0		
Peak of pressurization pressure	a	170	176	169	170	169	167	2	1.7
	b	195	195	193	194	192	192	3	1.9
	c	217	215	215	210	215	215	4	2.1
Shape of injection rate	a	Rectangle	Rectangle	Approximate rectangle	Approximate slope shape	Slope shape	Boot-shaped		
	b	Rectangle	Rectangle	Approximate rectangle	Approximate slope shape	Slope shape	Boot-shaped		
	c	Rectangle	Rectangle	Approximate rectangle	Approximate slope shape	Slope shape	Boot-shaped		

7.4.1.4 Effect of the Control Time Series on Pressurization Pressure and Fuel Injection Law

Under the condition of determining the rail pressure and the duration of the injection, the effect of the control time series on the pressure lifting and the injection rate forming of each system is compared at a rail pressure of 100 MPa and an injection control pulse duration of 2 ms, giving the control time series $\Delta T_{\text{int}} = +1.0, +0.0, -0.5, -1.0$ ms.

The results of the A, B, and C systems are shown in Figures 7.44, 7.45, and 7.46, respectively. Each system realizes the change of the fuel injection rate from the rectangle to the boot shape by adjusting the timing of the control. The direction of the arrow indicates the direction of ΔT_{int} change from small to large.

7.4.1.5 Test of System High-Pressure Oil Consumption

The high-pressure oil consumption of the DPCRS mainly includes the injector injection, the injector control oil return, the leakage from the fuel injector, and the supercharger controlling the oil return, in which the injector controls the oil return and the leakage oil of the fuel injector converge to an oil path, so it is generally called the injector return oil.

In the test condition, the pressures of the rail cavity are 100 MPa and 60 MPa, the pulse width of the injection control signal is 2 ms, and the test time is 30 seconds. The test results are shown in Tables 7.3 and 7.4.

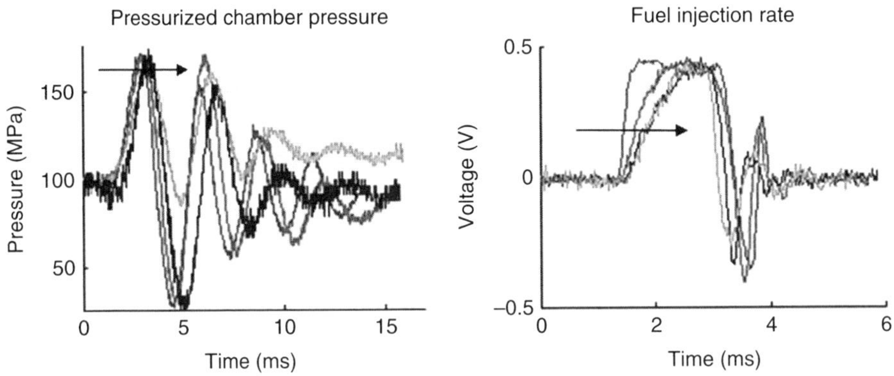

Figure 7.44 Comparison of the pressure and fuel injection rate in system A with different control timings.

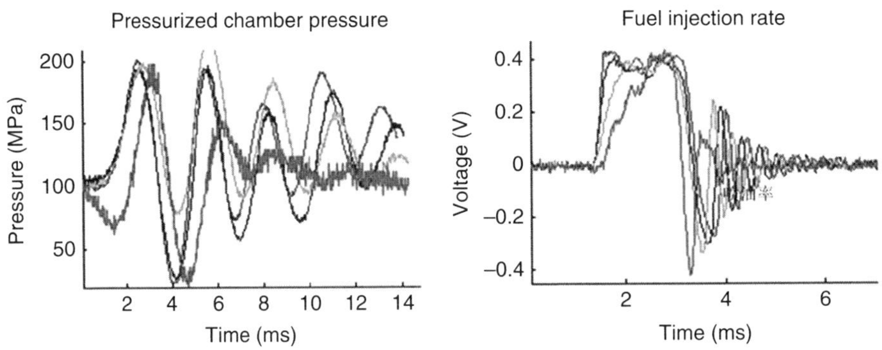

Figure 7.45 Comparison of the pressure and fuel injection rate in system B with different control timings.

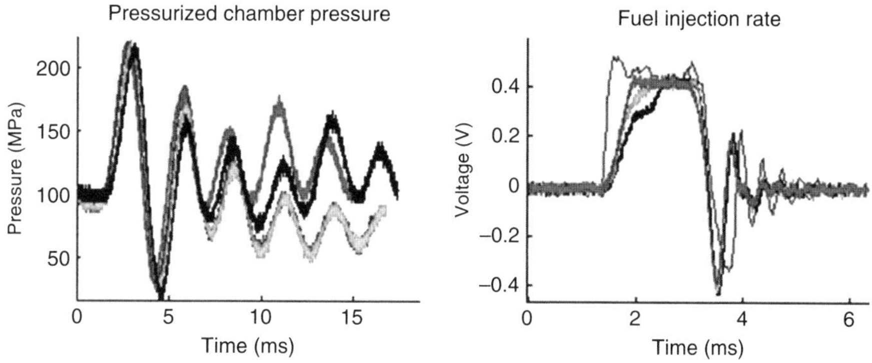

Figure 7.46 Comparison of the pressure and fuel injection rate in system C with different control timings.

Table 7.3 Oil consumption and percentage of 100 MPa rail pressure.

ΔT_{int} (ms)	Pressurized oil return (%) (ml)	Injector oil return (%) (ml)	Injector injection (%) (ml)	Total oil (ml)
+2.0	42(41.2%)	19(18.6%)	41(40.2%)	102
+1.5	38(40%)	19(20%)	38(40%)	95
+1.0	39(40.6%)	19(19.8%)	37(39.6%)	96
+0.5	34(36.2%)	21(22.3%)	39(41.5%)	94
0.0	28(32.2%)	18(20.7%)	41(47.1%)	87
−0.5	42(40%)	21(20%)	42(40%)	105
−1.0	39(40.6%)	19(19.8%)	38(39.6%)	96
−1.5	34(35.8%)	21(22.1%)	40(42.1%)	95

Table 7.4 Oil consumption and percentage of 60 MPa rail pressure.

ΔT_{int} (ms)	Pressurized oil return (%) (ml)	Injector oil return (%) (ml)	Injector injection (%) (ml)	Total oil (ml)
+2.0	14(23.3%)	15(25%)	31(51.7)	60
+1.5	2(4.3%)	15(31.9%)	30(63.8%)	47
+1.0	4(8.7%)	13(28.3%)	29(63%)	46
+0.5	3(6.4%)	14(29.8%)	30(63.8%)	47
0.0	5(10.4%)	15(31.3%)	28(58.3%)	48
−0.5	18(29.3)	14(22.6%)	30(48.1%)	62
−1.0	13(24.5%)	13(24.5%)	27(51%)	53
−1.5	6(12.8%)	13(27.7%)	28(59.5%)	47

The test data show that when the rail pressure is 100 MPa, the amount of oil return controlled by the supercharger is about 35% of the total oil consumption of the system (min. 32.2%, max. 41.2%). The effect of pressurization control on the amount of oil return and the fuel injection rate of the injector is significantly affected by the supercharging time. This is mainly because the supercharging time is different and the pressure rise value is different, which causes the different injection pressures. When the rail pressure is 60 MPa, the amount of oil return and the injection oil amount, compared with the 100 MPa rail pressure, are only a decrease in oil injection with the decrease in pressure. The change range of the pressure control return oil is larger (min. 6.4%, max. 29.3%).

7.4.2 Test on Spray Characteristics of the Dual Pressure Common Rail System

7.4.2.1 Spray Photography Test Platform

The spray photographic test device is shown in Figure 7.47. The whole system consists of a DPCRS, a simulated cylinder, a flash lamp, a camera, and a control system. The test equipment list is shown in Table 7.5.

The injector is fixed by a fixed fixture to maintain the same angle of installation as the engine. The flash lamp is arranged on the side of the nozzle. The inside of the simulated cylinder and the back of the nozzle are painted black. An SCM is used to control the

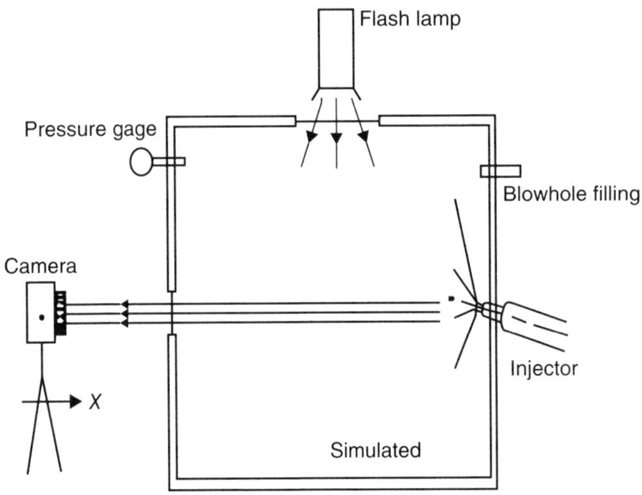

Figure 7.47 Sketch map of the spray photography system.

Table 7.5 Main test equipment for spray photography.

Device name	Model	Quantity
Dual pressure common rail system		1
Camera	Canon PowerShot A630	1
Flash lamp	Yinyan BY-30STZ	1
SCM	16f877a	1
Desktop	With serial communication	1

timing of the fuel injection and flash, so that the oil bundles at different times in the injection process can be photographed.

Simulated cylinder. The simulated cylinder can be inflated and pressurized, and the pressure in the cylinder is displayed through the pressure gauge. It simulates the high-pressure state at the end of the actual cylinder compression stroke, so that the spray can be produced closer to the real state. Both flashes and photographs are done through toughened glass. In the test, the maximum pressure of the simulated cylinder can reach 10 MPa.

Camera. The photo time can be controlled by controlling the flash.

The continuous injection of high-pressure oil in the simulated cylinder will make the whole shooting space full of oil mist and affect the quality of shooting. Therefore, we can take pictures with a controllable single shot. When the oil is foggy, we can use the compressed air.

The relative timing of the flashlight trigger time and the injection time of the injector is controlled by the 16f877A SCM.

Before shooting, the ECU of the common rail system is used to control the startuntil the pressure of the rail chamber reaches the predetermined value. When shooting, the camera is triggered first, then the fuel injection instruction is sent out, and the flash is triggered at the end. After the photo is taken, the quality of the photo should be confirmed, then the flash time is adjusted, and the next photo is taken. The timing difference between the blowout and the flash can be adjusted in 0.1 ms.

The timing of spray photography is shown in Figure 7.48. When the camera gets the trigger signal, at any time in the ΔT3 (ΔT3 < 2 seconds), the light enters the cylinder at any time and is exposed to the image. In photography, in addition to the flash, make sure that the cylinder is in an absolute dark state. The shorter the exposure duration of the flash, that is the smaller the ΔT2, the clearer and the more accurate the image is. The exposure duration of the selected flash is less than 0.1 ms, which fully meets the requirements of high-speed photography of fuel spray.

Td = 0 is the start time of the jet, and a series of images obtained at different times of the injection process by increasing Td can be used to study the development of the transient fuel spray. From the test results of the fuel injection law, the mean variance of

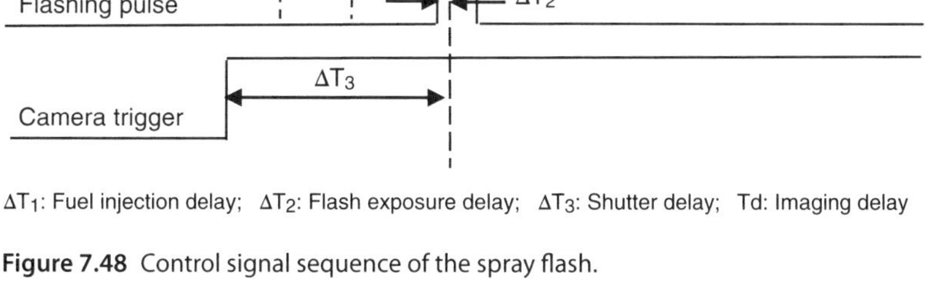

Figure 7.48 Control signal sequence of the spray flash.

Table 7.6 Injection characteristic test scheme.

	Injection pressure (MPa)					
	Base pressure	High pressure	Injection quantity (mm^3)	Shape of Fuel injection rate	ΔT_{int} (ms)	Injection duration (ms)
Scheme 1	100	170	107.4	Rectangle	+0.2	2
	100	169	97.3	Sloping	0	2
	100	167	87.1	Shoe-shape	−0.2	2
Scheme 2	100	100	107.4	Rectangle	+0.2	2
	100	100	107.4	Sloping	0	2.3
	100	100	107.4	Shoe-shape	−0.2	2.7
Scheme 3	100	100	87.1	Shoe-shape	−0.2	2
	80	80	87.1	Shoe-shape	−0.2	2.5
	64	64	87.1	Shoe-shape	−0.2	3.1

the oil quantity can be controlled within a very small range. Although the collection of images is not a whole record of the same spray development change, for a stable fuel spray, each image is a real reflection of the spray form at a specific time after the start of the injection. It can also be proved by experimental shooting at the same time for different injections.

7.4.2.2 Effect of the Fuel Injection Law on Fuel Injection Quantity

The test object is a DPCRS consisting of a scheme A supercharger.

The results of the test are shown in Table 7.6. When they have the same injection pulse width, the shape of the injection rate affects the amount of fuel injection: the maximum quantity for the rectangular shape, the second quantity for the slope shape, and the minimum for the boot shape. A boot-shaped injection needs a longer injection duration to obtain the same injection quantity. The reason is the effect of injection pressure on the amount of fuel injection.

7.4.2.3 Effect of the Injection Rate Shape on Spray Penetration and the Spray Cone Angle

According to the experiment in Table 7.6, the process of real shooting spray is shown in Figure 7.49.

After the liquid fuel is sprayed into the chamber space, a multiphase mixture field of liquid column, oil droplet, oil vapor, and air is formed – the spray field. For a diesel spray, the spray field is transient and extremely uneven in both dynamics and thermodynamics. In order to facilitate the study, the spray field is generally divided into multiple zones. Because the fuel spray is a two-phase mixture, starting from the coupling effect of the gas–liquid phase angle, the fuel spray field of literature according to the distance from the nozzle from far to near is divided into an extremely rarefied, a rarefied, a dense area, and a tossing flow of four zones. Figure 7.50 provides a definition of the spray structure and characteristic parameters of the spray hole under the conditions of an ambient temperature of 300 K and back pressure of 3 MPa. According to the axis Z direction, the fog structure can be divided into the following parts:

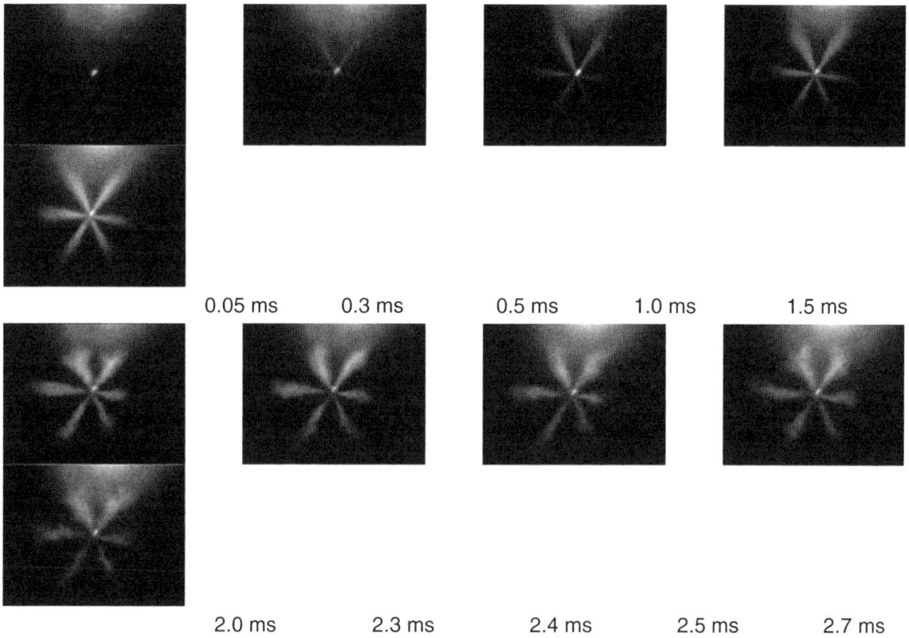

Figure 7.49 Development of the fog beam.

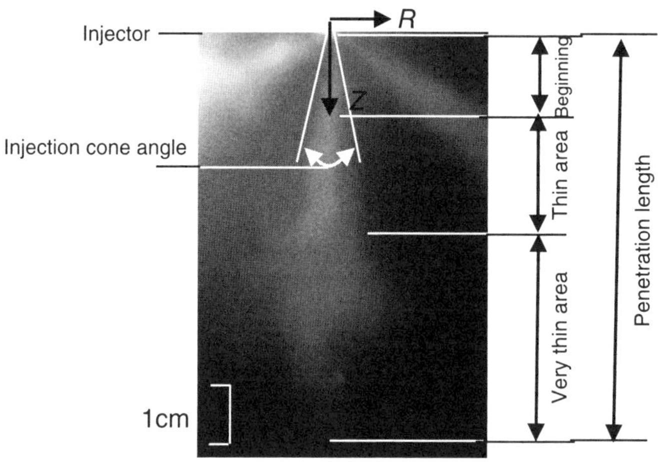

Figure 7.50 Injection cone angle and penetration definition.

(1) *Initial area.* This region consists of the liquid core region, the tossing flow area, and the dense area. In the vicinity of the liquid core region, the droplets of fuel begin to split. However, in a two-phase mixture, the volume fraction of the liquid is equal to that of the gas, or even more than the gas. Therefore, fuel cannot be dispersed in the gas to form a drop of oil, but appears in the form of a thin slice, filament, or gridding. The tossing flow is the first step of the atomization process and the intermediate state of the liquid core to the oil droplet. Between the rarefied and the

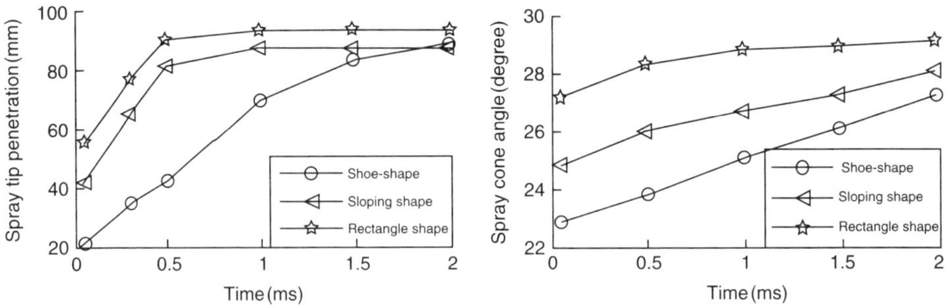

Figure 7.51 Spray tip penetration and conee angle affected by the injection rate.

tossed areas there is a small dense zone in which the droplets exist in a discrete state in a continuous gas field. Compared with the rarefied area, the distance between the droplets is much smaller and the interaction between the oil droplets cannot be ignored. The initial region changes obviously only at the beginning of the spray and the end of the spray. At other times, the region remains relatively stable and the change process is negligible.

(2) *Rarefied area.* This is also the main area of the spray, which develops with the development of the spray process, and its change is not negligible, but its shape remains almost the same as the cone.

(3) *Extremely rarefied area.* This belongs to the aerodynamical area. The fog beam is sucked and diffused by the resistance of ambient air. Air resistance pushes the fuel droplets at the front of the fog beam to the periphery of the fog, forming an inverted mushroom-shaped vortex ring structure in the front.

The cone angle of the spray is in the initial core zone of the fog, and the vertex of the spray cone angle is the center of the spray hole. The two edges are tangent to the edges of the initial core region. The length from the outlet of the spray hole to the tip of the fog is defined as the tip penetration distance of the fog beam.

Fuel injection is according to scheme 1 and shoot is its spray process. Through the image processing, the spray penetration distance and the spray cone angle at every working condition are obtained. The results are shown in Figure 7.51. A rectangle injection has a large penetration speed and reaches its maximum quickly. The slope shape has a small delay. Until the end of injection, the penetration speed for a boot-shaped injection reaches its maximum. From the experimental results, the spray cone angle is mainly influenced by the injection pressure, and the larger the opening pressure, the larger the spray cone angle is. In the same injection process, the injection pressure increases and the spray cone angle increases.

7.4.3 Experimental Research Conclusions

(1) The main performance indexes of the DPCRS are determined. These include the pressure of the electric control supercharger, the regulation of the law of fuel injection, the control of the fuel consumption, and the spray characteristics of the supercharger. The main test items are made up of the pressurized pressure test, the injection law test, the flow test, and spray photography. The performance test

platform and the spray photography platform for the DPCRS are set up according to the testing requirements.

(2) The results of the simulation are tested by the research results. The peak value of the test booster is up to 215 MPa, the peak value of the simulation supercharging is 230 MPa, and the test value is lower than the simulation value, which is mainly due to the insufficient machining precision of the couple parts. The continuous running of the system leads to a high oil temperature, which affects the effect of supercharging. The pressure rise rate of the simulation is higher than the test value, mainly because the theoretical response speed of the system is higher than the actual value. The simulation results show that changing the timing of turbocharging and injection control will change the shape of the injection law. This conclusion has been proved by the experimental results.

(3) The pressurization pressure and fuel injection rate of various systems are measured at different booster ratios. The experimental results show that the system has an obvious booster effect, and the turbocharging ratio measured in each system is lower than that of the original design. The change of the control timing of the solenoid valve and the supercharger solenoid valve in the system will have an obvious effect on the shape of the fuel injection rate. When ΔT_{int} goes from positive to negative, the shape of the injection rate is gradually transferred from the rectangle to the boot shape.

(4) The test data shows that when the rail pressure is 100 MPa, the amount of oil return controlled by the supercharger is about 35% of the total oil consumption of the system (min. 32.2%, max. 41.2%). The effect of pressurization control on the amount of oil return and the fuel injection rate of the injector is significantly affected by the supercharging time. This is mainly because the supercharging time is different and the pressure rise value is different, which causes the different injection pressures. When the rail pressure is 60 MPa, the amount of oil return and the injection oil amount, compared with the 100 MPa rail pressure, only shows a decrease of oil injection with a decrease of pressure. The change range of the pressure control return oil is larger (min. 6.4%, max. 29.3%).

(5) It is proved by the test that the shape of the injection rate has an obvious influence on the jet characteristics. When the rail pressures are the same, the pressurization pressures are the same, the injection pulse widths are the same, the rectangle injection has the largest quantity of injection, and the boot-shaped injection has the smallest quantity of injection. When the rail pressures are the same, the pressurization pressures are the same; in order to have the same quantity of injection, the boot-shaped injection needs more time.

(6) Results of the spray photography test of the DPCRS indicates that the rectangle injection has a large penetration speed and reaches a maximum quickly. The slope shape has a small delay. Until the end of the injection, the penetration speed for the boot-shaped injection reaches its maximum. From the experimental results, the spray cone angle is mainly influenced by the injection pressure, and the larger the opening pressure, the larger is the spray cone angle. In the same injection process, the injection pressure increases and the spray cone angle increases.

Index

Common Rail Fuel Injection Technology in Diesel Engines, First Edition.
Guangyao Ouyang, Shijie An, Zhenming Liu and Yuxue Li.

g

h